Conference Proceedings of the Society for Experimental Mechanics Series

Series Editor
Tom Proulx
Society for Experimental Mechanics, Inc.
Bethel, CT, USA

For further volumes:
http://www.springer.com/series/8922

H. Jerry Qi • Bonnie Antoun • Richard Hall • Hongbing Lu
Alex Arzoumanidis • Meredith Silberstein • Jevan Furmanski
Alireza Amirkhizi • Joamin Gonzalez-Gutierrez
Editors

Challenges in Mechanics of Time-Dependent Materials, Volume 2

Proceedings of the 2014 Annual Conference on Experimental and Applied Mechanics

Springer

Editors
H. Jerry Qi
Georgia Institute of Technology
Atlanta, GA, USA

Richard Hall
Air Force Research Laboratory
Wright-Patterson AFB, OH, USA

Alex Arzoumanidis
Psylotech, Inc.
Evanston, IL, USA

Jevan Furmanski
ExxonMobil
Los Alamos, NM, USA

Joamin Gonzalez-Gutierrez
University of Ljubljana
Ljubljana, Slovenia

Bonnie Antoun
Sandia National Laboratories
Livermore, CA, USA

Hongbing Lu
The Erik Johnson School of Engineering, EC-38
University of Texas-Dallas
Dallas, TX, USA

Meredith Silberstein
Cornell University
Ithaca, NY, USA

Alireza Amirkhizi
University of California
San Diego, La Jolla, CA, USA

ISSN 2191-5644 ISSN 2191-5652 (electronic)
ISBN 978-3-319-06979-1 ISBN 978-3-319-06980-7 (ebook)
DOI 10.1007/978-3-319-06980-7
Springer Cham Heidelberg New York Dordrecht London

Library of Congress Control Number: 2012952717

Printed on acid-free paper

Springer is part of Springer Science+Business Media (www.springer.com)

Preface

Challenges in Mechanics of Time-Dependent Materials, Volume 2: Proceedings of the 2014 Annual Conference on Experimental and Applied Mechanics represents one of the eight volumes of technical papers presented at the SEM 2013 SEM Annual Conference & Exposition on Experimental and Applied Mechanics, organized by the Society for Experimental Mechanics and held in Greenville, SC, June 2–5, 2014. The complete proceedings also include volumes on: *Dynamic Behavior of Materials; Advancement of Optical Methods in Experimental Mechanics; Mechanics of Biological Systems and Materials; MEMS and Nanotechnology; Composite, Hybrid, and Multifunctional Materials; Fracture, Fatigue, Failure and Damage Evolution; and Experimental and Applied Mechanics.*

Each collection presents early findings from experimental and computational investigations on an important area within Experimental Mechanics, the Mechanics of Time-Dependent Materials being one of these areas.

This track was organized to address constitutive, time (or rate)-dependent constitutive, and fracture/failure behavior of a broad range of materials systems, including prominent research in both experimental and applied mechanics. Papers concentrating on both modeling and experimental aspects of Time-Dependent Materials are included.

The track organizers thank the presenters, authors, and session chairs for their participation and contribution to this track. The support and assistance from the SEM staff is also greatly appreciated.

Atlanta, GA, USA — H. Jerry Qi
Livermore, CA, USA — Bonnie Antoun
Wright-Patterson AFB, OH, USA — Richard Hall
Dallas, TX, USA — Hongbing Lu
Evanston, IL, USA — Alex Arzoumanidis
Ithaca, NY, USA — Meredith Silberstein
Los Alamos, NM, USA — Jevan Furmanski
San Diego, La Jolla, CA, USA — Alireza Amirkhizi
Ljubljana, Slovenia — Joamin Gonzalez-Gutierrez

Contents

Chapter 1
Unimorph Shape Memory Polymer Actuators Incorporating Transverse Curvature in the Substrate

Jason T. Cantrell and Peter G. Ifju

Abstract Shape memory polymers (SMP) utilized in reconfigurable structures have the potential to be used in a variety of settings. This paper is primarily concerned with the use of Veriflex-S shape memory polymer and bi-directional carbon fiber in a unimorph actuator configuration. One of the major deficiencies of SMP unimorphs is the permanent set (unrecovered shape) after a single or multiple temperature cycle(s). The novel concept of incorporating transverse curvature in the composite substrate, similar to that of an extendable tape measurer, was proposed to improve the shape recovery. A set of experiments was designed to investigate the influence of transverse curvature, the relative widths of SMP and composite substrates, and shape memory polymer thickness on actuator recoverability after multiple thermomechanical cycles. Flat carbon fiber and shape memory polymer unimorph actuators were evaluated for performance versus actuators of increasing transverse curvature. Digital image correlation was implemented to quantify the out-of-plane deflection of the unimorph composite actuators (UCAs) during the actuation cycle. Experimental results indicate that an actuator with transverse curvature significantly reduces the residual deformation while increasing the shape memory recoverability which could be further tailored to enhance the performance of shape memory polymers in reconfigurable arrangements.

Keywords Shape memory polymer • Unimorph • Transverse curvature • Digital image correlation • Composite

Nomenclature

CF	Carbon fiber
DIC	Digital image correlation
MAV	Micro air vehicle
SMP	Shape memory polymer
T_g	Glass transition temperature
u, v, w	Lengthwise widthwise, and vertical displacements
UCA	Unimorph composite actuator
VIC	Visual image correlation
x, y, z	Lengthwise widthwise, and vertical coordinates

1.1 Introduction

Shape memory polymers (SMPs) are a category of smart material with the ability to change their shape upon the application of external stimuli such as temperature electricity, magnetism, or light. Classes of smart materials include piezoelectric, shape memory alloys, and shape memory polymers. Varieties of smart materials practical for various applications include shape memory alloys in orthodontic treatments, piezoelectric actuators for control of micro air vehicles, shape memory

J.T. Cantrell (✉) • P.G. Ifju
Mechanical and Aerospace Engineering Department, University of Florida, MAE Receiving, 134 MAE-C, Gainesville, FL 32611, USA
e-mail: jasontcantrell@gmail.com; ifju@ufl.edu

H.J. Qi et al. (eds.), *Challenges in Mechanics of Time-Dependent Materials, Volume 2: Proceedings of the 2014 Annual Conference on Experimental and Applied Mechanics*, Conference Proceedings of the Society for Experimental Mechanics Series,
DOI 10.1007/978-3-319-06980-7_1,

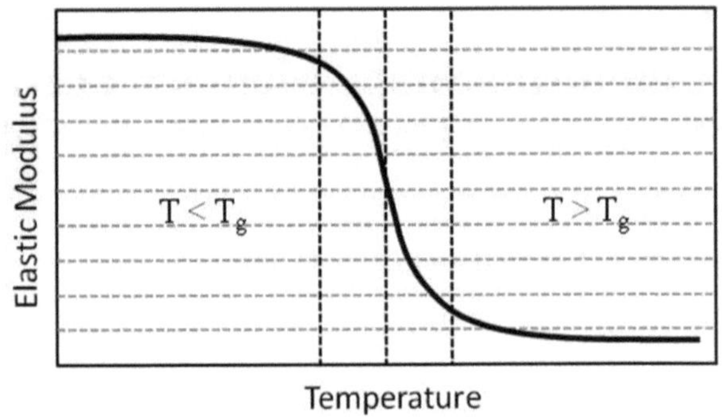

Fig. 1.1 Generalized plot of the elastic modulus (E) versus temperature for Veriflex SMP. The graph is divided into the plastic ($T < T_g$), transition, and rubbery ($T > T_g$) regions

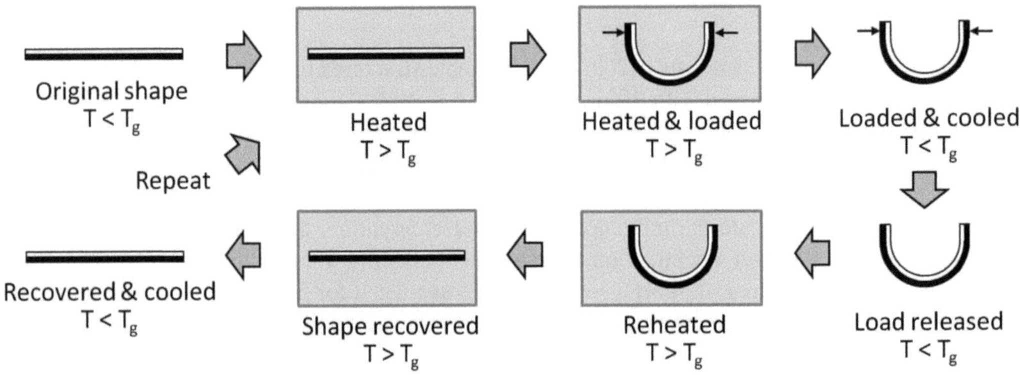

Fig. 1.2 Illustration of a shape memory cycle for recovery of thermally activated SMPs

polymers as cardiovascular stents, and a multitude of smart materials for the morphing of aircraft structures [1–5]. Veriflex-S, the SMP used during the studies in this paper, uses a thermal external stimulus to allow reconfiguration and recovery. This material's properties, and those of its higher temperature counterpart, Veriflex-E, have been studied extensively by researchers including Fulcher et al. [6–8], Nahid et al. [9], McClung et al. [10–14], Liu et al. [15], and Atli et al. [16]. The Veriflex SMPs have been utilized for notable applications including active disassembly for recycling, deployment of satellite solar panels, and deployable aircraft wings [17–19].

Veriflex can be divided into two categories of stiffness and material behavior: the high glassy modulus and low rubbery modulus [12, 13]. At temperatures below their glass transition temperature (T_g), the material is relatively stiff and has a high elastic modulus; however once the SMP is heated above T_g the modulus drops by several orders of magnitude. This transition from the glassy to the rubbery state is illustrated in Fig. 1.1. In the rubbery state, shape memory polymers can deform at levels up to 400 % and after cooling below T_g maintain this new shape indefinitely [20]. The original shape can be recovered by heating the polymer above T_g again. The glassy state is classified as the temperatures lying 10 °C or more below the T_g, while the rubbery state is identified as temperatures lying 10 °C or greater above the T_g [21]. The area in between the glassy and rubbery state is classified as the transition region in which the elastic modulus transitions rapidly.

SMPs can change their shape from their original cast shape (flat beams in this study) to a deformed shape and return to the original shape when exposed to elevated temperatures. An illustration of an ideal shape memory thermomechanical cycle is shown in Fig. 1.2. The SMP begins in its original shape at a high modulus below T_g and then heat is applied to the sample causing the modulus to fall into the rubbery state. Once in the rubbery state the sample is bent into the desired deformed shape (a U-shaped configuration for this study) and then allowed to cool below T_g locking the current deformed shape. The sample can be stored at this configuration to await the reapplication of heat. After heating the sample will release and return to the unconstrained original form. The sample is then cooled and would ideally return to 100 % of the original shape seen before the heating cycle. However, in reality, the Veriflex SMP can achieve a final shape that is only close to the original shape. Various researchers have studied this behavior and determined that the recoverability of SMP can vary between 65 and 95 % of the original shape after repeated cycling depending upon testing conditions [12, 18, 22, 23].

Despite these hindrances SMPs are still advantageous over other shape memory materials due to the fact that they are low cost low density, and highly deformable among other benefits [24, 25]. Shape memory materials are valued for their potential use in adaptive structures in applications such as micro air vehicles (MAVs) and morphing aircraft [26]. The University of Florida has worked with adaptive structures and MAVs extensively, adopting both active adaptation with piezoelectric actuators and passive adaptation with flexible membrane wings [2, 3, 27–31]. Ifju et al. developed a bendable load stiffened MAV wing that is compliant in the downward direction for storing the aircraft, but uses the wing curvature to avoid buckling due to flight loads [32]. Using this knowledge of MAVs and morphing wing structures, a plan of study was devised for a multipurpose morphing actuator to determine if the same bendable composite technology used in MAV wings could assist in increasing the recoverability of the Veriflex-S SMP. In order to properly understand the overall performance of the SMP in a unimorph composite actuator (UCA) configuration, extensive digital image correlation (DIC) testing was required to determine the residual deformation present. A UCA was described as an element capable of bi-stable configuration when supplied with an external stimulus consisting of one active layer (SMP) to which the stimulus is applied and one inactive layer (carbon fiber laminate) that supports the active layer. A simple flat carbon fiber (CF) beam with SMP adhered to its surface was compared against a transversely curved CF beam with curvature similar to that of the MAV wing discussed previously. Additionally, a more detailed survey investigating the influence of other variables present was also documented via DIC. The details of the UCA analysis and the experimental procedure are explained in the subsequent sections. During the course of experimenting with flat generic unimorph actuators, research indicated that by incorporating a transverse curvature (similar to an extendable tape measure) in the composite layer one can vastly improve the shape recovery of an SMP unimorph actuator. The following paper will cover the findings from the investigation of generic SMP unimorphs.

1.2 Unimorph Composite Actuator Experimental Procedure

1.2.1 Unimorph Composite Actuator (UCA) Fabrication

Each unimorph composite actuator consists of a layer of SMP bonded to a graphite/epoxy substrate. Both flat carbon fiber composite unimorphs and unimorphs incorporating transverse curvature followed the same fabrication methodology. A single layer of [±45°] oriented plain weave, bi-directional carbon fiber was cut and placed on a Teflon covered plate or curved tooling board. The entire assembly was covered in an additional layer of Teflon, vacuum bagged, and cured at 130 °C for 4 h. After curing, the carbon fiber was cut down to the appropriate size, then a Veriflex-S shape memory polymer panel was bonded to it using Araldite 2011 two-part epoxy. After the epoxy cured the actuators were coated with a base coat of flat white spray paint then speckled for DIC using flat black spray paint.

1.2.2 Digital Image Correlation (DIC) Set-Up

The primary objective of this research was to determine the deformation and shape of the composite beam samples. This was done through the use of the DIC system a non-contact, full-field shape and deformation technique developed at the University of South Carolina [33, 34]. The system uses two Point Grey Research 5-megapixel grayscale cameras to simultaneously capture images of the random speckle pattern applied to the samples. The cameras are calibrated via a high contrast dot pattern of known diameter and spacing. In these set-ups, it was a 9 × 9 grid of points with a separation of 10 mm. Once calibrated, the system is ready to photograph the composite beam and determine deflection as a function of time. Reference images of the beams were initially taken after the samples were painted. Subsequent images were taken before starting each testing cycle. These images were contrasted against images taken over the hour observation time to determine the deflection as the sample cooled. Images are captured via VIC Snap 2009 and processed via VIC-3D 2009 to determine deformations. Figure 1.3 shows the digital image correlation experimental set-up to measure the remaining deformation on the generic UCA specimens.

1.2.3 Environmental Chamber Set-Up

The UCAs were placed in a Sun Systems Model EC12 environmental chamber and was used to regulate the temperature to the desired point above the shape memory polymer glass transition temperature. The temperature was monitored via a thermocouple inside of the chamber and confirmed via a Fluke 561 series infrared thermometer. Beam samples were placed on a Teflon plate within the chamber to allow for full expansion under elevated temperature conditions.

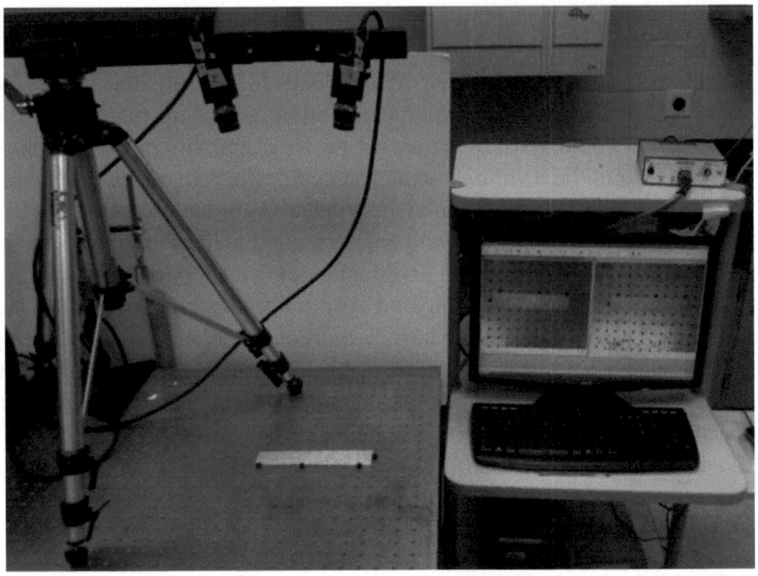

Fig. 1.3 Experimental set-up for DIC analysis of the UCAs

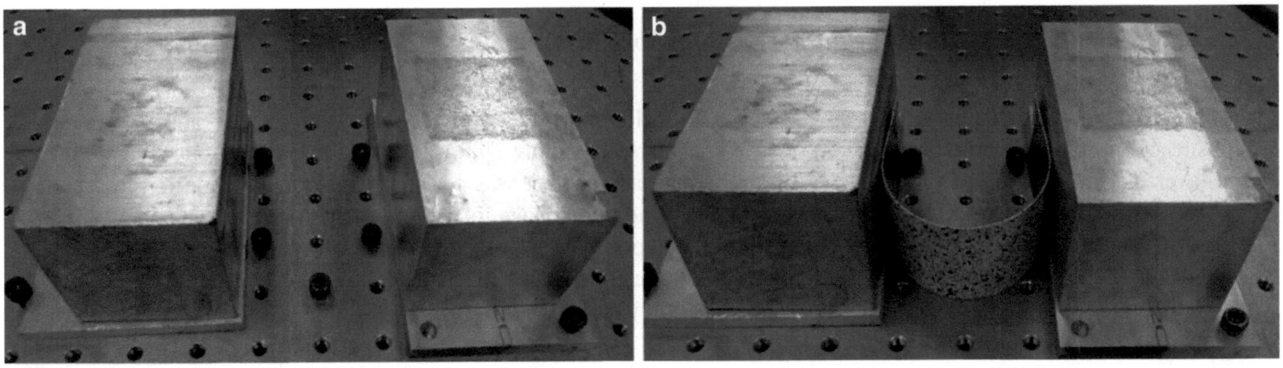

Fig. 1.4 (a) Sample container without a sample. (b) Sample container holding a curved carbon fiber beam

1.2.4 UCA Sample Holder Set-Up

Once samples were removed from the environmental chamber they were folded into a U-shaped configuration as shown in Fig. 1.4, and stored in a tabletop retainer to ensure equivalent loading conditions for all actuators. This apparatus consisted of five 1/4-20 bolts in a U configuration secured to the table in order to constrain the samples from folding inwardly, and two metal blocks spaced 60 mm apart to constrain the samples in the outward direction.

1.2.5 Procedure to Measure UCA Recoverability

Step-by-Step Procedure to Measure Shape Recovery of the UCA Using DIC

The procedure for measuring the out-of-plane residual deformation with DIC after a temperature cycle is enumerated below.

Step 1. After applying a speckle pattern to the sample take an initial (reference) image of the UCA using the DIC set-up.
Step 2. Place the undeformed UCA in the environmental chamber for 1 h at 85 °C.
Step 3. Bend the UCA beam into a U-shaped configuration and place it within the holder to cool for 1 h in the stored configuration.
Step 4. Return the sample to the environmental chamber set to 85 °C and allow the beam to hold for 1 h at temperature.
Step 5. Remove UCA from the oven to start recovery to original position.
Step 6. Monitor via DIC while the UCA cools to room temperature.

1.3 Unimorph Composite Actuator Results

Out-of-plane deflection (w) was the main focus of the UCA experiments. The goal was a beam with minimal residual deformation that when stored would hold the desired shape. Initial samples consisted of 200 mm long by 38 mm wide flat (zero curvature) and 63.5 mm radius of curvature carbon fiber samples. These samples consisted of a 12.7 mm wide and 1.6 mm thickness strip of shape memory polymer adhered via Araldite 2011 epoxy to the center of the carbon fiber beam. Figure 1.5 shows a mock-up of these variables on a concave curved carbon fiber beam. Figure 1.6 shows a schematic of the flat concave, and convex actuators used for testing.

Post-processing of the DIC data was required to properly determine the deflection for each UCA over time. Post-processing was done by extracting the XYZ coordinates and UVW displacements for the centerline of each sample at the desired timestamp. Next the data are extracted to an Excel file, the deformation (W) data sorted by timestamp, and shifted to the desired coordinate system via MATLAB. Once in the desired X–Z plane, the data are rotated to eliminate rigid body motion making sure to rotate the sample in the X or lengthwise direction to maintain the correct displacement directions. After rotation the data undergo a final vertical translation to the X-axis ensuring all images can be compared in the same coordinate system. This process is illustrated in Fig. 1.7.

Data for both the flat and concave UCA samples were collected in 2 min intervals for the entirety of the 30 min cool down time. The centerline shape was measured for the reference (before any temperature cycle) and at various times after the temperature cycle. To obtain the deformation the reference shape was subtracted from the shape after the temperature cycle. In order to properly control for any manufacturing defects, only the deflection from the original shape is covered in the subsequent results. Table 1.1 shows the maximum out-of-plane deflection for both the flat and concave samples while Figs. 1.8, 1.9, 1.10, and 1.11 show the centerline deformation along the longitudinal direction for both samples through

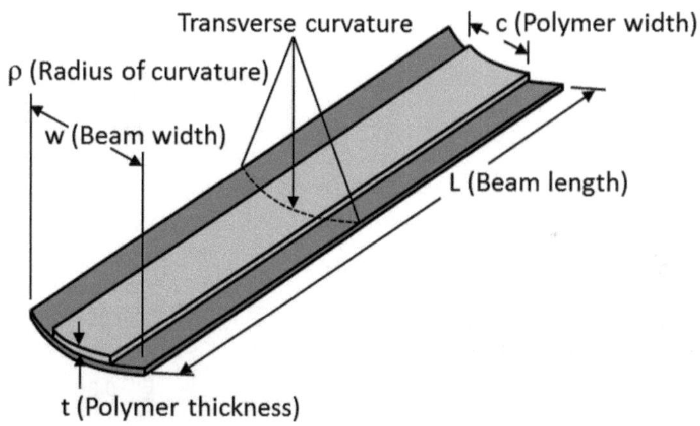

Fig. 1.5 Illustration of the variables present on a UCA

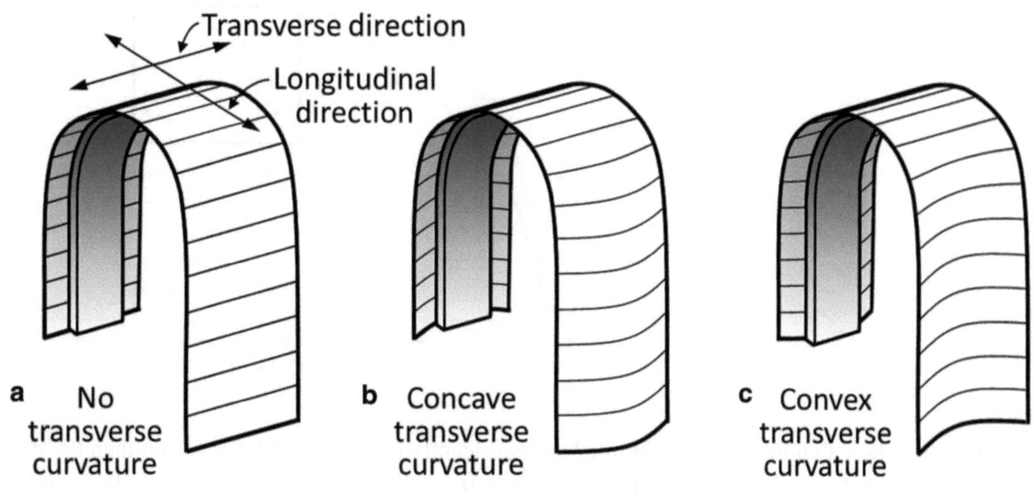

Fig. 1.6 Comparison of a UCA (a) without curvature (flat composite), (b) with concave transverse curvature, and (c) a UCA with convex transverse curvature

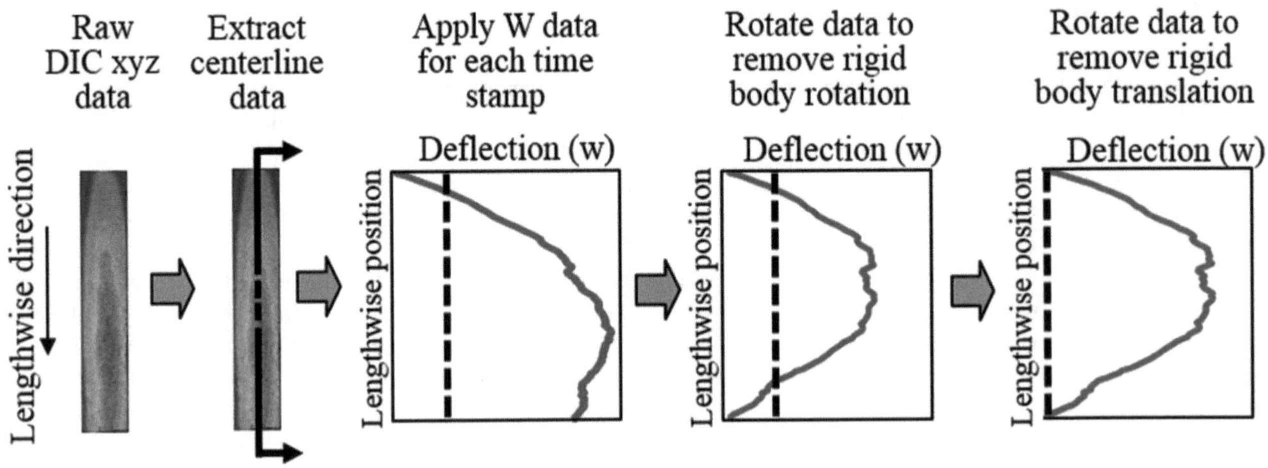

Fig. 1.7 Illustration of the process of converting the DIC data to the desired coordinate system and removing rigid body motion

Table 1.1 Maximum deflections for the UCA samples at each marked time

Time (min)	Max deflection concave sample (recovered-reference) (mm)	Max deflection flat sample (recovered-reference) (mm)
Reference	0.00	0.00
2	0.28	5.89
4	0.34	9.80
6	0.35	11.4
8	0.35	12.0
10	0.35	12.1
20	0.31	12.3
30	0.30	12.7

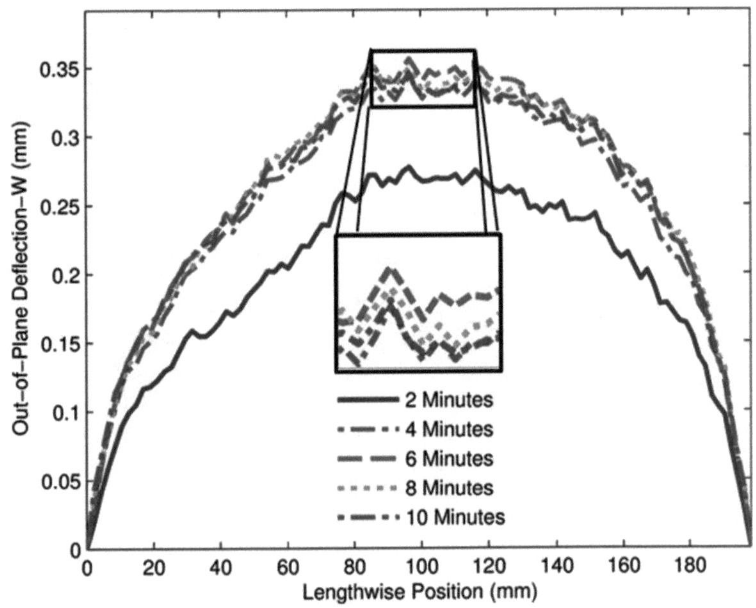

Fig. 1.8 Lengthwise versus out-of-plane deflection for the first 10 min of the 63.5 mm concave UCA cooling cycle

30 min. The data clearly show that the concave sample has significantly less residual deformation than the flat sample over the 30 min trial. The concave sample has a maximum variation from the original sample of only 0.35 mm while the flat sample has a maximum difference of 12.7 mm. The graphs show that the concave UCA reaches a peak deflection at approximately 6 min then relaxes a distance of 60 μm by the 30 min mark. The flat UCA does not reach equilibrium in

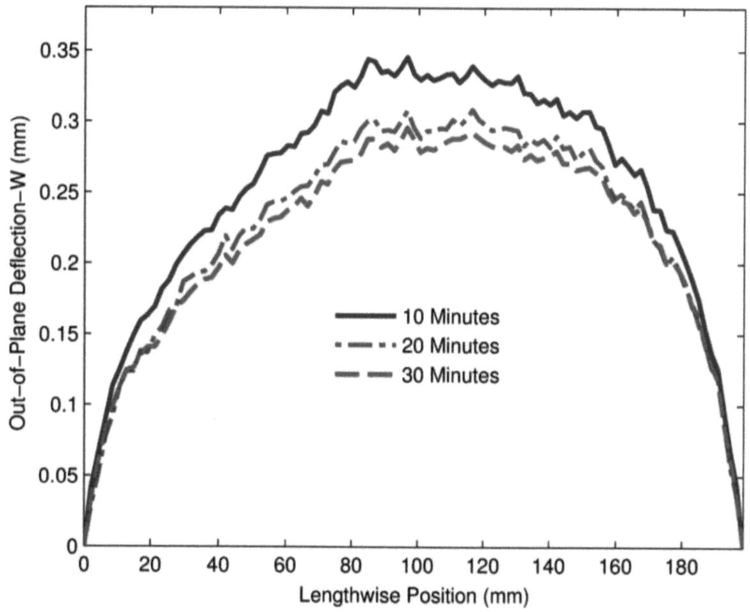

Fig. 1.9 Lengthwise position versus out-of-plane deflection for times 10–30 min of the 63.5 mm concave UCA cooling cycle

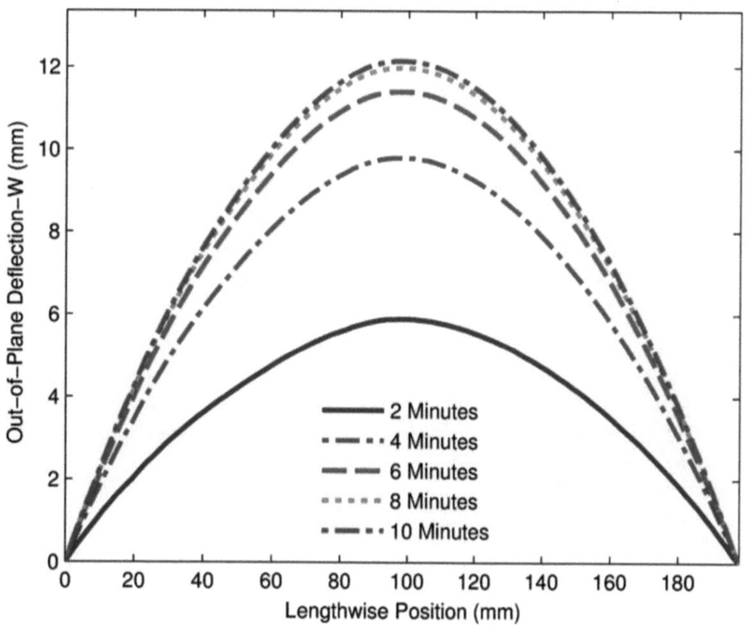

Fig. 1.10 Lengthwise position versus out-of-plane deflection for the first 10 min of the flat UCA cooling cycle

30 min as it continues to deflect until the 30 min mark. However, the data show that a majority of the deformation has already occurred after 6 min which was also true of the concave sample.

The preliminary test clearly shows that the concept of applying concave transverse curvature to a unimorph substrate substantially improves shape recovery. As such in order to explore further this concept, an additional curved UCA was created but instead of the conventional concave orientation (saddle configuration) it was created with convex orientation (trough configuration). The convex sample was constructed to determine the effect on the residual deformation in an alternate orientation. An example of a convex sample was shown previously in Fig. 1.6. The convex set of samples was monitored via DIC for 30 min during the cool down like the previously tested samples. Table 1.2 shows the maximum out-of-plane deviation with respect to time for the convex sample versus the original concave sample. The data show that while the original concave sample had more initial deformation the convex sample has the larger change in residual deflection. As stated previously, the concave sample deflects only 0.35 mm whereas the convex sample deflects 1.29 mm in the same time period. The convex sample behaves similarly to the concave sample with respect to relaxation. Both samples reach maximum deflection at approximately 6 min and decrease in deflection to some extent up to the 30 min. Figures 1.12 and 1.13

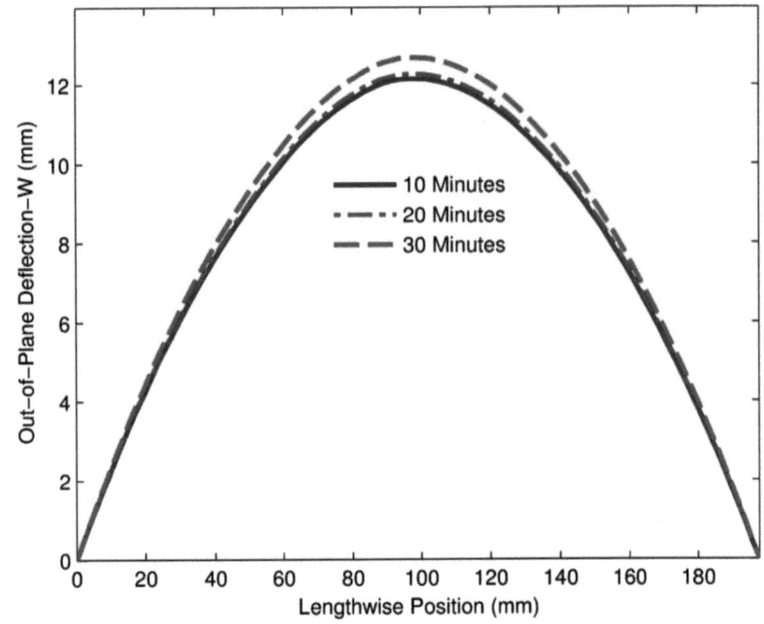

Fig. 1.11 Lengthwise position versus out-of-plane deflection for times 10–30 min of the flat UCA cooling cycle

Table 1.2 Maximum deflections for the UCA samples at each marked time

Time (min)	Max deflection concave sample (recovered-reference) (mm)	Max deflection convex sample (recovered-reference) (mm)
Reference	0.00	0.00
2	0.28	0.86
4	0.34	1.19
6	0.35	1.29
8	0.35	1.29
10	0.35	1.29
20	0.31	1.20
30	0.30	1.13

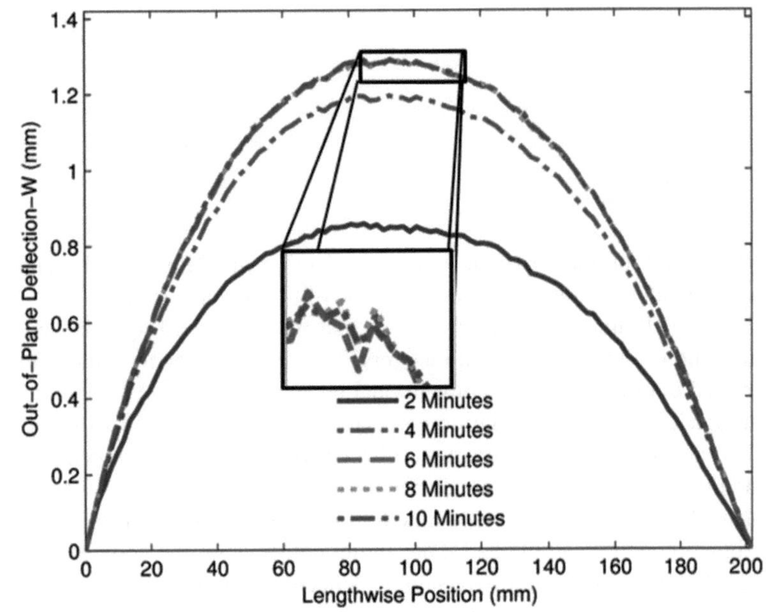

Fig. 1.12 Lengthwise position versus out-of-plane deflection for the first 10 min of the convex UCA cooling cycle

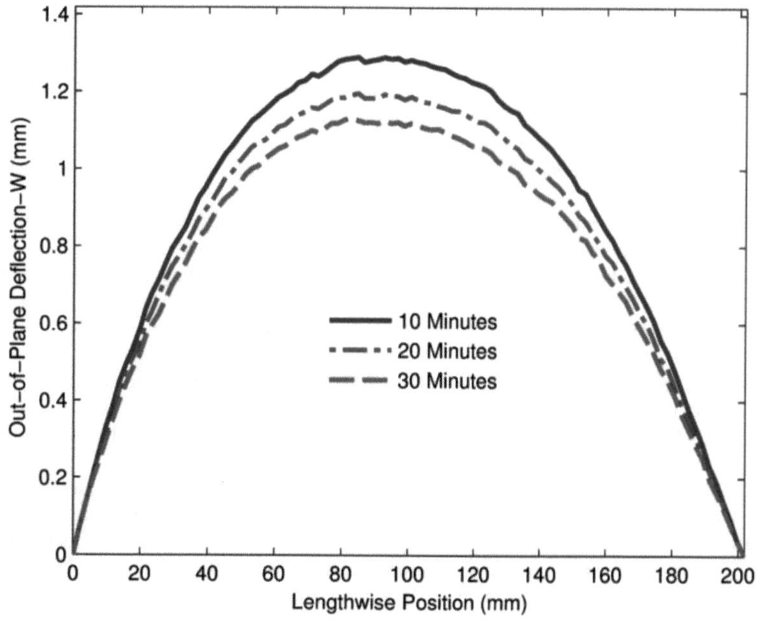

Fig. 1.13 Lengthwise position versus out-of-plane deflection for times 10–30 min of the convex UCA cooling cycle

Table 1.3 Repeatability test data for the concave curved UCA sample

Time (min)	Test 1 position (mm)	Test 2 position (mm)	Test 3 position (mm)	Test 4 position (mm)	Standard deviation (mm)	Coefficient of variation (%)
6	1.93	1.95	1.91	1.93	1.63E−2	0.85
60	1.81	1.80	1.78	1.81	1.41E−2	0.78

support these findings and show the centerline deformations for the convex sample. The results indicate the original concave configuration should be used for any further testing due to the minimal deflections seen under comparable conditions.

A final series of testing was done on a new concave curved sample to determine the repeatability of testing and any residual deformation as additional deflection cycles were performed on the UCA. A series of four consecutive tests were conducted and compared at the maximum out-of-plane position (Z + W) time of 6 min as well as at the end of the data collection period. Table 1.3 shows the data range was only 40 μm at 6 min and 30 μm at 60 min. Both values that are well within an acceptable range for repeatability.

1.4 Conclusion

A series of tests were conducted on carbon fiber and shape memory polymer composite actuators to determine the effect of radius of curvature on the residual deformation. Digital image correlation was employed to find the out-of-plane deformation and allowed for the study of the recovery behavior of these unimorph composite actuators. In the experiments conducted a unimorph composite actuator with a 63.5 mm concave transverse curvature was able to reduce residual deformation by two orders of magnitude compared to a flat unimorph composite actuator keeping all other variables constant. A unimorph with convex transverse curvature was only able to reduce residual deformation by one order of magnitude making a concave actuator the best option for future use. Unimorph composite actuators display repeatable actuation and storage cycles as they do not increase residual deformation with increasing number of cycles. These discoveries can facilitate the expanded use of shape memory polymers on a reconfigurable folding wing micro air vehicles as well as various other applications.

Future research will continue to develop the design space presented in this paper. Unimorph composite actuators with varying transverse curvature polymer thickness, substrate width, and polymer width will all be evaluated to determine the correlation between each variable and residual deformation.

References

1. Otsuka K, Wayman CM (eds) (1998) Shape memory materials, 1st edn. Cambridge University Press, Cambridge, p 284
2. Lacroix BW, Ifju PG (2012) Utilization and performance enhancements of multiple piezoelectric actuators on micro air vehicles. In: AIAA aerospace sciences meeting, pp 1–14
3. Lacroix BW, Ifju PG (2013) Macro fiber composites and substrate materials for MAV wing morphing. In: Society for experimental mechanics, pp 1–13
4. Yakacki CM, Shandas R, Lanning C, Rech B, Eckstein A, Gall K (2007) Unconstrained recovery characterization of shape-memory polymer networks for cardiovascular applications. Biomaterials 28(14):2255–2263
5. Thill C, Etches J, Bond I, Potter K, Weaver P (2008) Morphing skins. Aeronaut J 112(3216):1–23
6. Fulcher JT, Karaca HE, Tandon GP, Foster DC, Lu YC (2011) Multiscale characterization of water-, oil-, and UV-conditioned shape-memory polymer under compression. In: Mechanics of time-dependent materials and processes in conventional and multifunctional materials, pp 97–103
7. Fulcher JT, Karaca HE, Tandon GP, Lu YC (2012) Thermomechanical and shape memory properties of thermosetting shape memory polymer under compressive loadings. J Appl Polym Sci
8. Fulcher JT (2011) Mechanical characterizations of environmentally conditioned shape memory polymers for reconfigurable aerospace structures. University of Kentucky
9. Nahid MNH, Wahab MA, Lian K (2011) Degradation of shape memory polymer due to water and diesel fuels. In: Mechanics of time-dependent materials and processes in conventional and multifunctional materials, pp 37–48
10. McClung AJW, Tandon GP, Goecke KE, Baur JW (2011) Non-contact technique for characterizing full-field surface deformation of shape memory polymers at elevated and room temperatures. Polym Test 30(1):140–149
11. McClung AJW, Ruggles-Wrenn MB (2009) Strain rate dependence and short-term relaxation behavior of a thermoset polymer at elevated temperature: experiment and modeling. J Press Vessel Technol 131(3):031405
12. McClung AJW, Tandon GP, Baur JW (2011) Fatigue cycling of shape memory polymer resin. In: Mechanics of time-dependent materials and processes in conventional and multifunctional materials, vol 3, pp 119–127
13. McClung AJW, Tandon GP, Baur JW (2011) Strain rate- and temperature-dependent tensile properties of an epoxy-based, thermosetting, shape memory polymer (Veriflex-E). Mech Time-Dependent Mater 16(2):205–221
14. McClung AJW, Tandon GP, Baur JW (2011) Deformation rate-, hold time-, and cycle-dependent shape-memory performance of Veriflex-E resin. Mech Time-Dependent Mater 17(1):39–52
15. Liu Y, Gall K, Dunn ML, Greenberg AR, Diani J (2006) Thermomechanics of shape memory polymers: uniaxial experiments and constitutive modeling. Int J Plast 22(2):279–313
16. Atli B, Gandhi F, Karst G (2008) Thermomechanical characterization of shape memory polymers. J Intell Mater Syst Struct 20(1):87–95
17. Carrell J, Tate D, Wang S, Zhang H-C (2011) Shape memory polymer snap-fits for active disassembly. J Clean Prod 19(17–18):2066–2074
18. Lan X, Liu Y, Lv H, Wang X, Leng J, Du S (2009) Fiber reinforced shape-memory polymer composite and its application in a deployable hinge. Smart Mater Struct 18(2):024002
19. Joo J, Smyers B, Beblo R, Reich G, Force A (2011) Load-bearing multi-functional structure with direct thermal harvesting for thermally activated reconfigurable wing design. In: International conference on composite materials, pp 1–6
20. Leng J, Lu H, Liu Y, Huang WM, Du S (2009) Shape-memory polymers — a class of novel smart materials. MRD Bull 34(11):848–855
21. Monkman G (2000) Advances in shape memory polymer actuation. Mechatronics 10(4–5):489–498
22. Schmidt C, Neuking K, Eggeler G (2008) Functional fatigue of shape memory polymers. Adv Eng Mater 10(10):922–927
23. Gall K, Mikulas M, Munshi N a, Beavers F, Tupper M (2000) Carbon fiber reinforced shape memory polymer composites. J Intell Mater Syst Struct 11(11):877–886
24. Liu C, Qin H, Mather PT (2007) Review of progress in shape-memory polymers. J Mater Chem 17(16):1543
25. Beloshenko V a, Varyukhin VN, Voznyak YV (2005) The shape memory effect in polymers. Russ Chem Rev 74(3):265–283
26. Vaia R, Baur J (2008) Materials science: adaptive composites. Science 319(5862):420–421
27. Cantrell JT, Lacroix BW, Ifju PG (2013) Passive roll compensation on micro air vehicles with perimeter reinforced membrane wings. Int J Micro Air Veh 5(3):163–177
28. Ifju PG, Jenkins DA, Waszak MR, Ettinger S, Lian Y, Shyy W (2002) Flexible-wing-based micro air vehicles, pp 1–13
29. Shyy W, Ifju P, Viieru D (2005) Membrane wing-based micro air vehicles. Appl Mech Rev 58(4):283
30. Stanford BK (2008) Aeroelastic analysis and optimization of membrane micro air vehicle wings. University of Florida
31. Albertani R (2005) Experimental aerodynamic and static elastic deformation characterization of low aspect ratio flexible fixed wings applied to micro aerial vehicles. University of Florida
32. Ifju P, Lee K, Albertani R, Mitryk SJ, Mary L, Boria FJ, Abd (2008) Bendable wing for micro air vehicle. 7,331,5462008
33. Sutton MA, Turner JL, Bruck HA, Chae TA (1991) Full-field representation of discretely sampled surface deformation for displacement and strain analysis. Exp Mech 31:168–177
34. Sutton MA (2008) Springer handbook of experimental solid mechanics. Department of Mechanical Engineering, The Johns Hopkins University, Baltimore, pp 565–600

Chapter 2
Yield Criterion for Polymeric Matrix Under Static and Dynamic Loading

B.T. Werner and I.M. Daniel

Abstract A polymeric matrix (3501-6) used in composite materials was characterized under multi-axial quasi-static and dynamic loading at varying strain rates. Tests were conducted under uniaxial compression, tension, pure shear and combinations of compression and shear. Quasi-static and intermediate strain rate tests were conducted in a servo-hydraulic testing machine. High strain rate tests were conducted using a split Hopkinson Pressure Bar system built for the purpose. This SHPB system was made of glass/epoxy composite (Garolite) bars having an impedance matching the test polymer closer than metals. The typical stress–strain behavior exhibits a linear elastic region up to a yields point, a nonlinear elastoplastic region up to an initial peak or critical stress, followed by a strain softening region up to a local minimum and finally, a strain hardening region up to ultimate failure. It was observed that under multi-axial loading, yielding is governed by one characteristic property, the yield strain under uniaxial tension. Furthermore, it was found that the yield point varied linearly with the logarithm of strain rate. A general three-dimensional elasto-viscoplastic model was formulated in strain space expressed in terms of an effective strain and its yield point. A unified yield criterion was proposed to describe the onset of yielding under any state of stress and at any strain rate.

Keywords Polymer-matrix • Multi-axial testing • Elastic–plastic behavior • Yield criteria • Strain rate effects

2.1 Introduction

Recent and ongoing research in fiber reinforced polymer composites has dealt with material characterization, constitutive behavior and failure prediction. The process of fabrication, testing and modeling of these composites is costly and time consuming and impedes the introduction of new materials. To facilitate and accelerate the process of introducing and evaluating new composite materials, it is important to develop/establish comprehensive and effective methods and procedures of constitutive characterization and modeling of structural laminates based on the properties of the constituent materials, e.g., fibers, polymers and the basic building block of the composite structure, the single ply or lamina.

Lamina characterization and modeling under multi-axial states of stress has shown that there are significant inelastic, nonlinear, viscoelastic and rate effects on the matrix dominated constitutive and failure behavior of these materials [1–3]. In the case of carbon/epoxy composites for example, the fiber itself shows little nonlinear behavior and no rate dependence in its mechanical response. This suggests that the matrix is the key element that controls the inelastic and nonlinear behavior of the composite and that its characterization and modeling are very important. Since the polymer matrix is basically isotropic, a much less costly evaluation of a composite can be achieved by characterizing the bulk matrix under multi-axial states of stress at various strain rates.

The constitutive and strain rate behavior of epoxies under various loading conditions has been studied by many researchers including the authors of this paper [4–14]. Some studies describe characterization of the resin at various strain

B.T. Werner
Sandia National Laboratories, Livermore, CA, USA
e-mail: bwerner@sandia.gov

I.M. Daniel (✉)
Northwestern University, Evanston, IL, USA
e-mail: imdaniel@northwestern.edu

H.J. Qi et al. (eds.), *Challenges in Mechanics of Time-Dependent Materials, Volume 2: Proceedings of the 2014 Annual Conference on Experimental and Applied Mechanics*, Conference Proceedings of the Society for Experimental Mechanics Series,
DOI 10.1007/978-3-319-06980-7_2,

rates [4–11]. Mayr et al. [12] and Behzadi and Jones [13] have focused on the yielding behavior of these polymeric materials including strain rate effects. The onset of yielding in the matrix of the composite may not be critical for quasi-static loading or even single impact, but is the limiting factor in fatigue life of a composite structure. Christensen has presented simplified and comprehensive approaches to the study of yield (failure) criteria for homogeneous materials with particular reference to polymers [14–16].

The objective of this study was to characterize the matrix resin under multi-axial loading at different strain rates and to develop a general yielding model that incorporates rate effects. It is based on experimental observations made in the process of developing a constitutive model for the polymer [17]. Emphasis was placed on development of a relatively simple yield model that would not require extensive testing for evaluation of a given polymeric matrix.

2.2 Material Characterization

The polymer matrix investigated is a high stiffness, high strength epoxy (3501-6) commonly used in composites. It has a highly crosslinked structure that provides stiffness and strength but also reduces its ductility and leads to fairly brittle behaviour. The B-staged resin (supplied by Applied Poleramic, Inc.) was chipped out, weighed, and placed in the mold. A small amount of acetone was added to reduce the resin viscosity and facilitate casting of the partially cured resin into the mold. The temperature was raised at a rate of 2 °C/min up to 120 °C and held at that level for 1 h. During this stage the resin viscosity was low and vacuum was drawn to remove any entrapped air and to boil off any acetone that might remain in the resin. For more complex casting geometries a small amount of solvent (acetone) was used to reduce the viscosity of the partially cured epoxy and facilitate casting into the mold. The material was cast into closed molds to produce thin (3 mm) plates for tensile coupons, thick blocks for prismatic compression coupons, and thin-wall cylinders for specimens to be tested under torsion and combinations of torsion and axial tension or compression.

The geometry and dimensions of the specimens used are shown in Fig. 2.1. Specimens for uniaxial tensile testing were thin dogbone coupons with a gage section of 12.7 × 50.8 mm machined from 3 mm thick sheets using a router with an abrasive bit. Uniaxial compressive tests were conducted on thick prismatic coupons 12.7 mm long with a cross section of 7.62 × 8.89 mm. Specimens with varying aspect ratios (from 0.125:1 to 2.7:1) were tested. It was found that the apparent stiffness reaches the plane strain value (C_{11}) when extrapolated to zero aspect ratio and reaches asymptotically the Young's modulus E of the material for larger aspect ratios (Fig. 2.2). The stiffness determined by extrapolation to zero aspect ratio, C_{11}, was used along with the Young's modulus to determine Poisson's ratio ν. Tests under pure shear and combinations of shear and normal tensile or compressive stress were conducted on thin-wall cylindrical specimens. The cylinders were machined to produce a 1.27 mm wall thickness in the gage section and a fillet radius of 11.43 mm. The resulting dogboned cylinder was mounted on a steel post used to apply compression and torsion.

The young's modulus, Poisson's ratio and shear modulus of the polymer were obtained from the uniaxial compression tests using the relations

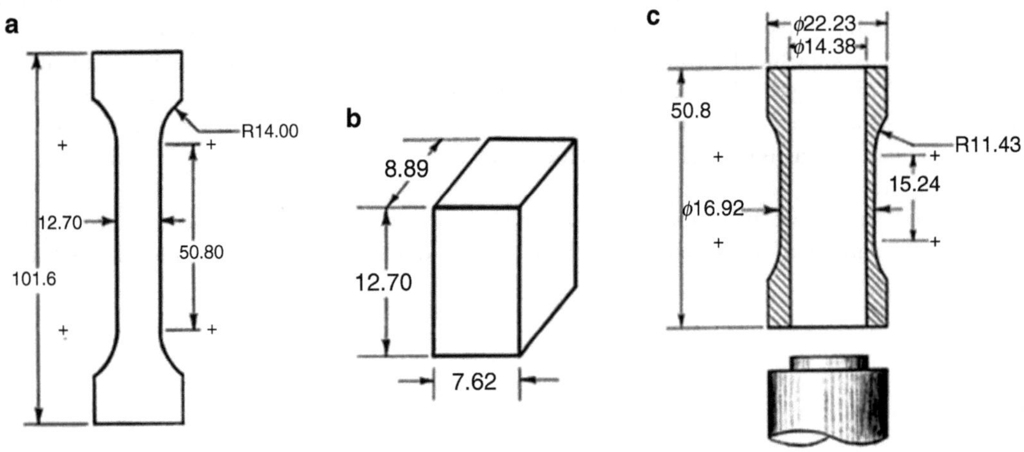

Fig. 2.1 Specimen geometries and dimensions. (**a**) Tensile specimen, (**b**) compressive specimen, (**c**) shear and combined stress specimen

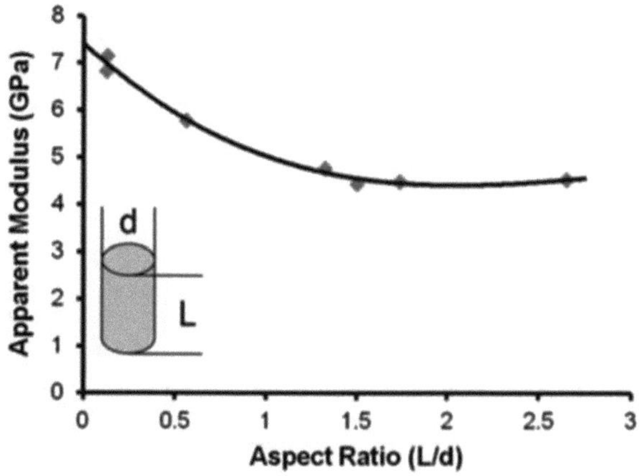

Fig. 2.2 Effect of aspect ratio on compressive response

$$C_{11} = \frac{E(1-\nu)}{(1+\nu)(1-2\nu)}$$
$$G = \frac{E}{2(1+\nu)} \tag{2.1}$$

Experiments were conducted at various strain rates ranging from 10^{-5} to 1,500 s^{-1}. Lower rate experiments at less than 1 s^{-1} were conducted in a servo-hydraulic testing machine. Higher rate testing was conducted in a Split Hopkinson (Kolsky) Pressure Bar (SHPB) system. Glass/epoxy composite (G-10) bars were used to minimize the impedance mismatch and reduce the noise-to-signal ratio.

A wave propagation analysis similar to that described by Daniel et al. [18, 19] was conducted to establish the limits of validity of dynamic testing in the SHPB system. The stresses at the ends of the specimen are not equal initially, but approach each other as the wave pulse is reflected back and forth within the specimen. The ratio of stresses at the two specimen ends was calculated and plotted versus number of wave transits through the specimen. The analysis shows that, for the selected configuration, the difference between stresses at the two ends is less than 10 % after four wave transits (one wave transit equals the time needed for the wave to propagate over one specimen length). This occurs at a time corresponding to less than 10 % of the pulse duration. Once an acceptable stress equilibrium state is reached, the stress strain curve for the material (epoxy) is obtained from the classical Hopkinson bar equations for stress, strain rate and strain in the specimen:

$$\sigma_s = \frac{EA_0}{2A_s}(\varepsilon_i + \varepsilon_r + \varepsilon_t) = \frac{EA_0}{A_s}\varepsilon_t$$
$$\dot{\varepsilon}_s = \frac{C_0}{L_s}(\varepsilon_i - \varepsilon_r - \varepsilon_t) = -\frac{2C_0}{L_s}\varepsilon_r \tag{2.2}$$
$$\varepsilon_s = \frac{C_0}{L_s}\int_0^t(\varepsilon_i - \varepsilon_r - \varepsilon_t)dt = -\frac{2C_0}{L_s}\int_0^t \varepsilon_r dt$$

where

$\varepsilon_i, \varepsilon_r, \varepsilon_t$ = incident, reflected and transmitted strain pulses in input and output bars
A_0, A_s = cross sectional areas of pressure bar and specimen
E = pressure bar modulus
L_s = specimen length
C_0 = wavespeed in pressure bar

Stress–strain curves were obtained over a wide range of strain rates. Figure 2.3 shows stress–strain curves under tension and pure shear loading at two strain rates. Figure 2.4 shows stress–strain curves under three states of biaxial compression and

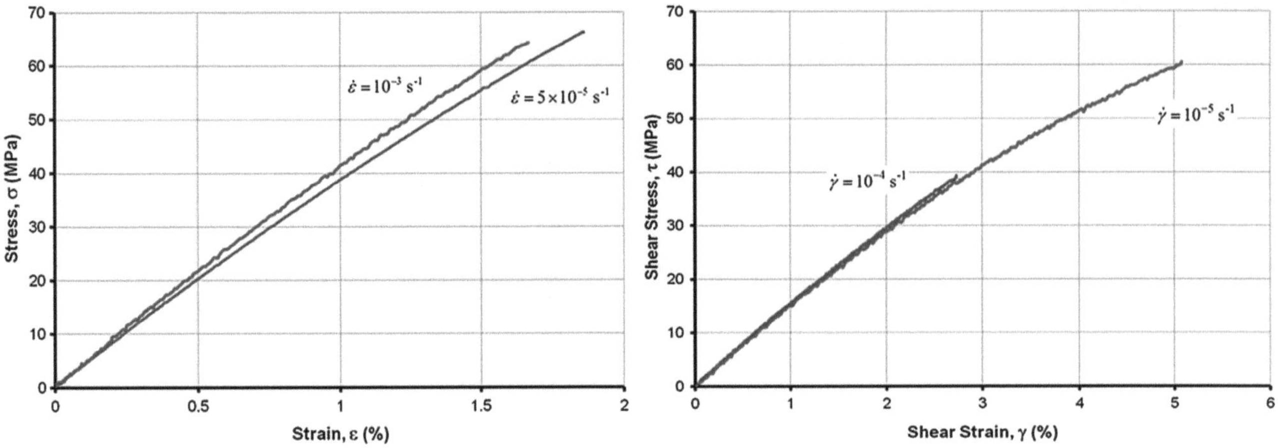

Fig. 2.3 Tension and shear stress–strain curves at two strain rates

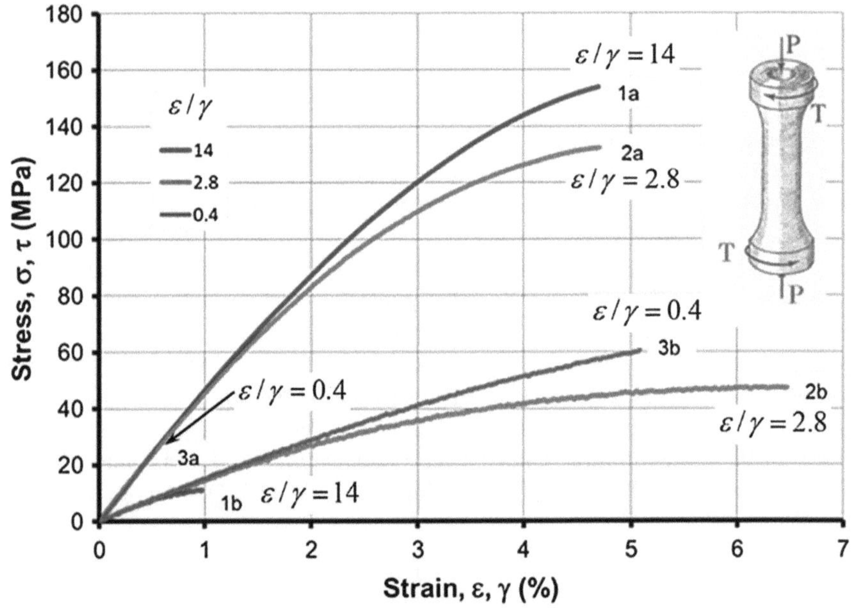

Fig. 2.4 Stress-stain curves under biaxial compression (**a**) and shear (**b**) at three stress ratios 1–3

shear and Fig. 2.5 shows similar curves under biaxial compression and shear at two strain rates. The compressive stress–strain curves shown in Fig. 2.6 have the same overall shape and share many key features. Each curve has a linear elastic region up to the yield point. This is followed by a plastic region up to a critical point of material instability, the "critical stress." Next, the resin reaches a minimum at the plateau point (local minimum) of the stress–strain curve and then it begins to harden. All curves share the same initial modulus. They only differ in the nonlinear behavior that shows pronounced rate dependence. The key features on the curves expand radially from the origin with increasing strain rate. This is highlighted in Fig. 2.6 where it can be seen that the yield point, the critical point, and the plateau point all fall along these radial lines. These characteristic properties vary linearly with the logarithm of strain rate as shown in Fig. 2.7. The resulting rate dependence can be expressed as

$$P(\dot{\varepsilon}) = P(\dot{\varepsilon}_0)\left[m\log_{10}\frac{\dot{\varepsilon}}{\dot{\varepsilon}_0} + 1\right] \tag{2.3}$$

where $P(\dot{\varepsilon})$ is the rate dependent property, $P(\dot{\varepsilon}_0)$ is that property at the reference strain rate, and m is the slope of the linear logarithmic relation. A similar relationship has been observed for the matrix dominated behavior of composites and for polymeric foams [3, 19]. In the present case, a reference strain rate of $10^{-4}\ s^{-1}$ was used and m was found to be 0.096.

Fig. 2.5 Stress–strain curves under biaxial compression and shear at two strain rates

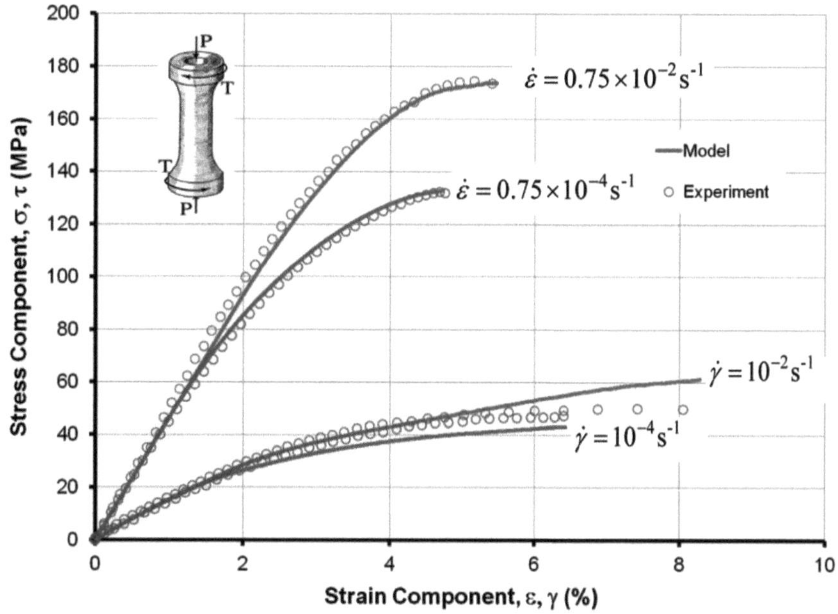

Fig. 2.6 Compressive stress–strain curves at various rates showing radial alignment of characteristic features

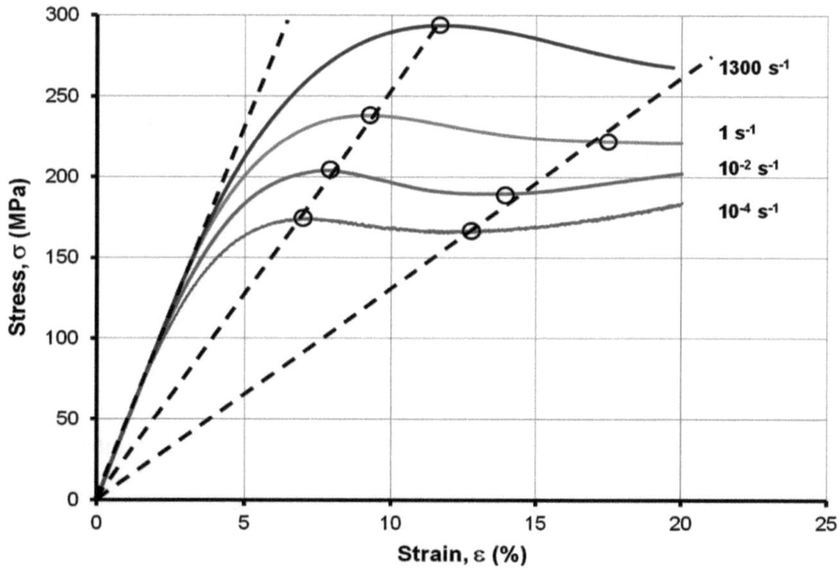

Fig. 2.7 Variation of normalized critical and yield stresses with strain rate

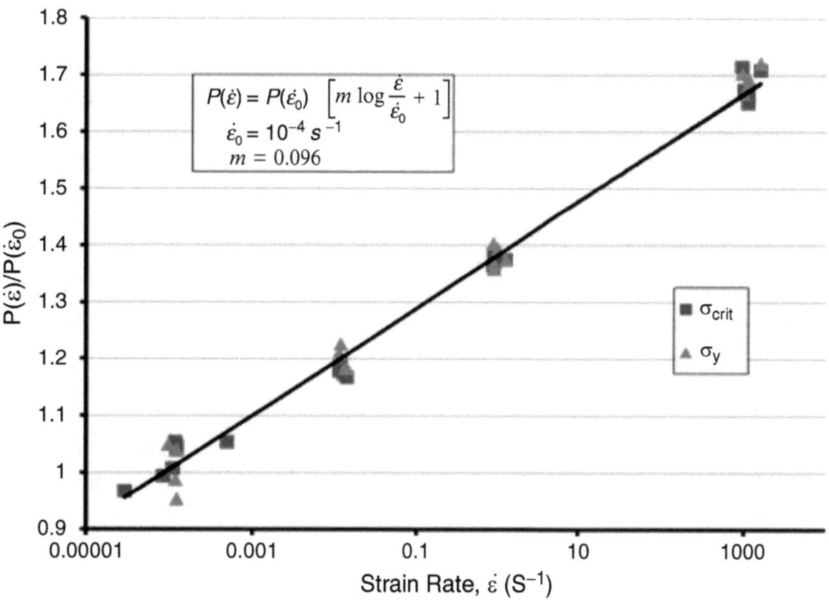

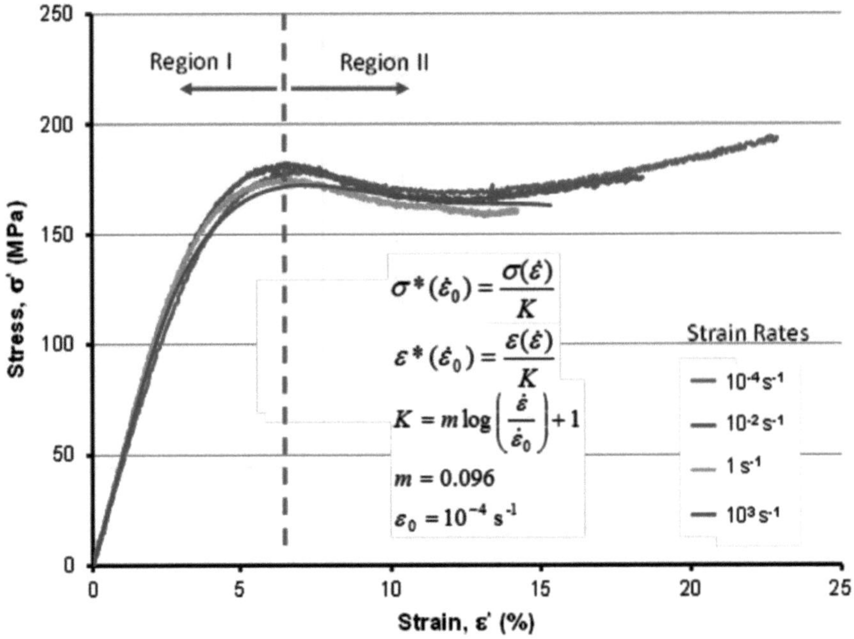

Fig. 2.8 Master compressive stress–strain curve at reference strain rate

Using the above relation, the stress and strain at any strain rate can be transformed into equivalent values at the reference strain rate by the following transformation relations

$$\varepsilon * (\dot{\varepsilon}_0) = \frac{\varepsilon(\dot{\varepsilon})}{K}$$

$$\sigma * (\dot{\varepsilon}_0) = \frac{\sigma(\dot{\varepsilon})}{K} \tag{2.4}$$

where $K = m\log\left(\frac{\dot{\varepsilon}}{\dot{\varepsilon}_0}\right) + 1$

The stress–strain curves of Fig. 2.5 can thus be transformed into one master curve at the reference strain rate as shown in Fig. 2.8.

2.3 Yield Criterion

The yield strains under uniaxial quasi-static tension, compression and shear are identified in the corresponding stress–strain curves of Figs. 2.3 and 2.6. The states of strain for these cases are represented by the Mohr circles of Fig. 2.9. Under uniaxial tension, the maximum normal strain is the yield strain of approximately $\varepsilon_t^y = 0.006$ and the maximum compressive strain is $-\nu\varepsilon_t^y = -0.35 \times 0.0066 = -0.002$. Under uniaxial compression (from Fig. 2.6) the maximum compressive strain at yield is $\varepsilon_c^y = -0.0165$ and the corresponding transverse tensile strain is $-\nu\varepsilon_c^y = 0.006 = \varepsilon_t^y$. Similarly, from the shear stress–strain curve of Fig. 2.3, it is seen that the shear strain at yield is $\gamma^y/2 \approx 0.006 = \varepsilon_t^y$. Thus, yield strains and stresses under uniaxial tension, compression and shear are related.

These experimental observations lead to a hypothesis that yielding behavior of a quasi-brittle polymer could be described in terms of only one parameter, the tensile yield strain or the tensile yield stress

$$\sigma_t^y = \varepsilon_t^y E = F \tag{2.5}$$

This represents a further simplification to the two property yielding theory presented by Christensen [14, 15]. The pattern of yielding governed by the uniaxial tensile yield strain (stress) holds true even under biaxial loading conditions. Results from compression/shear biaxial stress–strain curves shown in Fig. 2.4 are shown in Fig. 2.10. The yield criterion

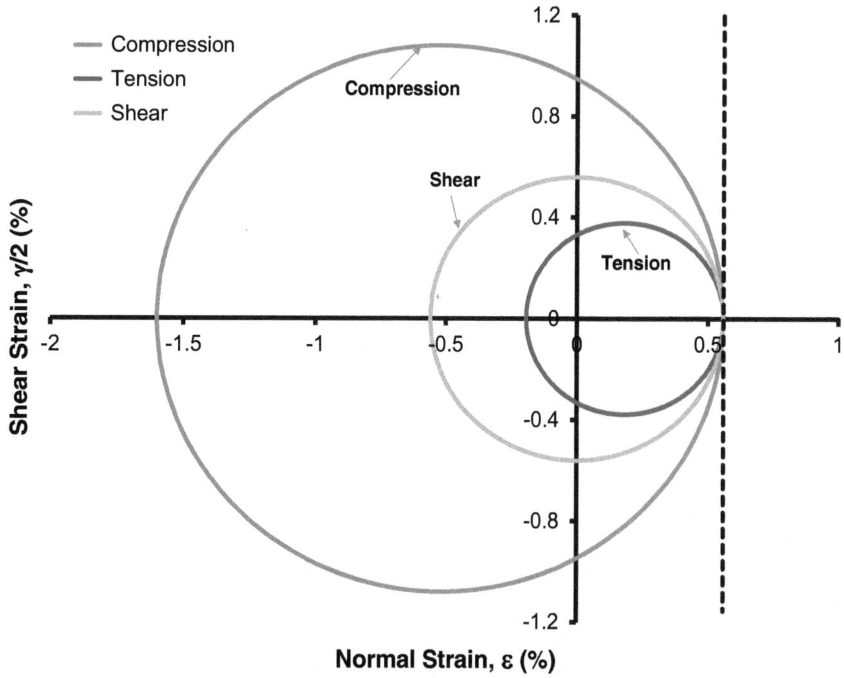

Fig. 2.9 Mohr circles for strain transformation under uniaxial tension, compression and shear

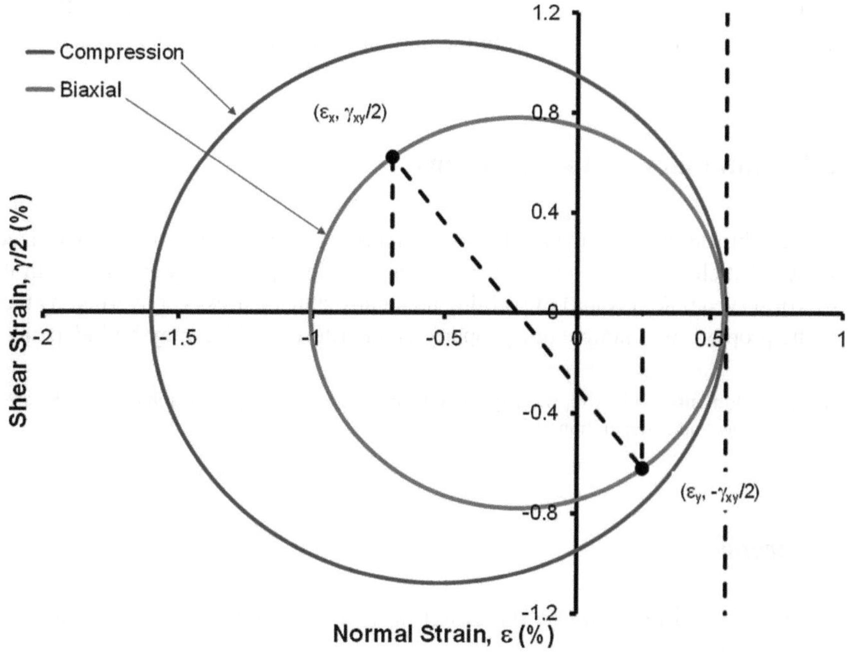

Fig. 2.10 Mohr circle based on normal and shear yield strains recorded in a biaxial test

was expressed in terms of the normal and shear strain components for a general biaxial state of stress noted on the Mohr circle of Fig. 2.10 as

$$\varepsilon_t^y = \frac{\varepsilon_x + \varepsilon_y}{2} + \sqrt{\left(\frac{\varepsilon_x - \varepsilon_y}{2}\right)^2 + \frac{\gamma_{xy}^2}{4}}$$

or

$$2\varepsilon_t^y = (1-\nu)\varepsilon_y + \sqrt{(1+\nu)^2\varepsilon_y^2 + \gamma_{xy}^2} \tag{2.6}$$

or, in terms of stresses

$$\varepsilon_t^y E = F = \sigma_1 - \nu\sigma_2 = (1-\nu)\sigma_y + (1+\nu)\sqrt{\sigma_y^2 + 4\tau_{xy}^2}$$

or

$$\nu\left(\frac{\sigma_y}{F}\right)^2 + (1-\nu)\left(\frac{\sigma_y}{F}\right) + (1+\nu)^2\left(\frac{\tau_{xy}}{F}\right)^2 = 1 \tag{2.7}$$

In the case of a general biaxial state of stress, the yield criterion would be

$$\frac{1}{F^2}\left[\nu\left(\sigma_x^2 + \sigma_y^2\right) - \left(1+\nu^2\right)\sigma_x\sigma_y\right] + (1-\nu)\frac{\sigma_x + \sigma_y}{F} + (1+\nu)^2\left(\frac{\tau_{xy}}{F}\right)^2 = 1 \tag{2.8}$$

provided the maximum principal stress is tensile.

This criterion accounts for strain rate effects without any modification since the characteristic properties of the material such as yield stresses and strains vary linearly with the logarithm of strain rate as depicted in Fig. 2.7 and described by Eqs. (2.3) and (2.4). This is true if normal and shear strain rates in a biaxial test are not significantly different from each other. In that case, the criteria of Eqs. (2.7) and (2.8) become independent of strain rate. Figure 2.5, for example, shows stress–strain curves from biaxial compression/shear tests at two different strain rates. In these tests, the normal strain rates (0.75×10^{-4} and 0.75×10^{-2} s^{-1}) were selected to be 75 % of the corresponding shear strain rates (10^{-4} and 10^{-2} s^{-1}).

The normalized ratios of stresses and uniaxial tension yield stress, F, are not changed with strain rate. The small difference in shear strain rate from the normal strain rate has a very small effect on the yield criterion of Eq. (2.8).

2.4 Summary and Conclusions

A quasi-brittle polymer material used as matrix in composites was characterized over a wide range of strain rates. It was shown that characteristic properties, such as the yield point, vary linearly with the logarithm of strain rate. Furthermore, it was observed that the onset of yielding under any state of stress is governed by the tensile yield stress or yield strain. This led to the proposal of a simple one property yield criterion which may be independent, or at least not sensitive to strain rate.

Acknowledgments This work was supported by the Office of Naval Research (ONR). We are grateful to Dr. Y. D. S. Rajapakse of ONR for his encouragement and cooperation.

References

1. Cho J-M, Fenner JS, Werner BT, Daniel IM (2010) A constitutive model for fiber reinforced polymer composites. J Comput Mater 44(26):3133–3150
2. Daniel IM, Cho J-M, Werner BT, Fenner JS (2011) Characterization and constitutive modeling of composite materials under static and dynamic loading. AIAA J 49(8):1658–1682
3. Daniel IM, Werner BT, Fenner JS (2011) Strain-rate-dependent failure criteria for composites. Compos Sci Technol 71(3):357–364
4. Arruda EM, Boyce MC, Jayachandran R (1995) Effects of strain-rate, temperature and thermomechanical coupling on the finite strain deformation of glassy-polymers. Mech Mater 19(2–3):193–212
5. Liang YM, Liechti KM (1996) On the large deformation and localization behavior of an epoxy resin under multiaxial stress states. Int J Solids Struct 33(10):1479–1500
6. Chen W, Zhou B (1998) Constitutive behavior of Epon 828/T-403 at various strain rates. Mech Time-Dependent Mater 2(2):103–111
7. Buckley CP, Harding J, Hou JP, Ruiz C, Trojanowski A (2001) Deformation of thermosetting resins at impact rates of strain. Part I: experimental study. J Mech Phys Solids 49(7):1517–1538
8. Gilat A, Goldberg RK, Roberts GD (2003) High strain rate response of epoxy in tensile and shear loading. J Phys IV 110:123–127
9. Gilat A, Goldberg RK, Roberts GD (2007) Strain rate sensitivity of epoxy resin in tensile and shear loading. J Aerosp Eng 20(2):75–89
10. Littell JD, Ruggeri CR, Goldberg RK et al (2008) Measurement of epoxy resin tension, compression, and shear stress–strain curves over a wide range of strain rates using small test specimens. J Aerosp Eng 21(3):162–173

11. Naik NK, Gadipatri R, Thoram NM et al (2010) Shear properties of epoxy under high strain rate loading. Polym Eng Sci 50(4):780–788
12. Mayr AE, Cook WD, Edward GH (1998) Yielding behaviour in model epoxy thermosets – I. Effect of strain rate and composition. Polymer 39(16):3719–3724
13. Behzadi S, Jones FR (2005) Yielding behavior of model epoxy matrices for fiber reinforced composites: effect of strain rate and temperature. J Macromol Sci: Phys 44(6):993–1005
14. Christensen RM (2004) A two-property yield, failure (fracture) criterion for homogeneous, isotropic materials. ASME J Eng Mater Technol 126:45–52
15. Christensen RM (2006) A comparative evaluation of three isotropic, two property failure theories. ASME J Appl Mech 73:852–859
16. Christensen RM (2007) A comprehensive theory of yielding and failure for isotropic materials. ASME J Eng Mater Technol 129(2):173–181
17. Werner BT, Daniel IM (2012) Characterization and modeling of polymeric matrix under static and dynamic loading. In: Proceedings of SEM XII international congress & exposition on experimental & applied mechanics
18. Daniel IM, Rao S (2000) Dynamic mechanical properties and failure mechanisms of PVC foams. In: Dynamic failure in composite materials and structures, ASME mechanical engineering congress and exposition; AMD-243, pp 37–48
19. Daniel IM, Cho JM, Werner BT (2013) Characterization and modeling of strain-rate-dependent behavior of polymeric foams. Composites Part A 45:70–78

Chapter 3
Investigating Uncertainty in SHPB Modeling and Characterization of Soft Materials

Christopher Czech, Aaron J. Ward, Hangjie Liao, and Weinong W. Chen

Abstract The increasing use of soft materials in industrial, commercial and military applications has necessitated a more thorough understanding of their visco-hyperelastic, reactive, and other non-linear properties. Additionally, testing and design methods for components that employ these materials have required innovation. Large-scale computational modeling has become an effective tool to mitigate the increased cost that accompanies the added complexity in testing and design, but modeling error in the forms of inaccuracy and uncertainty must be appropriately accounted for to effectively reduce both design-stage and validation-stage testing costs.

In this work, several models were built to simulate the split Hopkison pressure bar (SHPB) compression of plasticized polyvinyl chloride (PVC), butyl rubber (BR) and vulcanized rubber (VR) samples across a range of medium- to high-strain rates. Using these analyses, hyper-viscoelastic constitutive models were fit to experimental data for a number of samples at each strain rate, and effective material properties were determined for each curve. The model calibration values were also used to generate statistics to compare the utility of different fitting methods for soft materials.

Keywords Hyperelasticity • Viscoelasticity • Optimization • FEM • SHPB • Model calibration

Nomenclature

A	Visco-hyperelastic energy calibration value
α	Ogden strain energy calibration value
E	Green strain tensor
$\dot{E}$	Green strain rate tensor
f	Weighting parameter in Blatz–Ko strain energy
F	Deformation gradient tensor
G_i	Prony-series calibration modulus
G^{el}	Elastic/tangent shear modulus
G^{ve}	Viscoelastic shear modulus
I	Invariant of the Cauchy–Green tensor
J	Jacobian
λ	Stretch
μ	Ogden strain energy calibration value

C. Czech (✉) • A.J. Ward
Corvid Technologies, Mooresville, NC 28117, USA
e-mail: christopher.d.czech@gmail.com

H. Liao • W.W. Chen
School of Aeronautics and Astronautics and School of Materials Engineering, Purdue University, West Lafayette, IN 47907, USA

H.J. Qi et al. (eds.), *Challenges in Mechanics of Time-Dependent Materials, Volume 2: Proceedings of the 2014 Annual Conference on Experimental and Applied Mechanics*, Conference Proceedings of the Society for Experimental Mechanics Series,
DOI 10.1007/978-3-319-06980-7_3,

ν Poisson's ratio
σ Cauchy stress
t, τ Time variable
T Prony-series relaxation time
ψ Strain energy function
ω Prony-series relaxation frequency

3.1 Introduction

The ability to understand and engineer more robust and complex systems using soft materials has required innovation and maturation of material models that are capable of predicting broader ranges of phenomena, including hyperelasticity, viscoelasticity, temperature dependence, and damage. One superposition of these phenomena, hyper-viscoelasticity, arises from a subset of problems in which polymers are subjected to strains at rates that are orders of magnitude larger than those of quasistatic (strain rates ~10^{-2} s^{-1}). Accordingly, material models have been developed that treat hyper-viscoelastic polymers as a material in a superposition of hyperelastic and viscoelastic states, with each effect attributing a portion to the Cauchy stress based on strain, strain rate, and the material loading history [1–3].

Experimental methods have analogously been devised to observe and quantify high-rate stress–strain phenomena, many of which utilize the Split Hopkinson Pressure Bar (SHPB) apparatus. (For a detailed description and history of the SHPB, see [4] and sources therein.) The modern SHPB, capable of delivering strain rates to materials in a range from 10 to ~10^8 s^{-1}, utilizes an open-loop feedback control system to measure the engineering stresses and strains on a sample. For soft materials, these measurements are complicated by difficulties in obtaining stress equilibrium during experimentation, a problem is derived from low wave speeds. Nonetheless, accurate methods have been devised to measure stress and strain for these materials.

Given analytical models and accurate measurement methods, a material analyst will calibrate the analytical model with the experimental data by determining a set of optimal model parameters for the analytical model that replicates the behavior of the material in one or a number of experiments. This requires the analyst to make a set of guiding design assumptions, some of which include applicability of analytical models, calibration parameter bounds, simulation methods, optimization methods, boundary conditions and which experimental data to calibrate the parameters to. Coupling these assumptions with a complex material model results in a large amount of systematic uncertainty in the modeling and design processes; often, these errors propagate to make model calibration infeasible or implementation of the model intractable for a large-scale finite element analysis (FEA).

In this work, a subset of the above causes of systematic uncertainty for soft materials modeling will be addressed to help assess and ease the model calibration process for implementation in a large, parallel-computation FEA code. In particular, the material characterization of three materials, vulcanized rubber (VR), butyl rubber (BR), and polyvinyl chloride (PVC) will be pursued using hyperelastic and viscoelastic material models. Medium- to high-rate (200–2,000 s^{-1}) experimental data from SHPB experiments will be used in the model calibration. Contrary to methods reported by other researchers [1, 3], no quasistatic or stress relaxation data will be used in the polymer characterization. Instead, the strain rate history provided from the SHPB experiments will be used, and the calibrations will be compared a similar calibration process that does not use the strain rate history, instead assuming a single strain rate throughout the deformation process.

The remainder of the paper is organized as follow. First, the SHPB experimental setup and method will be briefly reviewed, and typical input and output data for the SHPB measurements of soft materials will be described. As part of this, a summary of experimental data for the three calibrated materials will be provided. Next, the candidate material models will be described, followed by the calibration process employed. Then, numerical results will be presented, and the comparison between methods will be made. This is followed by research conclusions and on-going work.

3.2 Experimental Setup and Data

Given only legacy quasistatic data for the BR and PVC materials and no quasistatic data for the VR, medium- to high-strain rate (200–2,000 s^{-1}) data were collected for all three materials using a split Hopkinson pressure bar (SHPB) apparatus. The experimental configuration consists of three major components; the striker bar, incident bar, transmission bar. A sample is confined between the incident and transmitted bars. A basic schematic of a typical SHPB setup is shown in Fig. 3.1. Strain gauges are placed on the incident and transmission bars to measure the generated pulse and the reflected/transmitted pulses.

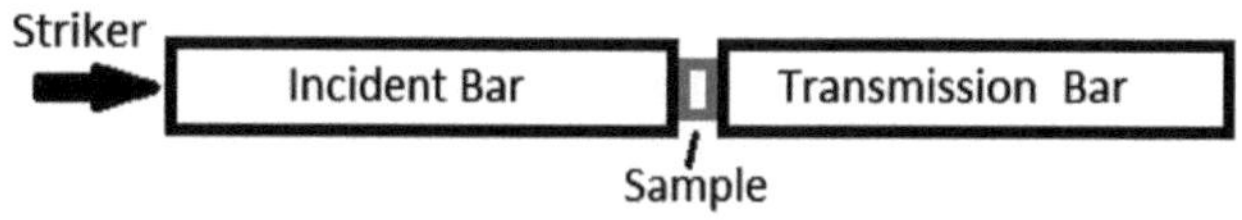

Fig. 3.1 Typical split Hopkinson pressure bar experimental setup

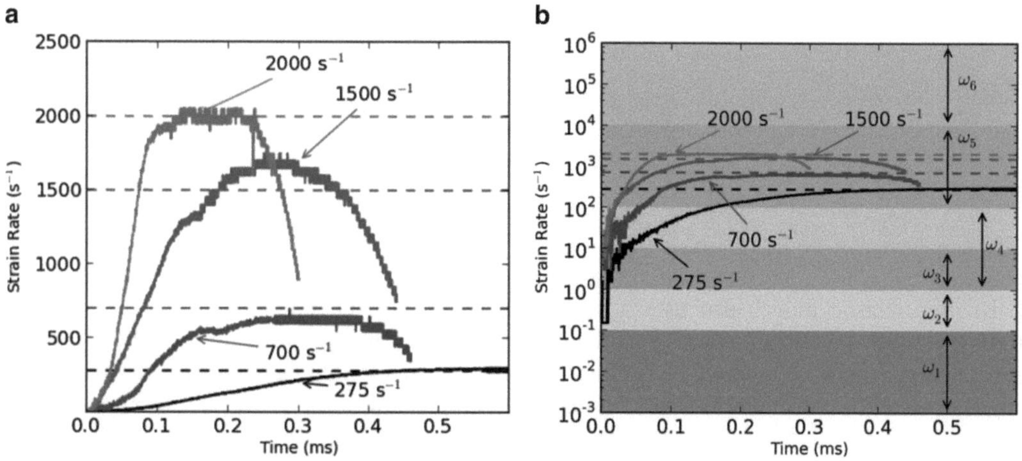

Fig. 3.2 *Solid lines* represent actual strain-rate pulses measured from SHPB experiments. *Dashed lines* represent approximate, constant strain rate histories. A linear plot (*left*, **a**) and a log-linear plot (*right*, **b**) are provided for the same rates. The *boxes* in (**b**) depict ranges of relaxation frequencies allocated for the Prony-series expansion of the viscoelastic behavior, where frequency ω_i is the inverse of the relaxation time T_i

Table 3.1 Summary of available stress–strain data at approximate strain rate histories

Butyl rubber (BR)		Polyvinyl chloride (PVC)		Vulcanized rubber (VR)	
Approx. strain rate (s^{-1})	Num. histories	Approx. strain rate (s^{-1})	Num. histories	Approx. strain rate (s^{-1})	Num. histories
200	4	275	3	250	4
700	5	700	5	700	3
1,400	5	1,500	5	1,400	5
2,000	4	2,000	5	1,800	5

To characterize materials at different rates, a striker bar with different initial velocities impacts the incident bar, propagating an elastic stress wave through the length of the bar. The elastic stress wave generates a reflected pulse that travels back into the incident bar; additionally, the stress wave passes through the sample, transmitting a pulse, measured in the transmission bar. When the sample is in stress equilibrium, the material response can be measured from the stress wave and reflected/transmitted pulses. A more in depth discussion of SHPB technique, assumptions and applications can be found in Chen and Song [4].

The low sound speeds observed in soft materials necessitates the use of a specialized pulse shaping technique for SHPB testing. These pulses are shaped in a way that maintains stress equilibrium in the sample by sculpting the incident pulse from a typical square wave to one which provides a gradual transition from a low strain rate to the desired upper limit. Experimental pulses created using this technique are shown in Fig. 3.2. Of particular note is the relatively large amount of experimental time spent loading a sample at lower strain rates for soft materials.

Table 3.1 summarizes the experimental data generated using the SPHB apparatus at different approximate strain rates.

3.3 Material Models

To generate an analytical stress for the material response given an input strain, the deformation and stress physics was separated two parts: hyperelastic and viscoelastic. This approach was taken to allow for a general treatment that could be applied to other models. Currently, the materials are assumed to be incompressible, deforming in uniaxial compression. These assumptions provide the ability to solve for the deformation gradient, strain invariants, and strain rate tensor can be directly [1, 3, 12, 13].

Hyperelastic material models are typically described by strain energy density functions, ψ, which is often a function of the principal stretches, λ_i, or strain invariants, I_i [5]. Given previous research, this study is limited to two strain energy density functions, Ogden [6] and Blatz–Ko [7, 8], but other forms could be examined [9].

The strain energy density form for an Odgen material [Eq. (3.1)] is written in terms of the principal stretch components. The Ogden model is particularly useful for its relations to other existing models, as it reduces to the Mooney–Rivlin when $m = 2$ and $\alpha_{1,2} = 2, -2$ and to the Neo–Hookean model when $m = 1$ and $\alpha_1 = 2$. This aspect gives it a broad range of applicability for little implementation cost. The Cauchy stress associated with the Ogden model is given in Eq. (3.2).

$$\psi_{Ogden} = \sum_{m=1}^{n} \left\{ \frac{\mu_m}{\alpha_m} \left[J^{-\alpha_m/3} \left(\lambda_1^{\alpha_m} + \lambda_2^{\alpha_m} + \lambda_3^{\alpha_m} - 3 \right) \right] \right\} \tag{3.1}$$

$$\sigma_{ii}^{Ogden} = \frac{1}{J} \left[\lambda_i \frac{\partial}{\partial \lambda_i} \psi_{Ogden} \right] \tag{3.2}$$

The Blatz–Ko hyperelastic model also uses a strain energy density function [Eq. (3.3)]. The Blatz–Ko model is formulated from the strain invariants which are calculated from the Cauchy–Green tensor. The Blatz–Ko model only requires three parameters; the shear modulus, G^{el}, Poisson's ratio, ν, and a weighting parameter. f. The weighting parameter as described by Blatz and Ko transitions the material response from a hard rubber to a foam response.

$$\psi_{Blatz-Ko} = \frac{G^{el} f}{2} \left[I_1 - 3 + \frac{1-2\nu}{\nu} \left(J \left(\frac{-2\nu}{1-2\nu} \right) - 1 \right) \right] + \frac{G^{el}(f-1)}{2} \left[\frac{I_2}{J^2} - 3 + \frac{1-2\nu}{\nu} \left(J \left(\frac{2\nu}{1-2\nu} \right) - 1 \right) \right] \tag{3.3}$$

$$\sigma_{ij}^{Blatz-Ko} = \frac{1}{J} F \left[2 \frac{\partial}{\partial C} \psi_{Blatz-Ko} \right] F^T \tag{3.4}$$

The viscoelastic stress modeled by Li and Lau [Eq. (3.5)] was selected as it has shown to produce reasonable results for polymers under high-rate loadings [1, 3]:

$$\sigma_{visco} = \frac{1}{J} F \left\{ \int_0^t [A_1 + A_2(I_2 - 3)] \left[\sum_{i=1}^{6} G_i e^{-(t-\tau)/T_i} \right] \dot{E}(\tau) d\tau \right\} F^T \tag{3.5}$$

The convolution integral was solved either through direct numerical integration or a state variables approach [2, 10, 11]. Both methods were evaluated, but the state variable approach, with better numerical efficiency with similar numerical error to that of the direct integration method, was used to produce all subsequent results. Under the conditions $A_1 = 1$ and $A_2 = 0$, the model reduces to the standard Prony series viscoelastic model.

Given a hyperelastic model and the viscoelastic correction model, the total Cauchy stress in the material is given as the sum of the hyperelastic and viscoelastic stresses:

$$\sigma = \sigma_{hyper} + \sigma_{visco} \tag{3.6}$$

3.4 Model Calibration

Given a lack of quasistatic or stress relaxation data and the additive nature of the hyperelastic and viscoelastic stresses, there is no way to distinguish the viscoelastic effects from the hyperelastic effects for a single stress–strain history. For example, a high-rate history measure at 2,000 s^{-1} could be modeled as entirely hyperelastic, with viscoelastic parameters yielding a null viscoelastic stress. While the calibration would be numerically accurate, the physical inaccuracy would be easily demonstrated by using the model to predict the stress–strain response at a different strain rate. However, by calibrating the total hyper-viscoelastic model over a series of stress–strain histories that represent a range of responses desired for model predictability, a better representation of the physics results may be obtained, potentially with compromise in model accuracy to experimental results for individual histories.

While this tradeoff seems immediately worthwhile, it raises a question regarding the individual strain rate histories applied to the model simulations for calibration. By assuming a constant strain rate for each curve during the calibration, computational efficiency is maintained with the sacrifice of model calibration error. Alternatively, by using the strain rate histories obtained from the experiment in the material simulations, the model is much less efficient during runtime due to incurred numerical integration costs, but the calibration error may be greatly reduced. This is complicated by the previously discussed specialized ramp loading used to test soft materials at higher rates.

To address the former issue, a least-squares fitting algorithm was applied to a model calibration of all available stress–strain histories for the three materials, where the least-squares response, R, was determined according to

$$R = \sum_{k=1}^{N}\sum_{m=1}^{n}\left(\frac{\left(\sigma_m^A(\mathbf{p}) - \sigma_m^E\right)^2}{n}\right)^{1/2} \tag{3.7}$$

where N is the number of stress–strain histories available for calibration, n is the number of individual points a provided stress–strain history for curve k, ${\sigma_m}^A$ is the Cauchy stress at engineering strain ${E_m}^E$ given by the model with parameters $\mathbf{p}$, and ${\sigma_m}^E$ is the experimental Cauchy stress at the same strain. The latter was investigated by calibrating models using each type of strain rate history: one constant history for each curve at the supposed rate, and one history for each curve at the observed rates during testing.

Each of the materials was calibrated to a hyperelastic model simultaneously with the viscoelastic model in an all-at-once, or AAO, optimization procedure. The BR was modeled using a three-term Ogden hyperelastic model; the VR was modeled using a Neo–Hookean hyperelastic model; and the PVC was calibrated using both the Blatz–Ko model and a three-term Ogden model in independent procedures. The viscoelastic relaxation times were divided into semi-independent ranges that provide for a spectrum of relaxation behaviors across varying timescales.

A sequential quadratic program (SQP) was used to solve the optimization problem involved in solving the least-squares calibration problem. To account for the possibility of premature convergence or non-convergence of the local SQP optimization problem, a stratified, Latin hypercube design of experiments was used to select initial design points from the feasible design space for each problem, and then a series of SQP problems was solved, one from each initial point. The solutions achieved at convergence of the SQP problems were used to conduct the study of the validity of the model calibration methods.

3.5 Results

The best resultant stress–strain curves from the model calibrations using both constant strain rate histories and observed strain rate histories are given in Figs. 3.3, 3.4, 3.5, and 3.6. The Ogden rubber calibrations to the BR data are given in Fig. 3.3, the Blatz–Ko rubber calibrations to the PVC are given in Fig. 3.4, the Ogden rubber calibrations to the PVC data are given in Fig. 3.5, and the Neo–Hookean model calibrations to the VR data are given in Fig. 3.6. In each case, the best calibration using a constant strain rate history is shown on the left, and the best calibration using the experimental strain rate history is shown on the right. Ogden rubber model calibration results to the butyl rubber data. The left plot depicts the best calibration using constant rate histories, and the right depicts the best calibration using the experimental rate histories.

In each plot, the experimental stress–strain data for all available tests are shown using solid lines, a single hyperelastic stress–strain curve is depicted with a solid-o line, and the total stress as given by Eq. (3.7) as a function of strain in shown with solid-square lines. There is one total-stress line provided for each experimental data set.

By comparing the plots from within any given material calibration scheme, it is clear that using the experimental strain rate history to calibrate the given model provides a better model calibration than using a constant strain rate history. This is to be expected, as a typical SHPB compression test for soft materials spends a significant portion of the experiment (up to 30 %) increasing to the desired strain rate. In addition to the better quality of calibration to the hyper-viscoelastic models, there is also evidence that the hyperelastic model can be more accurate calibrated using a time-dependent strain rate during calibration. For example, the Blatz–Ko hyperelastic stress values in Fig. 3.4b are much lower in magnitude than those in Fig. 3.4a. This trend is true of all of the best calibrations obtained for the four materials. Another consistent trend in the better distribution of stress–strain values across a spectrum of strain rates when the strain-rate histories were used for calibration. Of particular note is the failure of the constant rate methods to obtain model calibration values that accurately represent the experimental results.

A summary of numerical values for the model calibrations are given in Table 3.2. Response values given at algorithm termination for successfully converged attempts and summary statistics are provided, as well as values for two material parameters inferred from calibration parameters. The initial tangent shear moduli for Ogden and Neo–Hookean model calibrations are calculated according to [14]

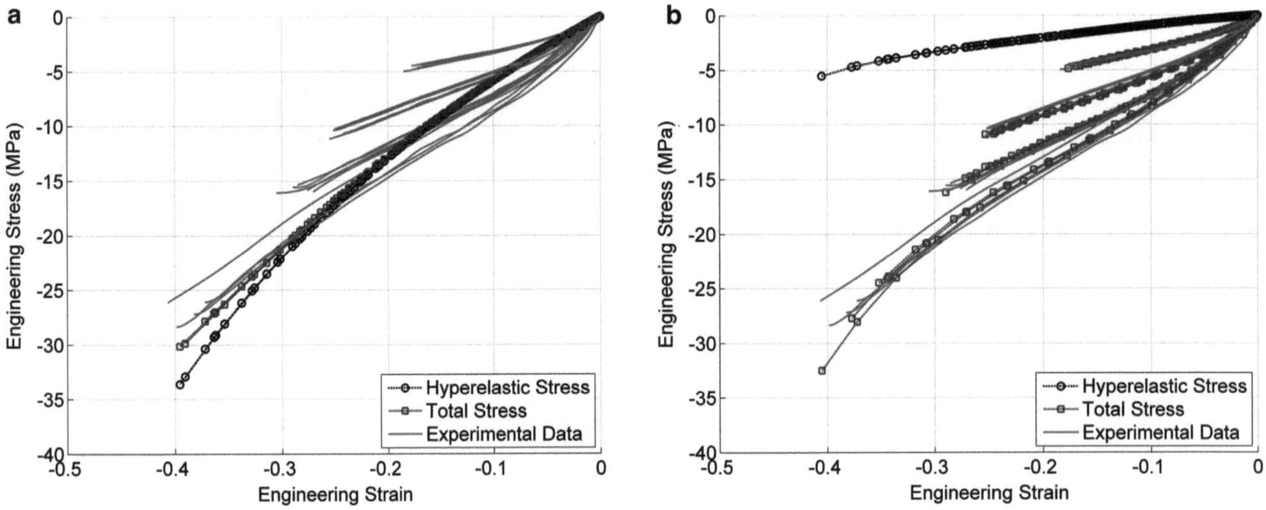

Fig. 3.3 Ogden rubber model calibration results to the butyl rubber data. The *left plot* (**a**) depicts the best calibration using constant rate histories, and the *right* (**b**) depicts the best calibration using the experimental rate histories

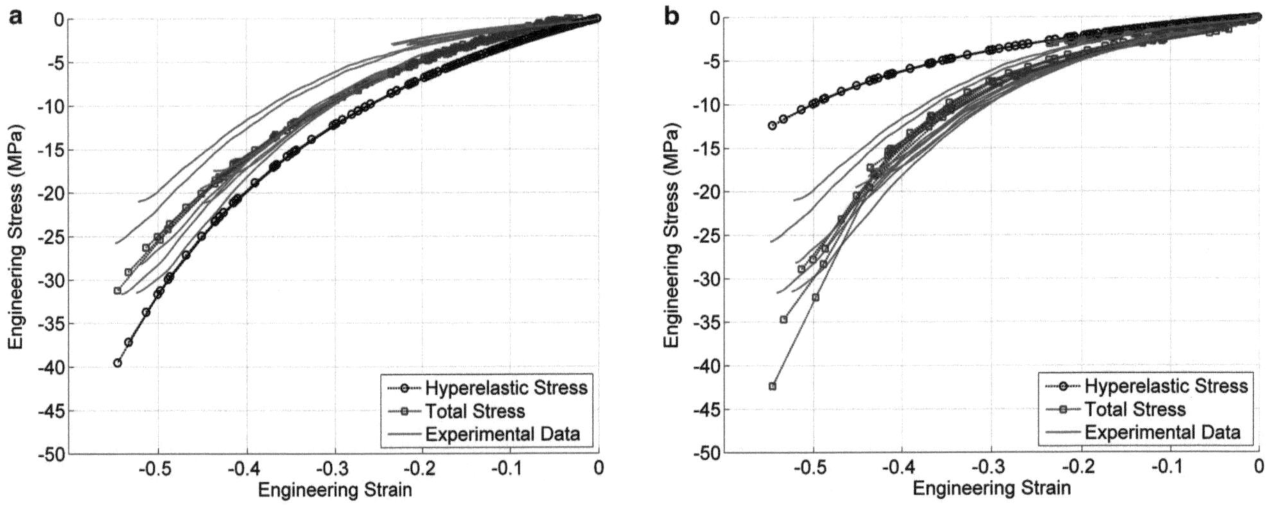

Fig. 3.4 Blatz–Ko rubber model calibration results to the PVC data. The *left plot* (**a**) depicts the best calibration using constant rate histories, and the *right* (**b**) depicts the best calibration using the experimental rate histories

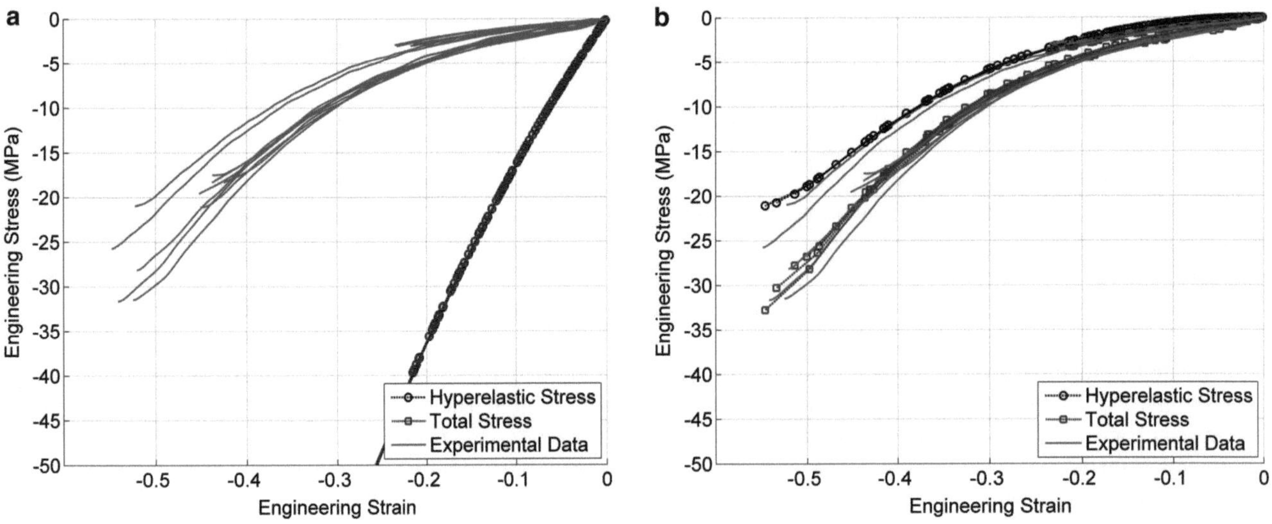

Fig. 3.5 Ogden rubber model calibration results to the PVC data. The *left plot* (**a**) depicts the best calibration using constant rate histories, and the *right* (**b**) depicts the best calibration using the experimental rate histories

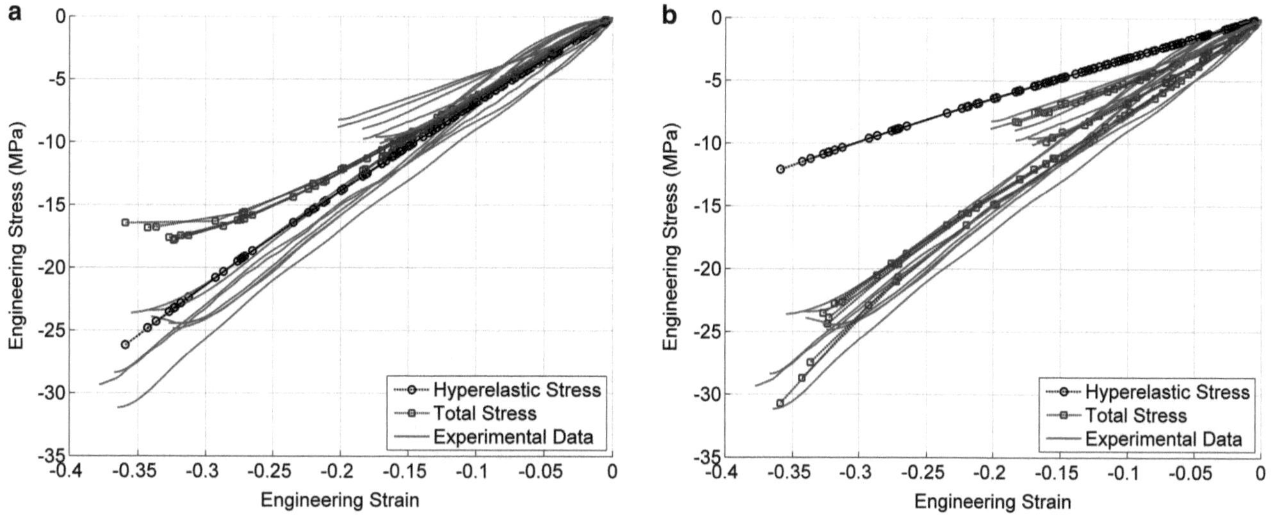

Fig. 3.6 Neo–Hookean model calibration results to the vulcanized rubber data. The *left plot* (**a**) depicts the best calibration using constant rate histories, and the *right* (**b**) depicts the best calibration using the experimental rate histories

Table 3.2 Summary of numerical results for model calibrations

Material/model	BR/Odgen		PVC/Blatz–Ko		PVC/Ogden		VR/Neo–Hookean	
SR method	Constant	Observed	Constant	Observed	Constant	Observed	Constant	Observed
Response value data at algorithm termination, R								
Num. successful runs	10	14	14	11	10	11	10	10
Minimum (MPa)	1056.800	245.719	767.073	596.490	573.498	384.273	864.151	561.500
Mean (MPa)	1437.390	268.915	842.320	658.379	744.618	487.929	966.674	626.598
99 % conf. int. (MPa)	252.679	22.899	27.722	87.688	160.238	66.873	7.334	91.588
Initial tangent shear modulus of hyperelastic curve, G^{el}								
Best calibration (MPa)	10.506	1.375	9.018	2.8242	1.171	2.68×10^{-3}	22.746	10.506
Minimum (MPa)	0.475	0.677	6.448	0.274	1.171	0.000	23.162	0.420
Mean (MPa)	4.722	1.264	7.864	2..651	1.760	0.578	24.105	9.368
99 % conf. int. (MPa)	4.415	0.167	0.992	1.047	1.539	0.522	0.343	5.061
Sum of viscoelastic shear moduli, G_0^{ve}								
Best calibration (MPa)	64.760	22.094	19.1816	40.000	9.1502	33.900	0.060	9.402
Minimum (MPa)	0.000	13.133	0.000	3.785	0.000	2.697	0.000	6.324
Maximum (MPa)	64.760	37.057	112.169	168.781	37.273	33.900	10.000	33.247
Mean (MPa)	15.766	20.725	17.383	53.203	12.949	17.102	1.444	13.533
99 % conf. int. (MPa)	25.393	4.854	27.512	40.161	12.209	8.744	3.148	7.562

$$G^{el} = \frac{1}{2}\sum_{m=1}^{n} \mu_m \, \alpha_m \tag{3.8}$$

For Blatz–Ko calibrations, G_{el} is a model parameter. The other material parameter calculated is the sum of the viscoelastic shear moduli. Given a standard Prony series expansion of the viscoelastic modulus, [2]

$$G(t) = G^{el} + G^{ve}(t) = G^{el} + \sum_{i=1}^{6} G_i e^{-t/\tau_i} \tag{3.9}$$

Because the exponential in the sum in Eq. (3.9) of the Prony series is bounded in the interval [0, 1], the maximum effect the viscoelastic shear modulus have corresponds to time $t = 0$, or

$$G^{ve}(0) := G_0^{ve} = \sum_{i=1}^{6} G_i \tag{3.10}$$

For each value, a 99 % confidence interval calculated using a student's-t distribution method was calculated, assuming the calibrations followed a small-number statistical normal distribution.

The results in Table 3.2 confirm much of what is evident from the plots in Figs. 3.3, 3.4, 3.5, and 3.6. In each case, the best response value is achieved by using the experimental strain rate history, and the mean values of the responses using this method are much lower when compared to the constant rate method. The 99 % confidence intervals calculated for each response data set do not intersect, providing statistical evidence that the observed input strain rate method provides more accurate calibrations to medium-and high-rate experimental compression testing data for these models. For the best available calibrations, the initial tangent shear modulus is generally much lower when using the non-constant method, while the initial viscoelastic shear modulus does not appear to follow any trend. However, for all materials, the minimum achieved initial viscoelastic shear modulus when using the constant rate method is 0 MPa, indicating a potentially faulty calibration method. For both the initial tangent and viscoelastic shear moduli, the confidence intervals overlap with varying widths, indicating the complexity of the calibration process and the large amount of uncertainty that is involved.

One important point to consider is that no quasistatic or stress–relaxation data were used to fit the models to the compressions tests. The quality of calibrations without these data when using for these cases using the time-dependent strain rates observed during experimentation suggests quasistatic and stress–relaxation data are not necessary to accurately populate the hyper-viscoelastic stress–strain model.

To further investigate, initial tangent shear moduli, G^{el}, obtained from quastistatic testing (0.01–0.1 s^{-1}) stress–strain legacy testing obtained for the BR and PVC. For the butyl rubber, initial moduli with a mean of 1.075 MPa and a 95 % student's-t confidence interval of 0.637 MPa were reported. Likewise, for PVC, the initial moduli with a mean of 0.352 MPa and a 95 % student's-t confidence interval of 0.260 MPa were reported. Ogden model fits for vulcanized rubber are reported in the original work by Treloar [15]. Two initial tangent moduli, 2.959 and 4.017 MPa, were calculated using Eq. (3.8).

Comparison plots of these moduli for all three materials are provided in Fig. 3.7. From the box plots, it is apparent that initial tangent shear moduli obtained from the two separate testing methods are consistent. There is one exception: the Blatz–Ko models calibrations to the PVC data suggest that the Ogden model should be used. Previously presented numerical and visual results in Table 3.2 and Fig. 3.4 are consistent with this conclusion.

3.6 Conclusions and Future Work

In this work, a series of medium- to high-rate stress–strain compression data and corresponding strain-rate data were provided from a SHPB testing apparatus. Given these tests, constant rate and experimentally-observed, time-dependent strain rate sets of model calibrations were performed using Ogden rubber, Neo–Hookean and Blatz–Ko rubber analytical models. Justified by mean least-squares response values, 99 % confidence intervals and corresponding visual results, the time-dependent calibration method was determined to outperform the constant rate method for these material-model combinations.

Additionally, as none of these model calibrations involved using quasistatic or stress–relaxation data, the quality of the calibrations using the time-dependent strain rates demonstrates that quasistatic and stress–relaxation data are unnecessary to perform model calibration for these material-model combinations, considering applicability of the model is over these ranges. It is also suggestive that similar methods may apply for most soft material-hyperelastic model combinations, given the varieties of materials and models used in this study. This is believed to be due to the relatively large amount of time utilized for increasing strain rates from zero to target ranges in soft materials rate testing, necessary to maintain stress equilibrium. This is advantageous for two reasons: the lessened burden of mechanical testing using a separate apparatus helps to reduce costs of experimentation, and the subsequent costs of simulating these quasistatic experiments is also lessened.

One major path forward for this research currently being pursued is large-scale implementation of similar hyper-viscoelastic models in a large, parallel, high-fidelity finite element analysis and computational physics simulation program. Such models have been shown to be difficult to implement in the past due to complications caused by compressibility and complex materials dynamics presented in large-scale simulations. Given this, a calibration method has been built that directly simulates the SHPB experiments utilizing similar hyper-viscoelastic material models where the assumption of incompressibility has no longer been assumed.

Figure 3.7 depicts the pressure states of a typical testing sample before and during compressive loading. The sample is modeled to the scale of a typical test sample, as stress-wave propagation can affect the total stress state (and SHPB compression test) of the material sample. Similar fitting methods as those presented here are currently being undertaken in an effort to stabilize the material modeling methods and properly calibrate the models for candidate materials. The same SHPB

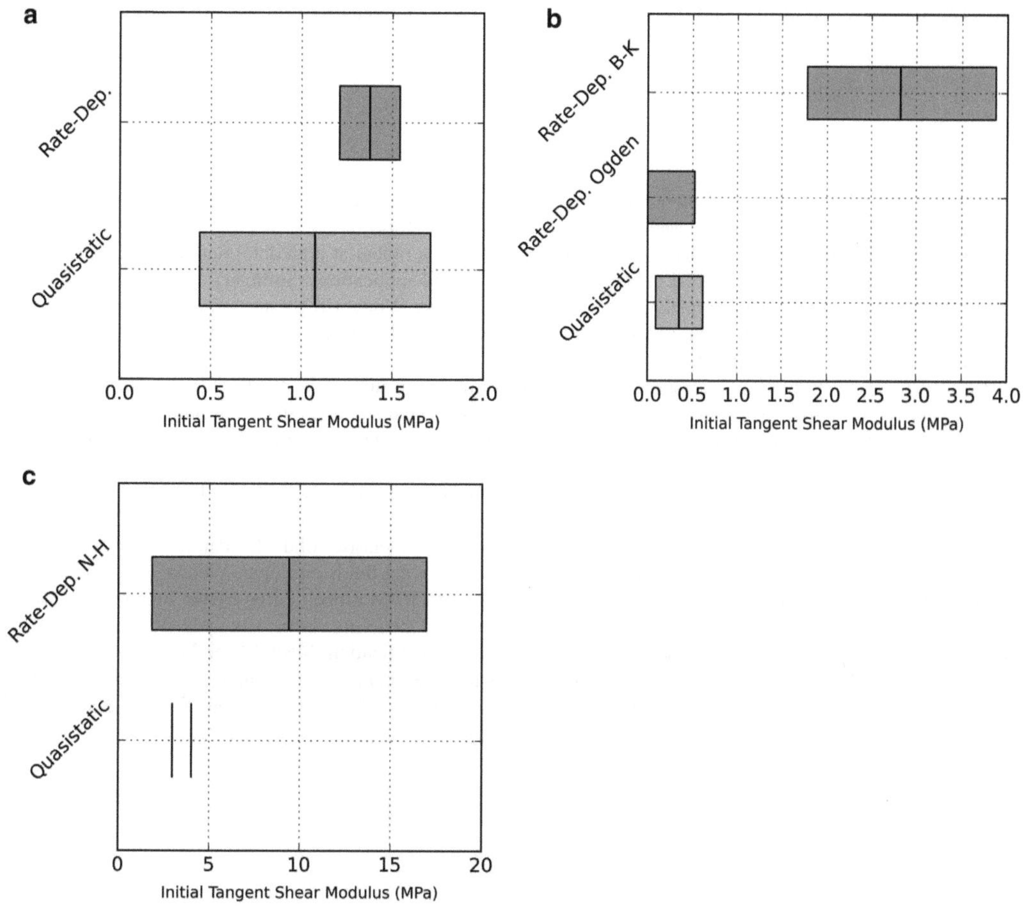

Fig. 3.7 Quasistatic data mean and 95 % confidence intervals of initial tangent shear moduli, and medium- to high-rate data mean and 99 % confidence intervals for three material-model sets: *top-left*, (**a**) butyl rubber, *top-right*, (**b**) PVC, and *bottom-left*, (**c**) vulcanized rubber. Only two values (displayed) were reported for the vulcanized rubber; statistics were not generated

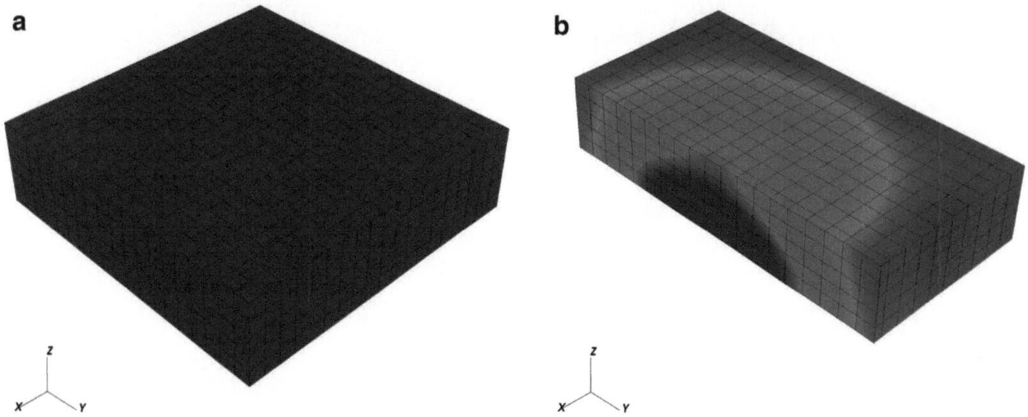

Fig. 3.8 *Left* (**a**) Depiction of a material sample before undergoing compression in an SHPB test simulation. *Right* (**b**) Sliced-view depiction of the pressure state of a the sample during compression in an SHPB test

data sets as those presented in this work are being used for model calibrations. Similar material models are being used, with the addition of a bulk-compressive term similar to those described in [5], as incompressibility assumptions are no longer valid for dynamic material modeling. Instead, the bulk compression term generates a small amount of compressibility within the material model (see Fig. 3.8).

Acknowledgment The authors would like to thank Anne Purtell at USMC Program Executive Office – Land Systems for funding this research.

References

1. Li C, Lua J (2009) A hyper-viscoelastic constitutive model for polyurea. Mater Lett 63(11):877–880
2. Marques SPC, Creus GJ (2012) Computational viscoelasticity. Springer, New York
3. Doman DA, Cronin DS, Salisbury CP (2006) Characterization of polyurethane rubber at high deformation rates. Exp Mech 46(3):367–376
4. Chen WW, Song B (2011) Split Hopkinson (Kolsky) bar – design, testing and applications. Springer
5. Crisfield MA (1991) Non-linear finite element analysis of solids and structures. Wiley, Chichester
6. Ogden RW (1972) Large deformation isotropic elasticity – on the correlation of theory and experiment for incompressible rubberlike solids. Proc R Soc Lond Math Phys Sci 326(1567):565–584
7. Blatz PJ, Ko WL (1962) Application of finite elastic theory to the deformation of rubbery materials. J Rheol 6(1):223
8. Ko WL (1963) Application of finite elastic theory to the behavior of rubber-like materials. PhD, California Institute of Technology
9. Martins PALS, Natal Jorge RM, Ferreira AJM (2006) A comparative study of several material models for prediction of hyperelastic properties: application to silicone-rubber and soft tissues. Strain 42:132–147
10. Onat ET (1970) Representation of inelastic mechanical behavior by means of state variables. In: Boley BA (ed) Thermoinelasticity. Springer, Vienna, pp 213–225
11. Onat ET (1968) The notion of state and its implications in thermodynamics of inelastic solids. In: Parkus PH, Sedov PLI (eds) Irreversible aspects of continuum mechanics and transfer of physical characteristics in moving fluids. Springer, Vienna, pp 292–314
12. Song B, Chen W (2003) One-dimensional dynamic compressive behavior of EPDM rubber. J Eng Mater Technol 125(3):294–301
13. Amin AFMS, Alam MS, Okui Y (2002) An improved hyperelasticity relation in modeling viscoelasticity response of natural and high damping rubbers in compression: experiments, parameter identification and numerical verification. Mech Mater 34(2):75–95
14. Ogden RW, Saccomandi G, Sgura I (2004) Fitting hyperelastic models to experimental data. Comput Mech 34(6):484–502
15. Treloar LRG (1944) Stress–strain data for vulcanised rubber under various types of deformation. Trans Faraday Soc 40:59–70

Chapter 4
Diffusion of Chemically Reacting Fluids through Nonlinear Elastic Solids and 1D Stabilized Solutions

Richard Hall, H. Gajendran, and A. Masud

Abstract This paper summarizes a 1D adaptation (Hall et al., Math Mech Solids, 2014) of the reactive fluid–solid mixture theory of Hall and Rajagopal (Math Mech Solids 17(2):131–164, 2012), which considers an anisotropic viscous fluid diffusing and chemically reacting with an anisotropic elastic solid. The present implementation introduces a stabilized mixed finite element method for advection–diffusion–reaction phenomena, which is applied to 1D isothermal problems involving Fickian diffusion, oxidation of PMR-15 polyimide resin, and slurry infiltration. The energy and entropy production relations are captured via a Lagrange multiplier that results from imposing the constraint of maximum rate of entropy production, reducing the primary PDEs to the balance equations of mass and linear momentum for the fluid and the solid, together with an equation for the Lagrange multiplier. The Fickian diffusion application considers a hyperbolic first-order system with a boundary discontinuity and stable approach to the usual parabolic model. Results of the oxidation modeling of Tandon et al. (Polym Degrad Stab 91(8):1861–1869, 2006) are recovered by employing the reaction kinetics model and properties assumed there, while providing in addition the individual constituent kinematic and kinetic behaviors, thus adding rich interpretive detail in comparison to the original treatment (Tandon et al., Polym Degrad Stab 91 (8):1861–1869, 2006); two adjustable parameters describing coupled chemomechanical and purely chemical dissipation are added. The slurry infiltration application simulates the imposed mass deposition process and consequent effects on the kinematic and kinetic behaviors of the constituents.

Keywords Mixture • Reaction • Dissipation • Stabilized • Slurry

4.1 General Mixture Theory

The equations of mass and linear momentum balance for the diffusion of a chemically reacting fluid through a finitely deforming thermoelastic solid are given as follows [1]:

$$\text{Balance of mass :}\quad \frac{D^\alpha \rho^\alpha}{dt} + \rho^\alpha \mathrm{div}\boldsymbol{\nu}^\alpha = \frac{\partial \rho^\alpha}{\partial t} + \mathrm{div}(\rho^\alpha \boldsymbol{\nu}^\alpha) = m^\alpha \tag{4.1}$$

$$\text{Balance of linear momentum :}\quad \rho^\alpha \frac{D^\alpha \boldsymbol{\nu}^\alpha}{dt} = \mathrm{div}(\mathbf{T}^\alpha)^T + \rho^\alpha \mathbf{b} + \mathbf{I}^\alpha \tag{4.2}$$

where, ρ^α is the mass concentration and m^α is the rate of mass transferred by chemical reaction, to constituent α, per unit mixture volume; $\boldsymbol{\nu}^\alpha$ is the velocity of constituent α and $\mathbf{T}^\alpha$ is its partial Cauchy stress, while $\mathbf{I}^\alpha$ and $\mathbf{b}$ are the interactive force per unit mixture volume on constituent α and the body force per unit mass.

R. Hall (✉)
Materials and Manufacturing Directorate, Air Force Research Laboratory, AFRL/RXBC Bldg 654, 2941, WPAFB, Hobson Way, OH 45433-7750, USA

H. Gajendran • A. Masud
Department of Civil and Environmental Engineering, University of Illinois at Urbana-Champaign, 205 North Mathews Ave., Urbana, IL 61801-2352, USA

H.J. Qi et al. (eds.), *Challenges in Mechanics of Time-Dependent Materials, Volume 2: Proceedings of the 2014 Annual Conference on Experimental and Applied Mechanics*, Conference Proceedings of the Society for Experimental Mechanics Series,
DOI 10.1007/978-3-319-06980-7_4,

The balance of energy and assumption of maximized rate of entropy production, together with Newton's third law, lead to the following relations for the partial stresses on the solid and fluid, $\mathbf{T}^s$ and $\mathbf{T}^f$; the interactive force $\mathbf{I}^f$ on the fluid, the constituent entropy η^α, and the rate of fluid mass conversion, m^f, all per unit mixture volume; and the heat flux $\boldsymbol{q}$, per unit mixture area:

$$\begin{aligned}\mathbf{I}^f = {} & g^f \frac{\rho^s}{\rho} \nabla \rho^f - g^s \frac{\rho^f}{\rho} \nabla \rho^s - \nabla \left[\frac{\rho^s \rho^f}{\rho} \left(\psi^f - \psi^s \right) \right] \\ & - (\nabla \theta) \frac{\rho^s \rho^f}{\rho} \left(\eta^f - \eta^s \right) - m^f \left(\boldsymbol{\nu}^f - \boldsymbol{\nu}^s \right) - \mu \, \mathbf{A}^v \left(\boldsymbol{\nu}^f - \boldsymbol{\nu}^s \right)\end{aligned} \tag{4.3}$$

$$\mathbf{T}^s = \rho \mathbf{F}^s \left(\frac{\partial \psi}{\partial \mathbf{F}^s} \right)^T - \rho^s \left(g^s + \frac{\rho^f}{\rho} \left(\psi^f - \psi^s \right) \right) \mathbf{I} \tag{4.4}$$

$$\mathbf{T}^f = -\rho^f \left(g^f + \frac{\rho^s}{\rho} \left(\psi^s - \psi^f \right) \right) \mathbf{I} + \mu \, \mathbf{A}^L \cdot \mathbf{D}^f \tag{4.5}$$

$$\eta^\alpha = -\frac{\partial \psi^\alpha}{\partial \theta} - \frac{\mu}{\rho} c_\theta^\alpha \dot{\theta} \tag{4.6}$$

$$\frac{\mathbf{q}}{\theta} = -\mu \, \boldsymbol{l} \, \nabla \theta + \frac{\rho^s \rho^f}{\rho} \left(\eta^f - \eta^s \right) \left(\boldsymbol{\nu}^f - \boldsymbol{\nu}^s \right) \tag{4.7}$$

$$m^f = \frac{1}{\mu c_{\mathrm{m}}} \left[-\left(g^f - g^s \right) - \frac{1}{2} \left(\boldsymbol{\nu}^f - \boldsymbol{\nu}^s \right) \cdot \left(\boldsymbol{\nu}^f - \boldsymbol{\nu}^s \right) \right] \tag{4.8}$$

where the chemical potential g^α of constituent α is defined through

$$g^\alpha \equiv \rho \frac{\partial \psi}{\partial \rho^\alpha} \tag{4.9}$$

ρ, ψ and θ are the mixture density, mixture Helmholtz energy and temperature; while ψ^α are the constituent Helmholtz energies; Material parameters $c_\theta{}^\alpha$ and c_{m} are respectively associated with the constituent entropies, and with mass transfer, while $\boldsymbol{l}$ is the mixture thermal conductivity tensor; $\mathbf{F}^s$ is the solid deformation gradient; $\mathbf{A}^v$ and $\mathbf{A}^L$ are drag and viscosity coefficient tensors and $\mathbf{D}^f$ is the fluid rate of deformation tensor.

The rate of mass transfer to the fluid, m^f, is determined in coordination with the orientation average of the rate of reaction tensor $\dot{\Gamma}$. Because of the presence of only two constituents, the mass balance provides that the rate of mass converted to the solid is the one lost from the fluid:

$$m^s = -m^f \tag{4.10}$$

In the diffusion-dominated approximation (diffusion of the reactants is far more rapid than the reaction), the operator $\dot{\Pi}\{\mathbf{n}, \mathbf{X}^s, t\}$ provides the directional solid mass conversion rate in the direction -$\boldsymbol{n}$, per unit mixture volume, such that:

$$m^s = \frac{1}{4\pi} \int_{\alpha=0}^{4\pi} \dot{\Pi}\{\mathbf{n}[\alpha], \mathbf{X}^s, t\} \, d\alpha \tag{4.11}$$

where, $\boldsymbol{n}$ is the outward unit normal, $\mathbf{X}^s$ is the reference coordinate of the solid, α is the solid angle, and a second-order representation is assumed for the operator $\dot{\Pi}\{\mathbf{n}, \mathbf{X}^s, t\}$

$$\dot{\Pi}\{\mathbf{n}, \mathbf{X}^s, t\} \approx \mathbf{n} \cdot \dot{\Gamma} \left[\mathbf{X}^s, t \right] \mathbf{n} \tag{4.12}$$

with the tensor

$$\Gamma[\mathbf{X}^s,t]=\int_0^t \dot{\Gamma}[\mathbf{X}^s,\bar{t}]\,d\bar{t} \tag{4.13}$$

thus, providing an anisotropic measure of the extent of reaction of the solid.

The Lagrange multiplier μ arises from the constraint of maximized rate of entropy production [1]. Because η^α and m^f depend on μ, a cubic equation in μ results. To obtain a single-valued relation for μ, the following approximations are made:

1. We assume that the attributes of the Helmholtz free energy functions of the constituents and the mixture can be represented in terms of suitably condensed forms, $\psi^s=\psi^f=\psi$, $\eta^s=\eta^f=\eta$.
2. Slow diffusion permits neglect of the squared relative kinetic energy terms $((\boldsymbol{\nu}^f-\boldsymbol{\nu}^s)\cdot(\boldsymbol{\nu}^f-\boldsymbol{\nu}^s))^2$, which are assumed also negligible relative to the drag force.
3. We assume that the reaction is near enough to equilibrium to neglect the squared difference in the chemical potentials of the constituents, and the product of the chemical potential difference with the relative kinetic energy.

The Lagrange multiplier is thus reduced to the following single-valued function:

$$\mu=\frac{1}{2}+\frac{\frac{1}{4}\dot{\Gamma}^0_{IJ}\Gamma^0_{KL}\Gamma^0_{MN}E^s_{OP}K^0_{IJKLMNOP}}{\left\{\begin{array}{l}\mathbf{D}^f\cdot\mathbf{A}^L\cdot\mathbf{D}^f+c_\theta\dot{\theta}^2+\left(\boldsymbol{\nu}^f-\boldsymbol{\nu}^s\right)\cdot\mathbf{A}^\nu\left(\boldsymbol{\nu}^f-\boldsymbol{\nu}^s\right)+\\ \dot{\Gamma}^0\cdot\mathbf{A}^{0\Gamma}\cdot\dot{\Gamma}^0+\nabla\theta\cdot\mathbf{I}\nabla\theta+\frac{1}{2}\dot{\Gamma}^0_{AB}\Gamma^0_{CD}\Gamma^0_{EF}E^s_{GH}K^0_{ABCDEFGH}\end{array}\right\}} \tag{4.14}$$

$\mathbf{E}^s$ is the Lagrangian solid strain measure and Γ is referred to material coordinates. The tensor $\boldsymbol{K}^\mathbf{0}$ will have mostly zero-valued components. If reaction processes such as oxidation are considered, in which the reaction is several times faster in the fiber direction than the transverse directions thus promoting a unidirectional reaction assumption, and assuming transversely isotropic coupling to the strains, the term involving $\boldsymbol{K}^\mathbf{0}$ reduces to the following expression, involving four independent constants:

$$\dot{\Gamma}^0_{IJ}\Gamma^0_{KL}\Gamma^0_{MN}E^s_{OP}K^0_{IJKLMNOP}=\dot{\Gamma}^0_{11}\left(\Gamma^0_{11}\right)^2\left[K^0_1E^s{}_{11}+K^0_2(E^s{}_{22}+E^s{}_{33})+K^0_3(E^s{}_{12}+E^s{}_{31})+K^0_4E^s{}_{23}\right] \tag{4.15}$$

In the present work, the influence of the energy and entropy production relations are retained through the presence of the Lagrange multiplier, which is obtained via invoking the constraint of maximized rate of entropy production. The equations explicitly retained are the constituent momentum balances and the mass balance equation, which can be considered most strongly enforced. In accordance with the present study being isothermal, the traditional heat capacity measures of the constituents are lost through the assumption above that the constituent entropy functions can be replaced by an overall entropy function. In general for anisothermal processes, the Helmholtz and entropy functions of each constituent would be retained. It is interesting to note, however, that the present system of equations incorporates the rate of temperature in combination with a non-traditional overall material property c_θ (the density average of the $c_\theta{}^\alpha$ properties), which may provide a simplified approach to accounting for a class of homogenized anisothermal effects. The present paper however considers only isothermal conditions.

We consider the following Helmholtz free energy function that corresponds to the 1-D representation of a transversely isotropic thermoelastic solid permeated by a chemically reacting Newtonian fluid.

$$\psi=A^s+(B^s+c^s)(\theta-\theta^s)-\frac{c^s_1}{2}(\theta-\theta^s)^2-c^s_2\theta\ln\left(\frac{\theta}{\theta^s}\right)+\frac{1}{\rho_T}\left\{\overline{R}\theta\rho^f+k^f_2\rho^f\right\}$$
$$+\frac{\rho^s}{\rho}\frac{1}{\rho^s_T}\left\{\frac{1}{2}\lambda^s+\mu^s_T+\alpha^s+2(\mu^s_L-\mu^s_T)+\frac{1}{2}\beta^s\right\}\left(E^s_{11}\right)^2+\Lambda \tag{4.16}$$

$$\Lambda=\int\frac{\mu}{2}\left\{\left(\Gamma^0_{11}\right)^2\overline{K}_1E^s_{11}+2\dot{\Gamma}^0_{11}\overline{A}^{0\Gamma}\right\}d\Gamma^0_{11} \tag{4.17}$$

where Λ describes the coupling between the solid strain and the extent of reaction, consistent with the developments of [1]; λ^s, α^s, ${\mu_L}^s$, ${\mu_T}^s$, β^s are the transversely isotropic material constants, which in one dimension reduce to the elastic moduli of the solid. ${\rho_T}^s$, ρ_T are the true solid density and the true mixture density respectively. $\overline{R}$ is the ratio of the universal gas constant to the molecular weight of the fluid. $K_1^0{=}{-}\rho\overline{K}_1$ and $A^{0\Gamma}{=}{-}\rho\overline{A}^{0\Gamma}$ are defined for convenient manipulations involving Λ.

Remark For the case of slurry deposition process that is presented in Sect. 4.3.2, Γ_{11}^0 represents the extent of material deposition. For this case, the term Λ provides coupling between the solid strain and the extent of deposition of the suspended particles. We assume that this deposition function Γ_{11}^0 is in fact a function of the volume fraction of particles, which is considered a process parameter.

4.1.1 Modeling Assumptions and Methodology

In mixture theory where both solid and fluid co-occupy the domain and fluid moves relative to the deforming solid, it is natural to write the fluid balance laws in an Arbitrary Lagrangian Eulerian (ALE) framework [2–4]. For the class of problems considered in this work, the inertial effects on the solid are assumed to be negligible.

Remark In [1] an expression for the rate of mass conversion for fluid m^f is derived via maximization of the rate of dissipation constraint. However, in the present work we prescribe an oxidation rate given in [5] that is developed based on physical measurements. Likewise, in the slurry infiltration model we prescribe a rate of particle deposition as is given in [6]. Because of these postulated rates, the physics involved in the consistent derivation of mass conversion given in [1] is circumvented. Thus,

$$m^f = \dot{\Gamma}_{11}^0 \tag{4.18}$$

4.2 Weak Form and Development of Stabilized Method

Let g^α and ${w_1}^\alpha$ denote the weighting functions for the balance of mass and linear momentum for the corresponding constituent, respectively.

4.2.1 Weak Form of Equations for the Fluid

$$\left(g^f, \left.\frac{\partial \rho^f}{\partial t}\right|_Y\right) + \left(g^f, \left(v_1^f - v_1^s\right)\frac{\partial \rho^f}{\partial x}\right) + \left(g^f, \rho^f \frac{\partial v_1^f}{\partial x}\right) - \left(g^f, m^f\right) = 0 \tag{4.19}$$

$$\begin{aligned}\left(\frac{\partial w_1^f}{\partial x}, T_{11}^f\right) - \left(w_1^f, \rho^f b_1\right) - \left(w_1^f, I_1^f\right) + \left(w_1^f, \rho^f \left.\frac{\partial v_1^f}{\partial t}\right|_Y\right) + \\ \left(w_1^f, \rho^f\left(v_1^f - v_1^s\right)\frac{\partial v_1^f}{\partial x}\right) - \left(w_1^f, T_{11}^f n_1\right)_{\partial\Omega_h^f} = 0\end{aligned} \tag{4.20}$$

4.2.2 Weak Form of Equations for the Solid

$$\left(g^s, \frac{\partial \rho^s}{\partial t}\right) + \left(g^s, \rho^s \frac{\partial v_1^s}{\partial x}\right) + \left(g^s, v_1^s \frac{\partial \rho^s}{\partial x}\right) - \left(g^s, m^s\right) = 0 \tag{4.21}$$

$$\left(\frac{\partial w_1^s}{\partial x}, T_{11}^s\right) - \left(w_1^s, \rho^s b_1\right) - \left(w_1^s, I_1^s\right) - \left(w_1^s, T_{11}^s n_1\right)_{\partial\Omega_h^s} = 0 \tag{4.22}$$

where, $(\bullet, \bullet) = \int_\Omega (\bullet) d\Omega$ is the $L_2(\Omega)$ inner product, $\left.\frac{\partial(\cdot)}{\partial t}\right|_Y$ represents the time derivative in the ALE frame [3, 4] and $v_1{}^m$ is the fluid mesh velocity. It is important to note that as the solid domain deforms, the Lagrangian mesh that is tied to material points deforms together with it. Consequently, the mesh velocity v_1^m is set equal to v_1^s where, v_1^s is the velocity of the solid domain.

4.2.3 Fluid Sub-system: Residual-Based Stabilization

Our objective is to model the diffusion of a chemically reacting fluid through a nonlinear elastic solid, a phenomenon that is observed in the process modeling of composites, oxidation of resin/composites, and slurry infiltration in porous media, to name a few. In the modeling of these processes, fluid mass concentration is invariably specified at the inlet boundary. Since the strong form of mass balance of fluid given in Eq. (4.1) is a first order hyperbolic equation, any specified mass concentration boundary condition at the inlet that is different from the initial condition results in a discontinuous fluid concentration field. This discontinuity introduces spurious oscillations in the computed solution right at the beginning of the nonlinear iterative process that can lead to non-convergent and therefore non-physical solutions.

To address this issue, we consider the weak form of the balance of mass equation for the fluid that is written in an ALE form. We employ Variational Multiscale (VMS) ideas [7–10] and develop a stabilized weak form for Eq. (4.19). Underlying idea of VMS is an additive decomposition of the solution field into coarse and fine scale components as given below.

$$\rho^f = \hat{\rho}^f + \tilde{\rho}^f \tag{4.23}$$

$$g^f = \hat{g}^f + \tilde{g}^f \tag{4.24}$$

where, $\hat{\rho}^f, \tilde{\rho}^f$ represents the coarse-scale and fine-scale components of the density field and $\hat{g}^f, \tilde{g}^f$ represents the coarse-scale and fine-scale counterpart of the weighting function respectively.

Substituting Eqs. (4.23) and (4.24) in Eq. (4.19) and employing the linearity of the weighting function slot in Eq. (4.19), we obtain the coarse-scale problem and the fine-scale problem as given in Eqs. (4.25) and (4.26) respectively.

$$\hat{W}^{\rho^f} = \left(\hat{g}^f, \left.\frac{\partial(\hat{\rho}^f + \tilde{\rho}^f)}{\partial t}\right|_Y\right) + \left(\hat{g}^f, \frac{\partial(\hat{\rho}^f + \tilde{\rho}^f)v_1^f}{\partial x}\right) - \left(\hat{g}^f, v_1^s \frac{\partial(\hat{\rho}^f + \tilde{\rho}^f)}{\partial x}\right) - \left(\hat{g}^f, m^f\right) = 0 \tag{4.25}$$

$$\tilde{W}^{\rho^f} = \left(\tilde{g}^f, \left.\frac{\partial(\hat{\rho}^f + \tilde{\rho}^f)}{\partial t}\right|_Y\right) + \left(\tilde{g}^f, \frac{\partial(\hat{\rho}^f + \tilde{\rho}^f)v_1^f}{\partial x}\right) - \left(\tilde{g}^f, v_1^s \frac{\partial(\hat{\rho}^f + \tilde{\rho}^f)}{\partial x}\right) - \left(\tilde{g}^f, m^f\right) = 0 \tag{4.26}$$

It is important to note that both systems are nonlinear, and are also fully coupled in terms of the scales. The key idea at this point is to solve the fine-scale problem Eq. (4.26) locally, using analytical methods or numerical methods, and extract the fine-scale component, $\tilde{\rho}^f$. This can then be substituted in the corresponding coarse-scale problem given in Eq. (4.25), thereby eliminating the fine-scales, yet modeling their effects.

4.2.4 Solution of the Fine Scale Problem

We segregate the terms into coarse-scale and fine-scale terms and group all the terms containing coarse-scale density field.

$$\tilde{W}^{\rho^f} = \left(\tilde{g}^f, \left.\frac{\partial\tilde{\rho}^f}{\partial t}\right|_Y\right) + \left(\tilde{g}^f, \frac{\partial\tilde{\rho}^f v_1^f}{\partial x}\right) - \left(\tilde{g}^f, v_1^s \frac{\partial\tilde{\rho}^f}{\partial x}\right) + \left(\tilde{g}^f, \hat{R}\right) = 0 \tag{4.27}$$

where, $\hat{R}$ is the residual of the Euler–Lagrange equations of the coarse-scales over element interiors and is given as,

$$\hat{R} = \left.\frac{\partial \hat{\rho}^f}{\partial t}\right|_Y + \frac{\partial \hat{\rho}^f v_1^f}{\partial x} - v_1^s \frac{\partial \hat{\rho}^f}{\partial x} - m^f\left(\hat{\rho}^f\right) \tag{4.28}$$

In obtaining the above form of the fine scale problem, we have assumed that the fluid mass conversion rate is a function of the coarse-scale fluid density field only, $m^f\left(\hat{\rho}^f, \tilde{\rho}^f\right) \approx m^f\left(\hat{\rho}^f\right)$. Discretizing (4.27) in time and space results in a solution for the fine scale density field in terms of the coarse scale fields, expressed as a multiple of the residual. This solution confers stabilization properties resulting in the dampening of the oscillatory effects resulting from the boundary discontinuity. Details are given in [11].

4.3 Numerical Results

We present three test cases that investigate the stability and accuracy of the numerical method developed for the mixture theory model described in Sect. 4.3. The present developments have been carried out in the context of 1D finite element method, and extension to 3D case will be pursued in a subsequent paper.

4.3.1 Fick's Diffusion Problem

In this section we employ a reduced mixture model to solve Fick's diffusion problem. The transient Fick's diffusion equation can be derived from the mixture theory balance laws, Eqs. (4.1) and (4.2) based on the following simplifications: (a) solid is assumed to be rigid, (b) fluid is assumed ideal, (c) fluid inertial effects are neglected, and (d) fluid is assumed non-reactive.

Remark Solving for fluid velocity from the equation of motion and substituting into the equation of conservation of mass for the fluid, one can obtain Fick's diffusion equation. Since our full mixture model results in a first order system, in this work we have opted to solve the reduced system also in its first order form to help serve as a test case for evaluating our numerical method.

The unknown fields in this problem are the fluid concentration and fluid velocity that are solved with zero initial conditions. The one-dimensional domain of length 0.001 m is exposed to air at the left end of the domain where fluid concentration is assigned a value of 22.8863E−3 kg/m^3 and fluid velocity is constrained to be zero at the right end of the domain. The gas constant $\overline{R}$ and drag coefficient A^v are assigned values of 286.987 J/kg-K and 1.63E17 s^{-1}, respectively. We employ the Backward Euler scheme for time integration and run the problem for a total time of 30,000 s. A variable time step increment is used: the time steps employed during the first second is $\Delta t = 1\mathrm{E}-4$, and it is increased to $\Delta t = 0.1$ for the remaining steps.

It should be noted that the fluid mass balance equation is a first order hyperbolic equation for fluid concentration. For a non-zero fluid concentration boundary condition applied at the inflow, the standard Galerkin finite element method results in oscillations around the steep front thereby causing numerical instability. We employ the variational multiscale method to stabilize the formulation, and provide a comparison between the stabilized numerical result and the exact solution. Figure 4.1a, b shows performance of the new method for h-refinement wherein we have used linear Lagrange interpolation functions. These plots show the spatial profiles of the fluid concentration and velocity fields at 1,000, 10,000 and 30,000 s. It can be seen that as the number of elements is increased, computed solution converges to the exact solution which is a numerical validation of the consistency of the formulation. Improved convergence rate is obtained for quadratic elements, with similar stability properties.

4.3.2 Oxidation of PMR-15 Resin

Thermo-oxidative aging of polymer matrix composites (PMC's) in high temperature applications influences the life and performance of these materials. In this section, we present numerical results for the oxidation behavior of polyimide PMR-15 resin based on the oxidation reaction model developed in the works of Tandon et al. [5]. For the sake of completeness, we provide a brief description of the oxidation process in polymer. However, for a detailed description of the oxidation process and the reaction kinetics model, refer to [5, 12]. Oxidation front in polymer materials advances through a combination of diffusion and reaction mechanism. The exposed surface reacts with the diffusing air, depleting the amount of polymer

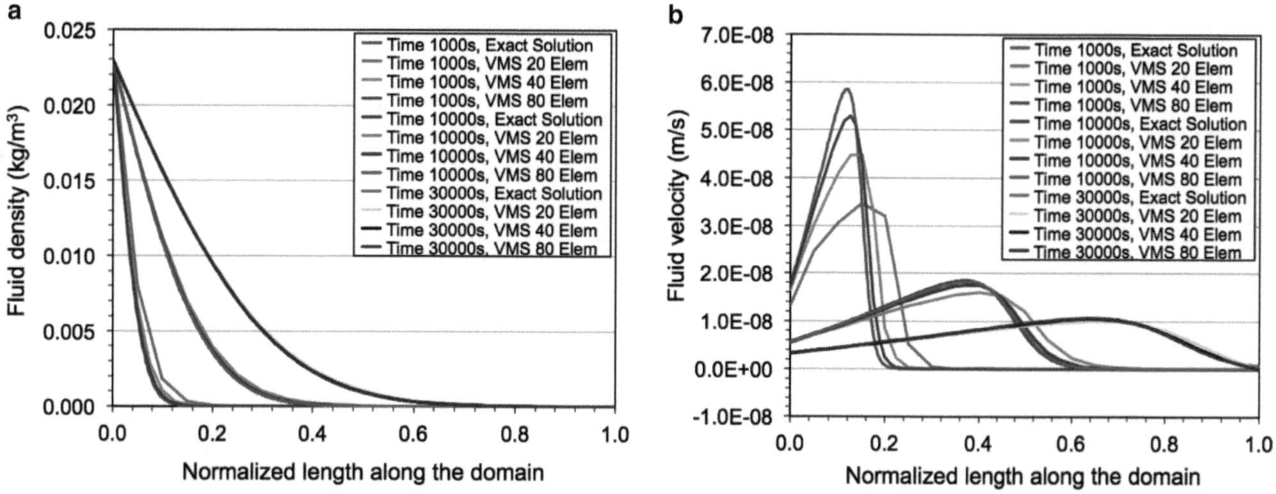

Fig. 4.1 Mesh refinement study at various time levels. (**a**) Fluid density: Linear Lagrange h-refinement. (**b**) Fluid velocity: Linear Lagrange h-refinement

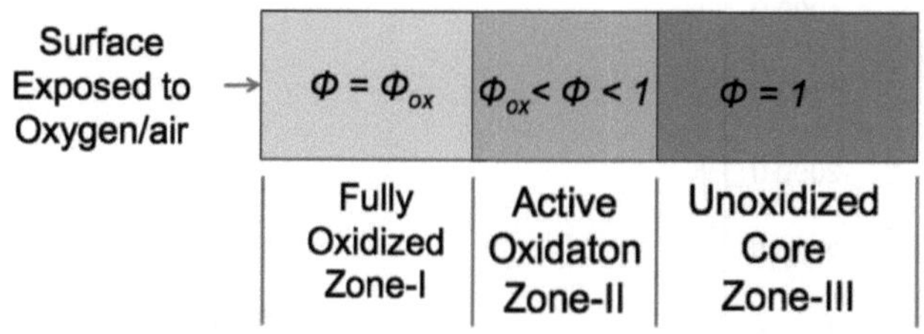

Fig. 4.2 Schematic representation of thermo-oxidation process

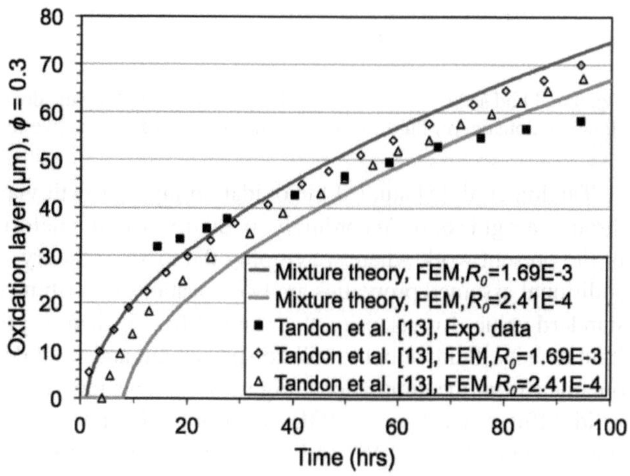

Fig. 4.3 Oxidation layer growth with time for various values of reaction rate

available to react in that region. Once this region is fully oxidized, it acts as a medium through which air/oxygen diffuses through and an active oxidation zone is formed ahead of the fully oxidized zone. Thus, at any instant of time, the oxidation process in polymers comprises of a fully oxidized zone, an active oxidation zone and a neat resin zone as shown in Fig. 4.2.

A parametric study was done for the oxidation layer growth with time and results are presented as shown in Fig. 4.3. The oxidation layer growth results shown in Fig. 4.3 are plotted for oxidative state parameter $\phi = 0.3$, a specific value within zone II [5, 12]. Figure 4.3 shows the variation in oxidation layer growth for different reaction rate parameters for a duration of 100 h. The solid line shows the results from the mixture theory, where it can be seen that the reaction rate of 2.41E−4 kg/m^3 s produces an oxidation layer growth of 66.9 μm as compared to 74.7 μm for the reaction rate of 1.69E−3 kg/m^3 s at the end of 100 h. The mixture theory results follow a similar trend in comparison with the Tandon et al. [5] numerical results.

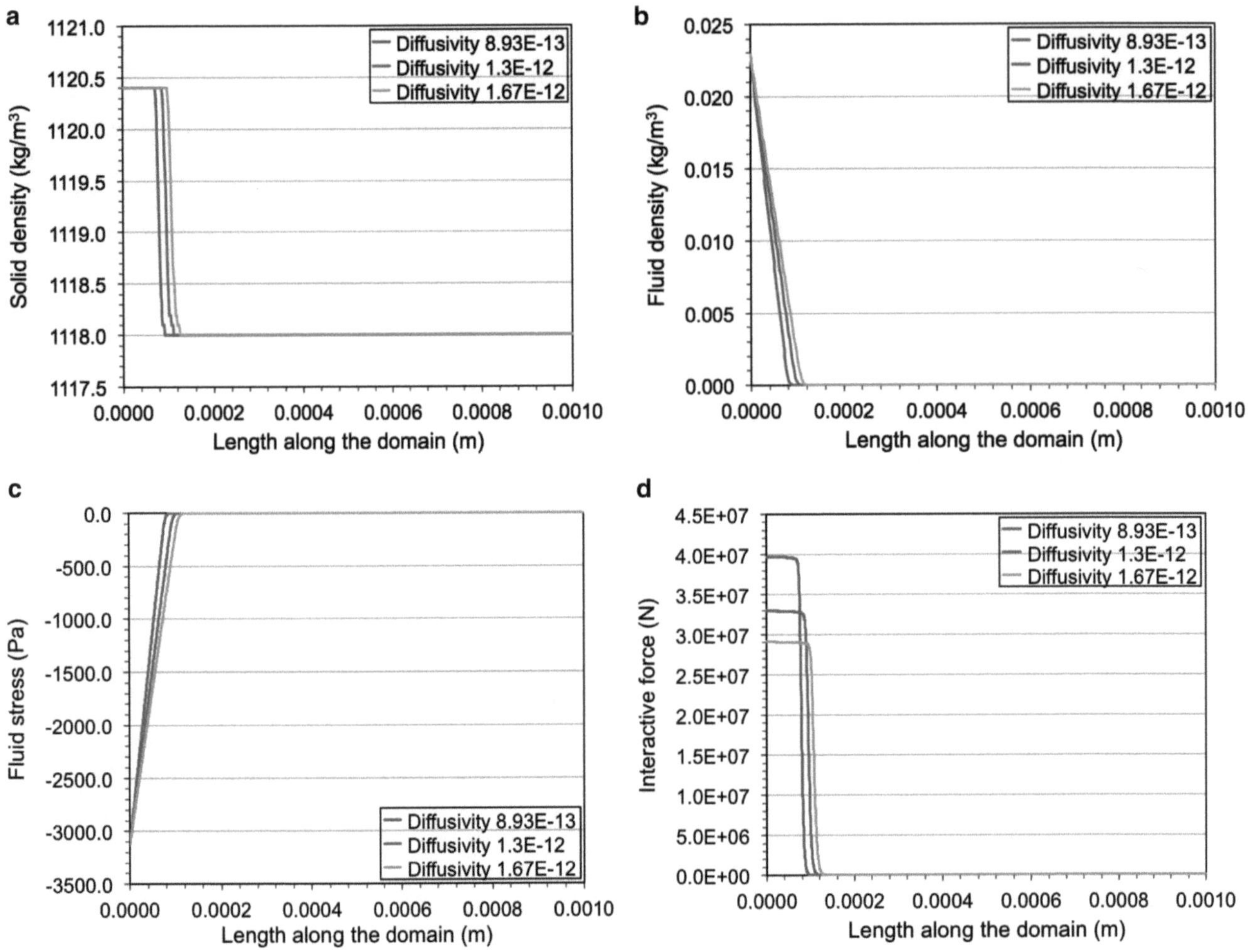

Fig. 4.4 Fluid and solid kinematic and force quantities along the domain at the end of 100 h. (**a**) Solid density along the domain, (**b**) fluid density along the domain, (**c**) fluid stress along the domain, (**d**) interactive force along the domain

Tandon et al. [5] studied the oxidation layer growth via diffusion reaction equation assuming an ideal fluid permeating through a rigid solid. Accordingly, in their model the deformation of the solid and viscous effects in the fluid are neglected. In the present work where we employ the mixture theory, a Newtonian fluid and an elastic solid are considered. The only additional assumed properties are two constants describing coupled chemomechanical and purely chemical dissipation, and standard values for viscosity of air and PMR-15 stiffness properties [11]. Since the unknown fields in the mixture model are fluid density, fluid velocity, solid displacement and solid density, therefore kinematic and the force measures can be readily obtained from the simulations. Figure 4.4 shows the variation of the fluid and the solid kinematic and force quantities for solid diffusivity values of 8.93E−13, 1.30E−12 and 1.67E−12 m^2/s. The plots shown are obtained for a saturated oxidation state value of $\phi_{ox} = 0.187$ and a reaction rate of 1.69E−3 kg/m^3 s. Figure 4.4a, b shows the variation of solid density and fluid density along the domain at the end of 100 h. Full oxidation of all presumed available sites results in a fixed solid density, as indicated in Fig. 4.4a.

Since there are only two constituents in the present model, loss of mass from one is the gain in mass for the other. Consequently, the density of the solid increases as shown in Fig. 4.4a wherein the apparent solid density has a higher value as compared to the neat resin region. This is rather contradictory to the experimental observations as the density of the PMR-15 resin is expected to decrease with increased levels of oxidation. (It does, however, correspond to initial weight gains in certain oxidizing materials systems before substantial mass loss to the environment occurs. The transfer of mass out of the material system is not explicitly addressed here.) If the two-constituent mixture model is extended to three-constituent model where the third constituent is allowed to evolve and also leave the domain, it can account for the experimentally observed weight loss in solid due to the oxidation process. Figure 4.4c shows that the variation in fluid stress is dominated by the hydrostatic pressure. Figure 4.4d shows the distribution of interactive force between the diffusing fluid and deforming solid. It can be seen that the interactive force becomes zero in the neat resin region where the fluid has not reached yet.

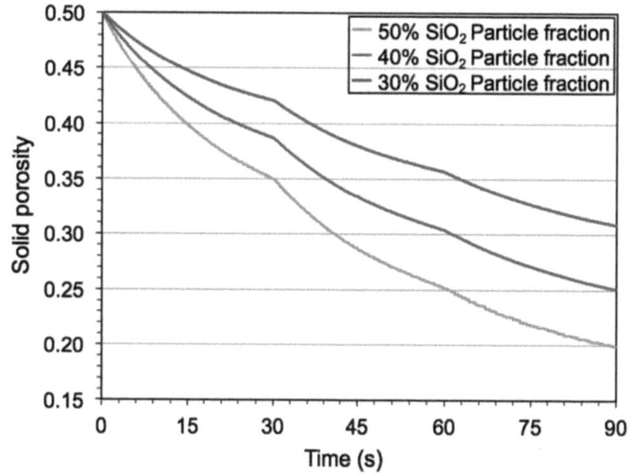

Fig. 4.5 Reduction in solid porosity with time for 30, 40 and 50 % SiO_2 particles in the slurry

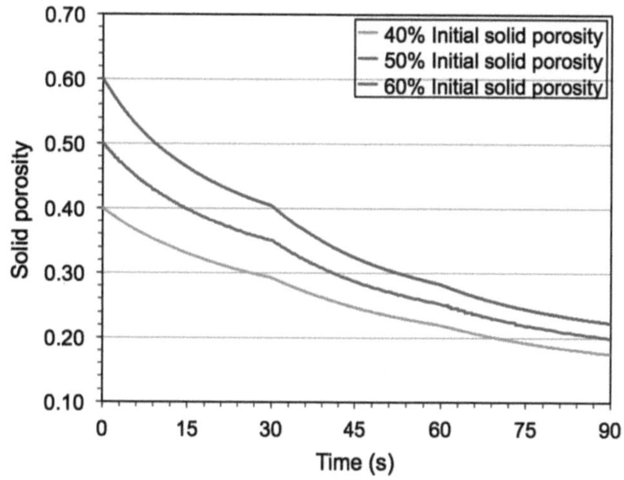

Fig. 4.6 Reduction in solid porosity with time for 40, 50 and 60 % initial solid porosity

4.3.3 Slurry Infiltration Problem

Slurry infiltration (SI) is an important step in the processing of ceramic matrix composites (CMC). In the slurry infiltration process, a viscous fluid that is laden with particles of various sizes, composition, and volume fraction is injected into a fiber preform, wherein fluid primarily serves as a medium that carries the suspended particles to the preform. This cycle is repeated several times till the density of the preform increases and its porosity reduces to some desired design value. Once slurry infiltration process is complete, a second process called melt infiltration is carried out with a viscous fluid that can chemically react with the preform as well as the deposited particles to make a composite with desired strength and density distribution [13].

In this section, we consider the SI and employ properties of a porous PMC as a surrogate model for CMC material. We assume that water based slurry has permeated the porous elastic solid and we model the process of deposition of suspension onto the fiber preform.

Below, we present the results for the case where the porous solid is subjected to three infiltration cycles of 30 s each, for a total of 90 s. At the end of each cycle, the particle mass fraction w is reset to the initial particle mass fraction in the slurry w_0. Figure 4.5 shows the variation of the solid porosity with time for 50 % porous solid and 30, 40 and 50 % SiO_2 particle volume fraction in the slurry. We see that as the particles get deposited, the porosity of the solid decreases. For all three different particle volume fractions in the slurry, this decrease in porosity is nonlinear, wherein the rate of reduction in porosity seems to be slowing down with time that is indicated by the relatively flatter portion of the curve at the end of each cycle. From the perspective of the physics of the problem this means that while there is more relative reduction in porosity during early infiltration cycles, because of the closure of pores that happens due to the solid mass buildup, the relative reduction in porosity in subsequent cycles also slows down. Figure 4.6 shows a similar trend in reduction in porosity with time for three different initial solid porosities that are infiltrated with 50 % particle slurry.

References

1. Hall R, Rajagopal K (2012) Diffusion of a fluid through an anisotropically chemically reacting thermoelastic body within the context of mixture theory. Math Mech Solids 17(2):131–164
2. Hughes TJ, Liu WK, Zimmermann TK (1981) Lagrangian–Eulerian finite element formulation for incompressible viscous flows. Comput Methods Appl Mech Eng 29(3):329–349
3. Khurram RA, Masud A (2006) A multiscale/stabilized formulation of the incompressible Navier–Stokes equations for moving boundary flows and fluid–structure interaction. Comput Mech 38(4–5):403–416
4. Calderer R, Masud A (2010) A multiscale stabilized ALE formulation for incompressible flows with moving boundaries. Comput Mech 46(1):185–197
5. Tandon G, Pochiraju K, Schoeppner G (2006) Modeling of oxidative development in PMR-15 resin. Polym Degrad Stab 91(8):1861–1869
6. Civan F (2011) Porous media transport phenomena. Wiley, Hoboken
7. Hughes TJR (1995) Multiscale phenomena: Green's functions, the Dirichlet-to-Neumann formulation, subgrid scale models, bubbles and the origins of stabilized methods. Comput Methods Appl Mech Eng 127(1):387–401
8. Masud A, Hughes TJR (1997) A space-time Galerkin/least-squares finite element formulation of the Navier–Stokes equations for moving domain problems. Comput Methods Appl Mech Eng 146(1):91–126
9. Masud A, Khurram R (2004) A multiscale/stabilized finite element method for the advection–diffusion equation. Comput Methods Appl Mech Eng 193(21):1997–2018
10. Masud A, Kwack J (2008) A stabilized mixed finite element method for the first-order form of advection–diffusion equation. Int J Numer Methods Fluids 57(9):1321–1348
11. Hall R, Gajendran H, Masud A (2014) Diffusion of chemically reacting fluids through nonlinear elastic solids: mixture model and stabilized methods. Math Mech Solids (in press)
12. Schoeppner G, Tandon G, Pochiraju K (2008) Predicting thermooxidative degradation and performance of high-temperature polymer matrix composites. In: Talreja R (ed) Multiscale modeling and simulation of composite materials and structures. Springer, New York, pp 359–462
13. Bansal NP, Boccaccini AR (2012) Ceramics and composites processing methods. Wiley.com, Hoboken

Chapter 5
Effect of Temperature on Mechanical Property Degradation of Polymeric Materials

Tong Cui, Yuh J. Chao, John W. Van Zee, and Chih-Hui Chien

Abstract Proton Electrolyte Membrane (PEM) fuel cell is a promising energy source because of its high efficiency and zero emission. One of the most important unresolved problems of PEM fuel cells today is the durability issue of its components. For example, the polymeric gasket material of PEM fuel cell must be durable enough to hold the liquid and gas inside the fuel cell channel, as its sealing force decreases gradually with time and also changes with temperature. Liquid Silicone Rubber (LSR) is commonly used as gasket or seal material in many industrial applications including PEM fuel cells. This paper discusses the compression stress relaxation of LSR under temperature cycling, which is to simulate the actual fuel cell operation. It is found that (a) in addition to stress relaxation, thermal expansion or contraction of the material contributes the most in the observed stress variation during temperature change, and (b) the stiffness of LSR appears to change according to temperature history, and (c) the Maxwell stress relaxation model can be used to predict the sealing force only after a correction of the change of material stiffness is implemented into the model.

Keywords Stress relaxation • Temperature cycling • Mechanical property degradation

5.1 Introduction

Stress relaxation of polymeric seals is a manifestation of viscoelastic behavior which can lead to leaking of gas inside fuel cells. Our previous work investigated effect of temperature and environment (i.e. water and acid solution) on the stress relaxation behavior. With the master curve constructed, it can be shifted to other temperatures, and thus estimate the stress relaxation at the temperature that can be obtained. However, this life prediction is only valid at constant operation temperature. In this paper, the cumulative effect of temperature change is studied. Most other work reported in the literature about stress relaxation/creep focused on constant temperature [1–3]. In a real application of the fuel cell, the temperature of fuel cell stack changes from start up to operation to shut down and thus the gasket/seal experiences thermal cycling. When temperature changes, sealing force is expected to change due to thermal expansion effects and accumulative effects of temperature change on material stiffness. Several researchers mentioned increasing stiffness of material after aging/stress relaxation. Derham noticed that load change during temperature change did not agree with the calculation value. He eventually found that material stiffness actually increased after stress relaxation [4]. Gordon et al. also pointed out that strain hardening was not lost when extra mechanical strain was added, and the time dependent strain hardening accumulated

T. Cui
University of South Carolina, Columbia, SC 29208, USA

Y.J. Chao (✉)
University of South Carolina, Columbia, SC 29208, USA

Tianjin University, Tianjin, China
e-mail: chao@sc.edu

J.W. Van Zee
University Alabama, Tuscaloosa, AL, USA

C.-H. Chien
National Sun Yat-Sen University, Kaohsiung, Taiwan

H.J. Qi et al. (eds.), *Challenges in Mechanics of Time-Dependent Materials, Volume 2: Proceedings of the 2014 Annual Conference on Experimental and Applied Mechanics*, Conference Proceedings of the Society for Experimental Mechanics Series,
DOI 10.1007/978-3-319-06980-7_5,

through initial creep relaxation and secondary torquing. This indicates that the stiffness of the gasket after initial dwell period actually increased [5, 6]. Hitendra et al also reported that the compressive load increased at the beginning of intermittent stress relaxation test because of the increasing stiffness of material during new chain formation [7]. It is worth noting that all of these researchers noticed the increasing material stiffness with different materials.

5.2 Experimental

The stress relaxation equipment was made by Elastocon AB (Sweden) and this particular model consists of three independent testing rigs. Please see the schematic of the rig [8]. The rig can be put in the oven, and temperature of the oven can be adjusted manually. Both the temperature and the resultant force of the specimen can be continuously recorded by a computer throughout the test.

The temperature cycling profiles includes both high and low temperature. The oven of the stress relaxation machine can only operate above 40 °C. In order to reach temperature below zero, a freezer with temperature controller is designed to hold the rig. So the rig can be switched from oven to freezer when needed. A regular freezer was bought and then three holes were drilled on the cover, and a basket inside the freezer can hold the lower part of the rig in the freezer.

In order to validate data collected by CSR machine, Instron is also used to do similar tests. The coefficient of thermal expansion of steel is 17×10^{-6} mm/mm °C, and it is much smaller than polymeric sample, which is 272×10^{-6} mm/mm °C. The oven associated with Instron is very large and can hold most part of the upper and lower steel rod, and the length of the steel rod inside the oven is hundred times bigger than the test sample. So the overall thermal expansion of the rod is much larger than the polymeric sample itself. This will introduce large error during the temperature cycling test.

In order to minimize error introduced by expansion of the arm of test machine, a fixture working with Instron is designed as shown in Fig. 5.1. Two ceramic rods each with 1 in. diameter are used to compress the standard polymer sample. The ceramic rods are surrounded by heating tape and insulation foam. A T type thermocouple is placed near the sample to monitor the temperature and provide feedback for temperature controller. The working temperature of the heating tape is up to 200 °C.

Material of the two ceramic rods is machinable high alumina ceramic. It has low thermal conductivity and coefficient of thermal expansion (CTE). The CTE of alumina ceramic is about 5×10^{-6} mm/mm °C, much lower than the CTE of liquid silicone rubber. Because the heating tape only surrounds the test sample and a short part of the ceramic, so the thermal expansion of the ceramic rod can be neglected compared with test polymeric sample.

Generally, the test sample is compressed with 25 % strain at room temperature. Then the temperature is adjusted periodically. It takes about several hours to change the temperature in the oven, and each constant temperature holds about several days. During the temperature cycling, temperature below zero is also included by switching rigs from oven to freezer when needed. It is designed to simulate the real operation temperature of fuel cell. Initially the test starts at room temperature, after several days, it is raised as high as 120 °C, and holds for several days before dropping back to room temperature. Then the rig is switched to the freezer and sample is cooled down to −10 °C. Similar cycles are repeated for a few times. Note that temperature change process is not very quick. Both heating and cooling takes about 2–3 h for the oven to reach the set temperature.

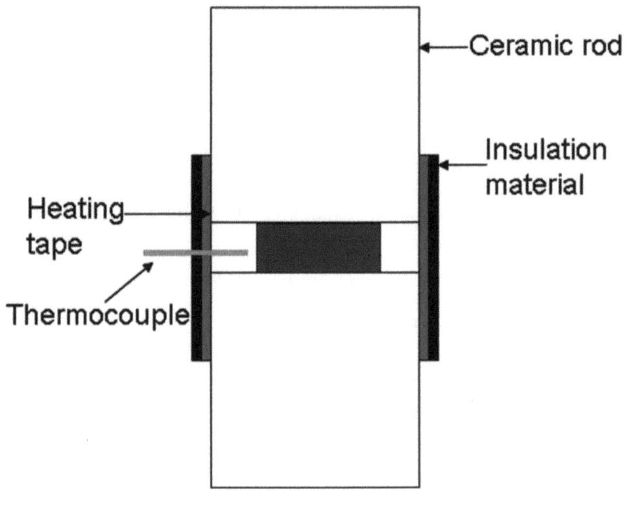

Fig. 5.1 Schematic of the micro oven

5.3 Results

5.3.1 Experimental Data and Model Prediction

Figure 5.2 shows the compressive load profile as the temperature changes. It can be seen that stress relaxation is significant only at the highest temperature (70 °C), and that at any other lower temperature, the stress is almost a constant due to extremely slow relaxation rate. Load change is very rapid due to temperature changes, but it is fully recoverable when temperature returns its initial value. Similar phenomena are found in other test as shown in Fig. 5.2.

In our previous work [8], master curves are constructed to predict the service life at any constant temperature. This section will discuss how to predict the stress relaxation behavior subjected to temperature cycling. Generalized Maxwell equation is widely used to model the stress relaxation behavior of polymers. The governing equation for a given Maxwell element is as following:

$$\frac{d\varepsilon_{total}}{dt} = \frac{\sigma}{\eta} + \frac{1}{k}\frac{d\sigma}{dt} \tag{5.1}$$

For stress relaxation at constant temperature, strain rate at left hand side of the above equation is always zero, since a constant strain is applied for stress relaxation test. However, for stress relaxation test subjected to thermal cycling, the total strain is always changing due to the effect of thermal expansion. The coefficient of thermal expansion of polymer is much higher than metal, so the thermal expansion effect of test sample makes it compressed by extra strain when temperature increases. Humfeld and Dillard proposed a modified Maxwell equation, which includes the nonzero thermal strain rate in the generalized Maxwell mode [9, 10]. After accounting the temperature shift factor, the following recursive solution can be reached:

$$\sigma_{i+1} = \sigma_i e^{\frac{-\Delta t}{a_T \tau}} + a_T \eta \dot{\varepsilon}\left(1 - \dot{e}^{\frac{-\Delta t}{a_T \tau}}\right) \tag{5.2}$$

Further derivation indicates that the recursive solution is the simple combination of stress relaxation and thermal stress. This model is used to calculate the stress relaxation subjected to temperature cycling. First, simple temperature cycling profiles are used to test the model, which is similar with the temperature cycles in reference [9]. Figure 5.2 shows the comparison between our experimental data and the model calculation. The temperature cycle range is from 70 to 40 °C. During the model calculation, three elements generalized Maxwell model is selected, which is sufficient to describe the stress relaxation behavior. Parameters needed in this model are generated from master curve, and MATLAB® program does the recursive calculation.

Figure 5.2 shows model calculation result compared with experimental data. It turns out that this model works very well at stress relaxation prediction, but the thermal expansion effect is largely underestimated. It can be seen from Fig. 5.2 that the

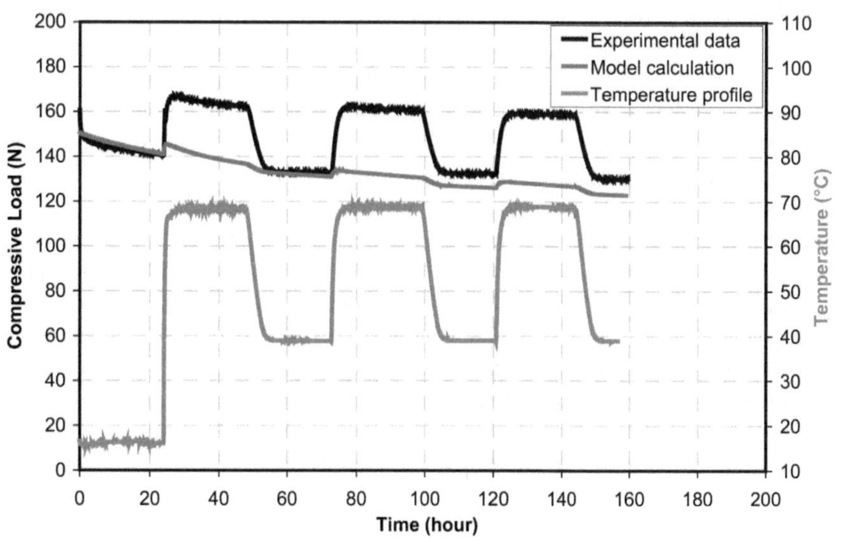

Fig. 5.2 Comparison between experimental data and generic model calculation for simple temperature cycling (cycling between 70 and 40 °C)

stress relaxation parts of the red curve are parallel with the blue curve. This indicates that if the thermal expansion effect was not underestimated, the model would work perfectly on predicting the stress relaxation behavior.

It is surprisingly found that at lower temperature, such as 40 °C where the stress relaxed very slowly, the calculated trend exactly follow the experimental data. This indicates that master curve can explain the stress relaxation behavior at different cycling temperatures. The recursive model consists of stress relaxation and thermal stress. All the parameters used in the model are generated from a master curve at a certain temperature. During temperature cycling, different parts of the master curve are shifted to different temperatures with the shift factor.

Thermal stress is usually calculated by the following equation:

$$\Delta\sigma = S\Delta\varepsilon \tag{5.3}$$

Where $\Delta\sigma$ is thermal stress, S is material stiffness and $\Delta\varepsilon$ is thermal strain.

Thermal strain can be expressed by coefficient of thermal expansion and temperature change:

$$\Delta\varepsilon = \alpha\Delta T \tag{5.4}$$

Where α is the coefficient of thermal expansion and ΔT is temperature change. Combining Eqs. (5.3) and (5.4) yields the following relation:

$$\Delta\sigma = S\alpha\Delta T \tag{5.5}$$

During the calculation of thermal stress, temperature change is specified by the experiment, coefficient of thermal expansion is a constant, and material stiffness is assumed to be a constant. In order to investigate the reason why the calculated thermal stress is much lower than the measured value, first it is suspected that the effective thermal expansion coefficient of the fixture is higher than the pure coefficient of the test sample. Due to thermal expansion, the gap between the two parallel plates may decrease, resulting in a much higher effective thermal expansion coefficient of the sample.

A new fixture is built later with two ceramic rods, and uses Instron to collect data about load change when temperature changes. It turns out that the data collected by Instron agrees very well with compression stress relaxation tester.

5.3.2 Material Property Change

Equation (5.5) shows that thermal stress depends on temperature change, coefficient of thermal expansion and material stiffness. However, with these parameters, the calculated thermal stress is far more less than the measured value from experiment. Material stiffness is to be examined, since temperature change is always prescribed by experiment.

A temperature profile test is performed as shown in Fig. 5.3. In this test, we first expose the sample to a high temperature and then continuously decrease the temperature. Note that in an old temperature region, thermal expansion and contraction has the same effect. The material stiffness is then determined from the data of this test. During temperature increases from 20 to 100 h, stress change in each small step as shown in Fig. 5.3 is divided by its temperature interval and CTE to get material stiffness according to Eq. (5.5). Table 5.1 summarizes the data derived from Fig. 5.3. Note that we use an average of stress and temperature for each step in Fig. 5.3 to correspond the material for each step in Table 5.1 as observed also in refs [4–6, 9, 10]. Material property change can explain why the thermal expansion effect is underestimated by the model proposed by Humfeld and Dillard. They used a constant material stiffness corresponding to the fresh material in their model [9, 10].

Note that during temperature increases or decreases, there is still stress relaxation, especially at a new temperature region. However, calculation shows that it is not significant at all compared with thermal expansion and contraction. So the stress changes during temperature change only account for thermal expansion or contraction.

Figure 5.4 shows that changed material stiffness is linearly dependent on both temperature and stress. However, temperature and stress are not independent, and additional experiments show that material stiffness is dependent on the specific stress level rather than temperature.

Figure 5.5 shows that material stiffness is linearly dependent on stress regardless of the strain level. Even though temperature can affect stress and then material stiffness as a consequence, the underlying factor that determines material stiffness is stress rather than temperature.

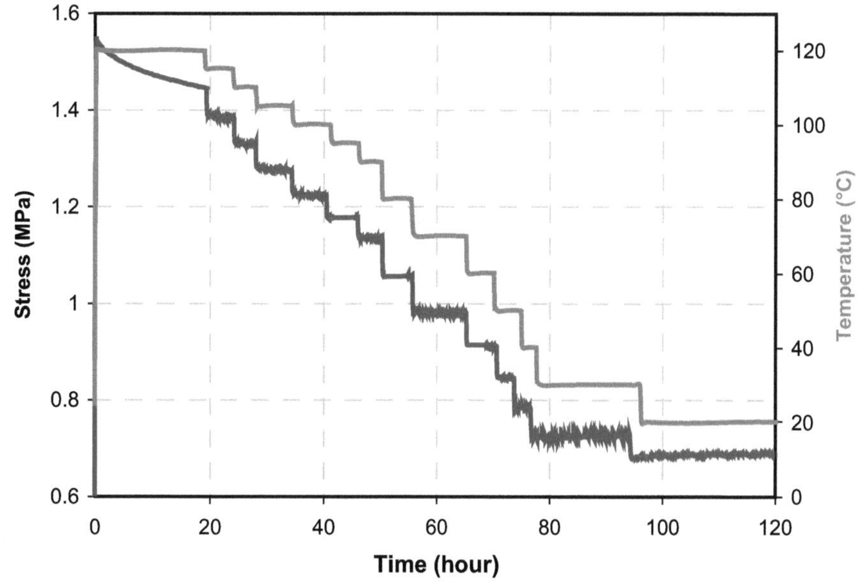

Fig. 5.3 Stress change due to temperature change (LSR, ΔT = 5 °C from 120 to 90 °C, ΔT = 10 °C from 90 to 20 °C, experiment condition: 25 % strain at t = 0 at room temperature and increases to 120 °C in about 10 min. Apparatus, Inston)

Table 5.1 A relation between material stiffness and stress and temperature

Thermal stress (MPa)	Temperature step (°C)	Average stress (MPa)	Average temperature (°C)	Material stiffness (MPa)
$\Delta\sigma$ (MPa)	ΔT (°C)	σ (MPa)	T (°C)	$S = \Delta\sigma/(\Delta T\alpha)$ (MPa)
0.057	5	1.42	117.5	41.6
0.053	5	1.36	112.5	38.8
0.049	5	1.31	107.5	36.0
0.048	5	1.25	102.5	35.2
0.046	5	1.2	97.5	33.8
0.045	5	1.16	92.5	32.7
0.079	10	1.09	85	29.1
0.074	10	1.02	75	27.2
0.069	10	0.95	65	25.5
0.067	10	0.88	55	24.5
0.059	10	0.81	45	21.8
0.061	10	0.75	35	22.4
0.052	10	0.7	25.5	19.2

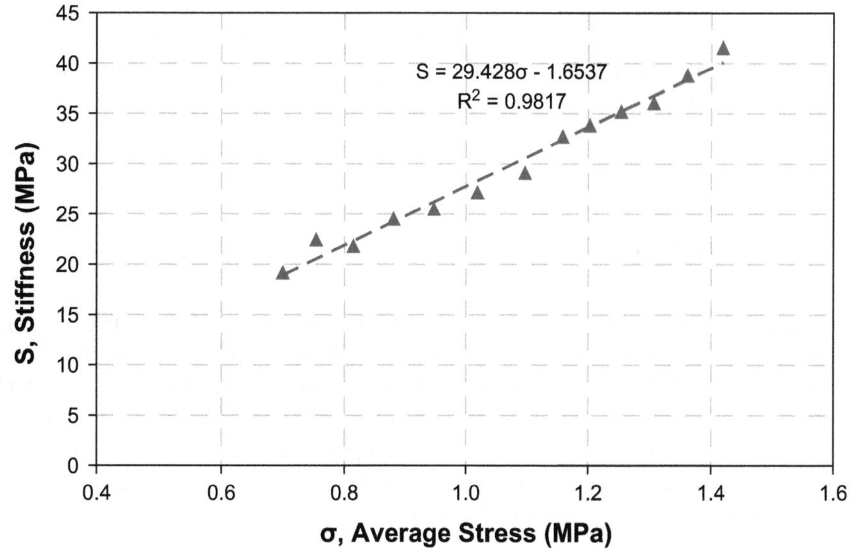

Fig. 5.4 Relation between stress and material stiffness for the data of Table 5.1

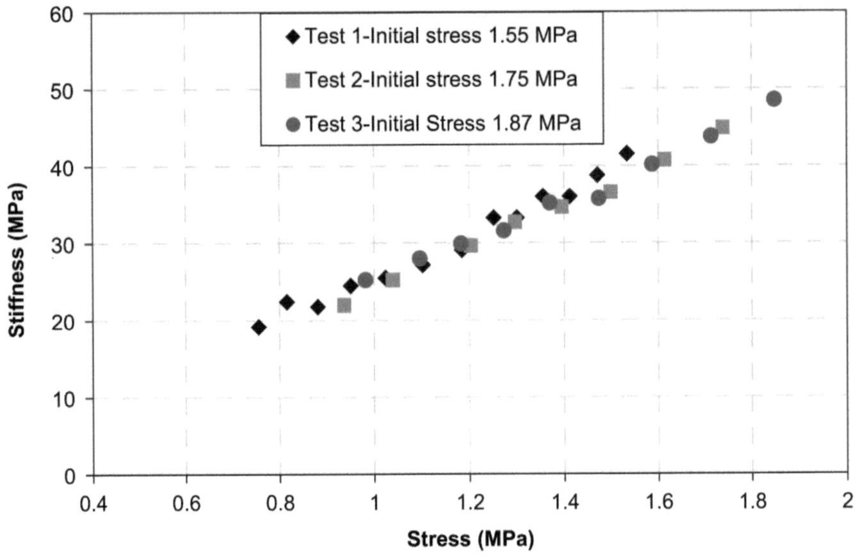

Fig. 5.5 A relation between stiffness and stress

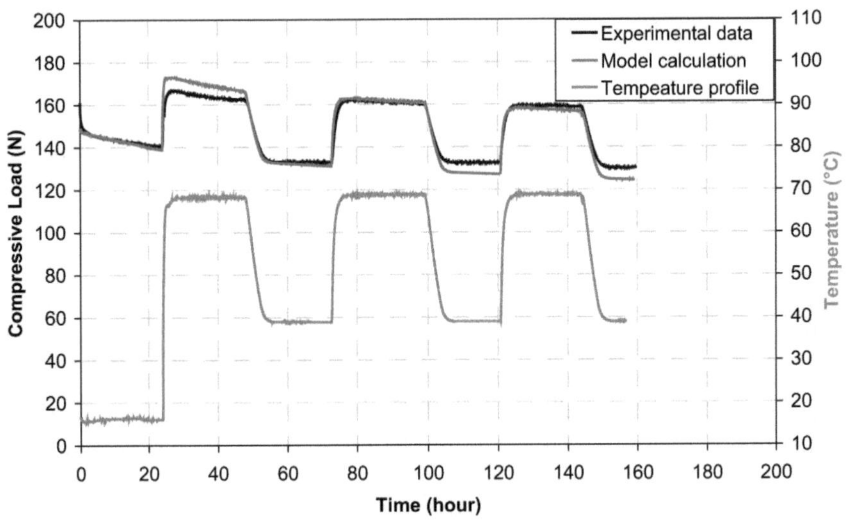

Fig. 5.6 Comparison between experimental data and modified model calculation for simple temperature cycling (between 70 and 40 °C)

After considering accumulative temperature effect on material stiffness, stress responses for different temperature profiles can be calculated. Figure 5.6 show the comparison between model calculation and experimental data, and they agree very well.

5.4 Conclusion

In this paper, stress relaxation behavior under temperature cycling is discussed. It is found that thermal expansion effect on compressive stress change is more pronounced than stress relaxation during thermal cycling. Modified generalized Maxwell model can predict stress relaxation at different temperatures but fails to calculate the thermal stress when temperature changes, in which a constant fresh material stiffness is always used. Experimental data shows that material stiffness increases when temperature changes, and temperature history exerts a large effect on material stiffness change. With the increased material stiffness, the solution to generalized Maxwell equation can successfully predict the stress relaxation behavior under different temperature profile.

Acknowledgements This study is sponsored by the US Department of Energy (DE-FC36-06G086041 and DE-FG36-08GO88116) and the NSF Industry/University Cooperative Research Center for Fuel Cells at the University of South Carolina. The author Y.J. Chao would also like to thank the financial support from the National Natural Science Foundation of China (51275338).

References

1. Liu D et al (2006) Relaxation of proton conductivity and stress in proton exchange membranes under strain. J Eng Mater Technol Trans ASME 128(4):503–508
2. Bernstein R, Gillen KT (2009) Predicting the lifetime of fluorosilicone o-rings. Polym Degrad Stab 94(12):2107–2113
3. Ronan S et al (2007) Long-term stress relaxation prediction for elastomers using the time-temperature, superposition method. Mater Des 28 (5):1513–1523
4. Derham CJ (1997) Transient effects influencing sealing force in elasomeric O ring seals. Plast Rubber Compos Proc Appl 26(3):130–136
5. Gordon AP, Drilling B, Williams K, Hebert S, Kammerer CC, Baldwin F (2011) Optimization of re-torque and relaxation parameters of the GUCP. In: ASME 2011 PVP conference, Baltimore
6. Gordon AP, Maricela De Santiago JW (2011) Analytical modeling of the mechanics of re-torque. In: ASME 2011 pressure vessel and piping conference (PVP 2011), Baltimore
7. Singh HK (2009) Lifetime prediction and durability of elastomeric seals for fuel cell application. The Virginia Polytechnic Institute and State University, Blacksburg, pp 21-22
8. Tong Cui CWL, Chien CH, Chao YJ, Van Zee JW (2011) Service life estimation of liquid silicone rubber seals in polymer electrolyte membrane fuel cell environment. J Power Sources 196(3):1216–1221
9. Humfeld GR, Dillard DA (1998) Residual stress development in adhesive joints subjected to thermal cycling. J Adhes 65(1–4):277–306
10. Dillard JG et al (1998) The durability of adhesively-bonded titanium: performance of plasma-sprayed polymeric coating pretreatments. J Adhes 65(1–4):217–238

Chapter 6
Small Strain Plasticity Behavior of 304L Stainless Steel in Glass-to-Metal Seal Applications

Bonnie R. Antoun, Robert S. Chambers, John M. Emery, and Rajan Tandon

Abstract Cracks in glass-to-metal seals can be a threat to the hermeticity of isolated electronic components. Design and manufacturing of the materials and processes can be tailored to minimize the residual stresses responsible for cracking. However, this requires high fidelity material modeling accounting for the plastic strains in the metals, mismatched thermal shrinkage and property changes experienced as the glass solidifies during cooling of the assembly in manufacturing. Small plastic strains of just a few percent are typical during processing of glass-to-metal seals and yet can generate substantial tensile stresses in the glass during elastic unloading in thermal cycling. Therefore, experimental methods were developed to obtain very accurate measurements of strain near and just beyond the proportional limit. Small strain tensile characterization experiments were conducted with varying levels and rates of strain ratcheting over the temperatures range of −50 to 550 °C, with particular attention near the glass transition temperature of 500 °C. Additional experiments were designed to quantify the effects of stress relaxation and reloading. The experimental techniques developed and resulting data will be presented. Details of constitutive modeling efforts and glass material experiments and modeling can be found in Chambers et al. (Characterization & modeling of materials in glass-to-metal seals: Part I. SAND14-0192. Sandia National Laboratories, January 2014).

Keywords Stainless steel • Plasticity • Small strain • Relaxation • Temperature

6.1 Material

The material used in this study was a 4 in. (101.6 mm) diameter bar of 304L VAR stainless steel, produced by Electralloy. The chemical composition is shown in Table 6.1 and the average grain size was ASTM 5 with the largest grain size in the material equal to ASTM 4 (100 μm). Although mechanical characterization of alloys usually includes tensile testing to large strains and failure (e.g. [2]), such as shown in Fig. 6.1 [3], the focus of this work is in the small strain regime of just a few percent strain [1]. Tensile specimens 4.25 in. (107.95 mm) in total length were removed from the bar stock material by electric discharge machining (EDM), and were machined to have an overall straight gage section of 1.5 in. (38.1 mm) and gage diameter of 0.35 in. (8.89 mm), with 5/8 in (15.875 mm) threaded ends that couple to the tensile test fixtures.

To mimic the material condition in glass-to-metal seals, a representative glass-ceramic sealing cycle was chosen and all machined 304L VAR specimens were subjected to this thermal process prior to testing. The prescribed cycle was as follows: heat to 990 °C at 25 °C/min, hold for 10–12 min at 990 °C, cool to 800 °C at 10 °C/min, ramp from 800 to 482 °C at 25 °C/min, hold at 482 °C for 45 min and finally ramp to 75 °C at 15 °C/min. However, the thermal history traces provided from the heat treatment facility showed that the specimens were actually were held at 990 °C for about 40 min rather than the prescribed 10–12 min. Whether this makes a significant difference on the resulting microstructure or mechanical response is currently under study by the authors.

B.R. Antoun (✉)
Sandia National Laboratories, Livermore, CA 94551, USA
e-mail: brantou@sandia.gov

R.S. Chambers • J.M. Emery • R. Tandon
Sandia National Laboratories, Albuquerque, NM 87185, USA

H.J. Qi et al. (eds.), *Challenges in Mechanics of Time-Dependent Materials, Volume 2: Proceedings of the 2014 Annual Conference on Experimental and Applied Mechanics*, Conference Proceedings of the Society for Experimental Mechanics Series,
DOI 10.1007/978-3-319-06980-7_6, © The Society for Experimental Mechanics, Inc. 2015

Table 6.1 Chemical composition (wt%) of 304L VAR stainless steel material

Al	B	C	Co	Cr
0.009	0.0014	0.022	0.055	19.02
Cu	Fe	Mn	Mo	N
0.092	Balance	1.38	0.16	0.02
Ni	O	P	S	W
10.14	0.003	0.020	0.002	0.017
Si	Ti	V	Nb	Sn
0.63	0.0036	0.049	0.019	0.012

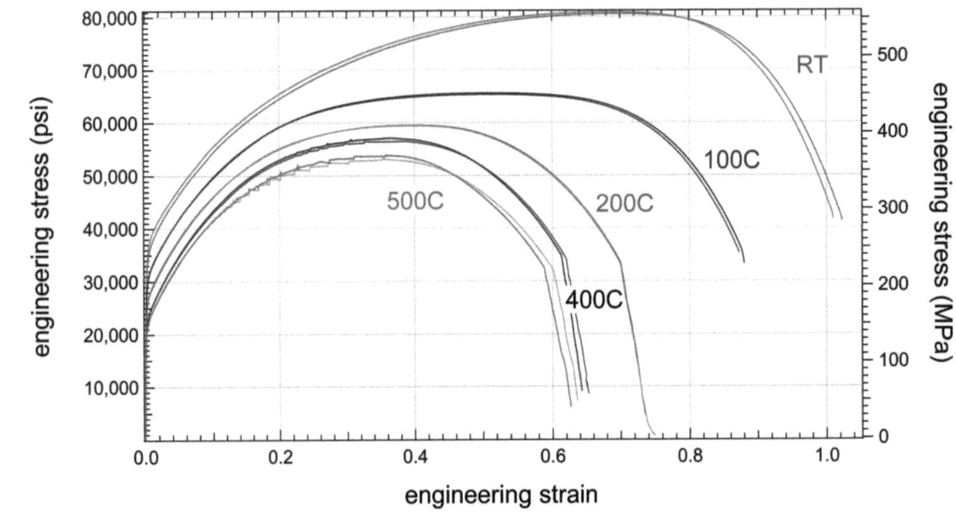

Fig. 6.1 Large strain behavior of 304L VAR stainless steel material similar to the one used in this study [3]

6.2 Experimental Equipment

304L stainless steel is used in many Sandia applications and components, but typically is chosen for its excellent ductility. Hence, experiments to characterize this alloy use techniques that allow measurement of very large deformation through failure, often at strain levels over 1. Using the same techniques and measurement methods for small strains does not result in acceptable results, however, due to signal noise and reduced accuracy of the strain measurements at very low levels. Therefore, this study required the development of an experimental apparatus to measure small strains from room temperature to 500 °C.

Early in the study, a few approaches were considered and tried to address the need for high accuracy measurements in elevated temperature environments. Figure 6.2 shows a photograph of the test system with both a small three zone furnace and a large environmental chamber. Furnace heating produces high temperature uniformity with little air movement that can adversely affect the extensometer but requires mounting the extensometer outside the furnace. This necessitates extended rods to contact the specimen. Heating within the environmental chamber introduces air flow but allows direct, close contact of the extensometer to the specimen. However, an extensometer that can withstand temperatures of 500 °C was required and had to be acquired. After trying both methods, the latter one was chosen based on better results such as less signal noise and better repeatability.

A small strain extensometer was purchased from Epsilon Technology, a model 3555-HT, which could be placed directly into the environmental chamber for temperatures up to 540 °C. This model uses a high temperature capacitive sensor that does not require cooling, but does require a dedicated, factory calibrated signal conditioner. The extensometer has a gage length of 0.5 in. (12.7 mm) and a total travel or extension of 0.02 in. (0.508 mm) for a maximum measureable strain of 0.04 or 4 %. The extensometer was purchased with a rigid, side-mounted cable so it could be fed straight out of a port hole on the side of the chamber to allow support of the cable and reduction of movement from air currents.

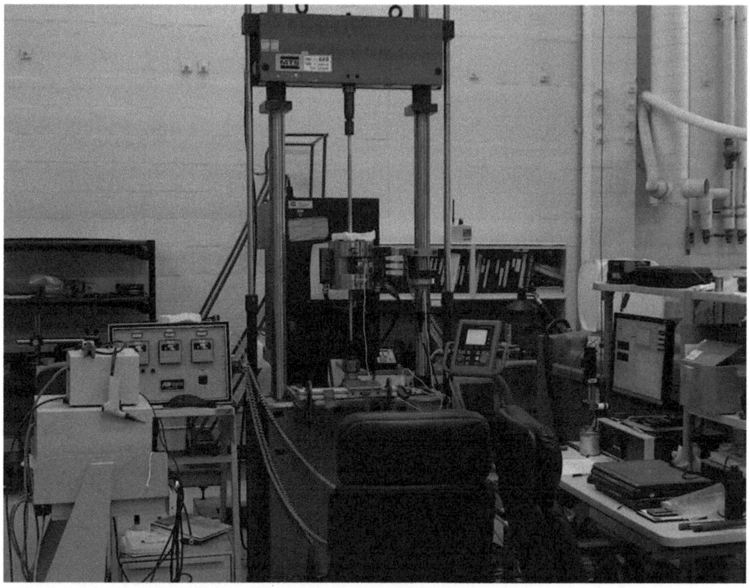

Fig. 6.2 MTS test frame used for all small strain characterization experiments

Fig. 6.3 Extensometer mounted on a specimen for testing

All experiments were conducted on a 20 Kip (89 kN) MTS 880 test frame, the environmental chamber can be seen in the back of the photograph in Fig. 6.2. A photograph of the extensometer mounted on a specimen is shown in Fig. 6.3. The MTS test system controller and software, Flextest GT and Flexware, were used to program the various experiments, typically in strain control using the extensometer signal or in load control for some of the strain ratcheting tests. Experiments conducted in strain control were conducted at 3.1E−5 s^{-1}, similar to the strain rate expected in glass-to-metal seals sealing operation.

6.3 Experimental Results for Tension and Stress Relaxation

Material characterization experiments were conducted at six temperatures, room temperature, 100, 200, 300, 400 and 500 °C. Primarily, tests were conducted in extensometer (strain) control at 3.1E−05 s^{-1}, but at room temperature, 300 and 500°C, additional tests were conducted at a slower strain rate of 3.1E−06 s^{-1}. A loading scheme was used to maximize the information that could be collected during each test. All specimens were loaded to a strain of 0.03 (3 %), then held in

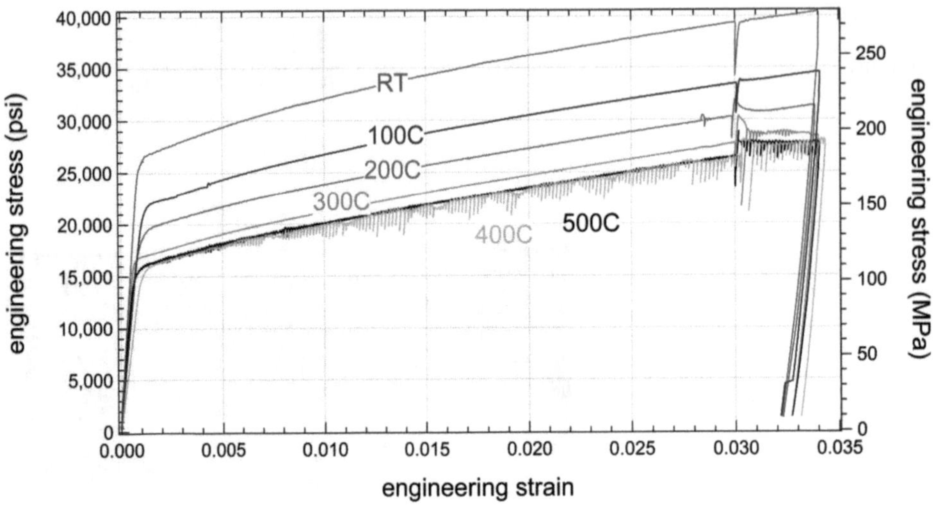

Fig. 6.4 Tensile response of glassing cycle treated 304L VAR stainless steel material

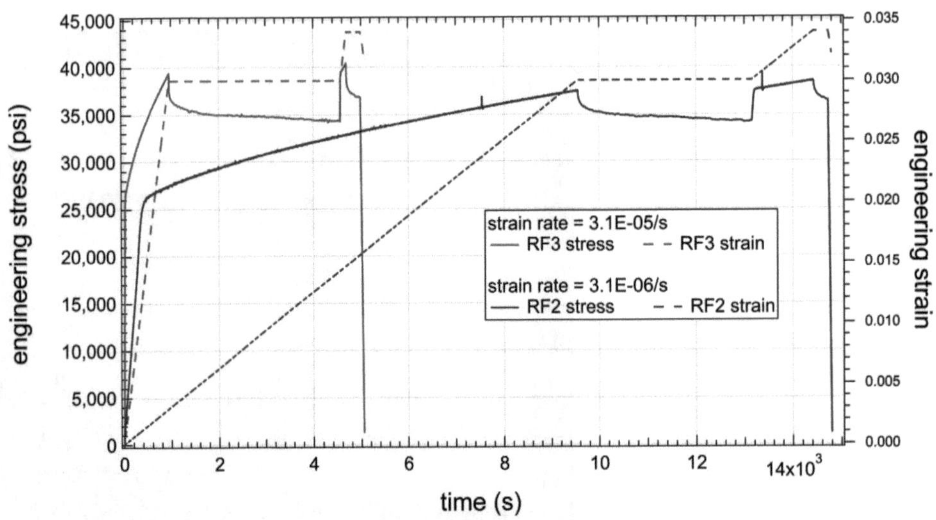

Fig. 6.5 Stress relaxation in 304L VAR stainless steel after tensile loading at 3.1E−05 and 3.1E−06 s^{-1} at room temperature

extensometer (strain) control for a period of 60 min while the load was allowed to relax. This was followed by a brief reloading at the starting strain rate to a strain of 0.034 (3.4 %) and another relaxation period of 5 min.

A graph showing stress versus strain for all of the temperatures is shown in Fig. 6.4, where the relaxation period followed by reloading at 0.03 strain is clear. The tensile curves generally drop with increasing temperature as expected. Note that dynamic strain aging (serrated yielding) occurs at 400 and 500 °C throughout the loading, with larger serrations measured at 400 °C. The result, at least in the low strain regime is that the 500 °C tensile curve lies just above the 400 °C tensile curve. Dynamic strain aging is also evident at 300 °C during the second loading portion, possibly activated by the stress relaxation period or by increasing strain levels.

The next several figures show stress (solid line, left axis) and strain (dashed line, right axis) versus time to illustrate the stress relaxation response of the material. Figure 6.5 shows the stress relaxation response at room temperature after loading at both strain rates to a strain of 3 %. Notice that substantial stress relaxation occurs even though relaxation started at a total strain of only 0.03 (3 %). Also, relaxation appears to have reached a steady state or constant slope, but has not leveled off which seems to indicate relaxation of stress could continue to occur, if allowed, for a very long period of time. Finally, due to the strain rate sensitivity of 304L stainless steel, including at room temperature, the higher strain rate test RF3 reaches a higher stress level than the lower strain rate test RF2. This is likely the primary reason that more stress relaxation occurs in

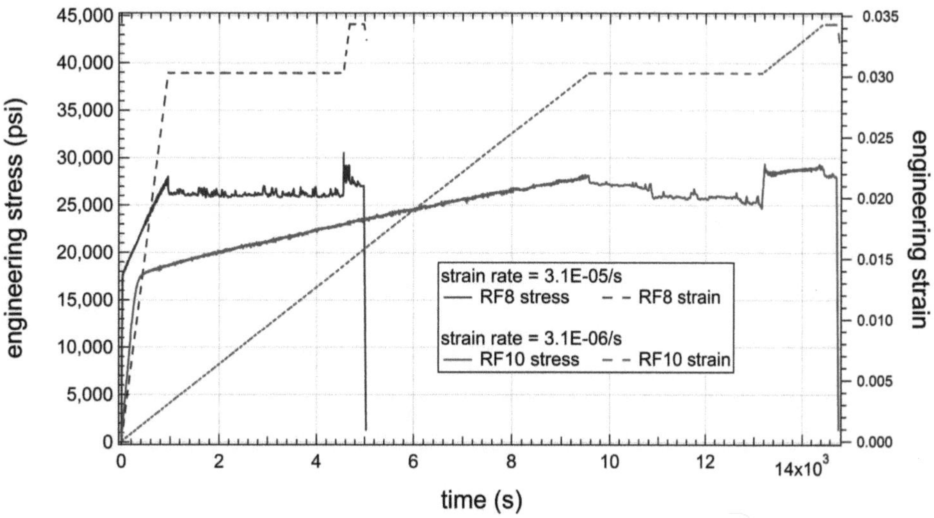

Fig. 6.6 Stress relaxation in 304L VAR stainless steel after tensile loading at 3.1E−05 and 3.1E−06 s^{-1} at 300 °C

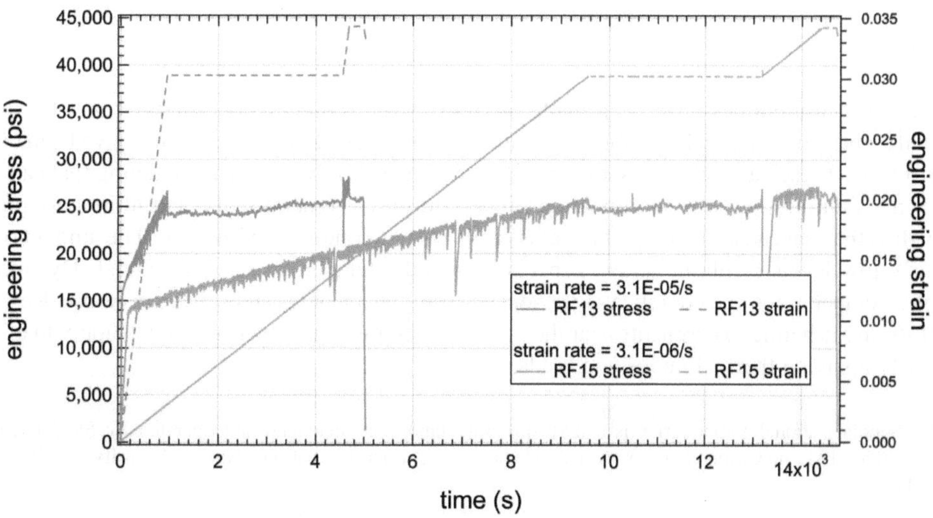

Fig. 6.7 Stress relaxation in 304L VAR stainless steel after tensile loading at 3.1E−05 and 3.1E−06 s^{-1} at 500 °C

RF3 than RF2. Figure 6.6 shows the stress relaxation response at 300 °C, again at both strain rates. An initial relaxation response is measured for RF8, the specimen loaded at the slower strain rate, but relaxation ends after a short amount of time and is quickly overshadowed by serrated loading due to dynamic strain aging of the material during the strain hold period. For RF10, loaded at the higher rate, dynamic strain aging is also interfering with the material stress relaxation. Figure 6.7 shows the stress relaxation response at 500 °C, for both strain rates. At this temperature, dynamic strain aging is completely entangled with relaxation. It appears that relaxation happens quickly. Finally, Fig. 6.8 shows the remaining temperatures, 100, 200, and 400 °C, that were loaded only at the higher strain rate. At 100 °C, specimen RF5 shows immediate relaxation followed by a transition to a steady state, slow relaxation and eventually relaxation appears to end. The reloading of RF5 appears to show a similar material response. At 200 °C, specimens RF6 and RF7 look similar to the 100 °C response, but the amount of relaxation is lower and steady state is reached sooner. At 400 °C, the results are similar to that described previously for 300 and 500 °C.

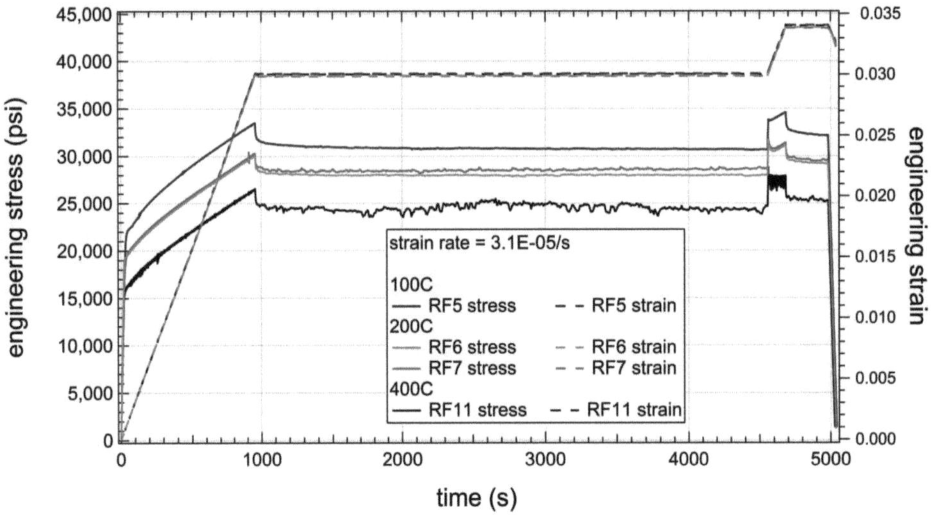

Fig. 6.8 Stress relaxation at 100, 200 and 400 °C in 304L VAR stainless steel after tensile loading at 3.1E−05 s^{-1}

6.4 Summary

Detailed experimental measurements have been made on 304L VAR stainless steel material that has been processed to mimic that found in a glass-to-metal seal. Combined tension and stress relaxation experiments have been completed over the range from 25 to 500 °C. These data are necessary for high fidelity constitutive material modeling that can account for small plastic strains, changes during the glass solidification process in glass-to-metal seals, and ultimately predict the residual stress state in the glass that can cause cracking of the glass and loss of hermeticity of the seal. The amount of stress relaxation that occurs at fairly low levels of plastic strain, especially at room temperature, is substantial. Dynamic strain aging is evident at the higher temperatures and complicates the tensile straining and stress relaxation response. Research continues in the areas of small strain ratcheting experiments near the proportional limit and experiments that apply loading and unloading while simultaneously applying thermal ramps and holds.

Acknowledgements Sandia National Laboratories is a multi-program laboratory managed and operated by Sandia Corporation, a wholly owned subsidiary of Lockheed Martin Corporation, for the U.S. Department of Energy's National Nuclear Security Administration under contract DE-AC04-94AL85000.

References

1. Chambers RS, Emery JM, Tandon R, Antoun BR, Stavig ME, Newton C (2014) Characterization & modeling of materials in glass-to-metal seals: Part I. SAND14-0192. Sandia National Laboratories (January 2014)
2. Antoun BR, Korellis JS, Song B (2007) Techniques for testing 304L stainless steel over a wide range of temperatures. In: Proceedings of the 2007 SEM annual conference on experimental & applied mechanics, Springfield, 4–6 June 2007
3. Antoun BR (2013) Sandia National Laboratories Internal Memo (2 August 2013)

Chapter 7
Observations of Rate-Dependent Fracture of Locally Weakened Interfaces in Adhesive Bonds

Youliang L. Guan, Shantanu Ranade, Ivan Vu, Donatus C. Ohanehi, Romesh C. Batra, John G. Dillard, and David A. Dillard

Abstract Studies of the fracture behavior of adhesive joints can provide scientific understanding of failure processes as well as properties required for engineering design purposes. The focus of the present paper is to discuss the role that locally weakened interfaces in adhesive bonds can have on the fracture behavior of double cantilever beam specimens loaded in both mode I conditions and in mixed (mode I/II) conditions in a dual actuator load frame that permits independent control of the applied loads. Locally weakened areas are created by several methods of contamination of one aluminum adherend, including physical vapor deposition of copper through a mask perforated with the desired size, spacing, and pattern. Results from this experimental study have provided evidence of the size of a weakened zone that is required to be detected by a growing cohesive crack for a commercial adhesive system. The detection size depends on the mode mixity applied, with opening shear conditions rendering detection of smaller weakened zones and closing shear conditions detecting only larger weakened zones. In addition, an interesting rate dependence will be described in which rapidly growing cracks are more likely to detect locally weakened zones than more slowly growing cracks for several systems studied. Possible mechanisms will be suggested.

Keywords Rate-dependent fracture • Weakened interfaces • Adhesive • Double cantilever beam (DCB) specimens • Aluminum/epoxy bonds

7.1 Introduction

Surface preparation is often important in developing strong adhesive bonds that will effectively resist environmental and loading challenges. Contaminants such as oil, dust, and weakly adhering layers can all affect bond strength and durability, though the significance can vary dramatically depending on the nature of the contaminant, its location and extent, and the robustness of the adhesive. For strength tests, such as the single lap joint, the presence of even relatively large defects within the central region of the overlap, where the stresses are relatively low, as suggested by shear lag and beam on elastic analyses [1], causes relatively little effect in adhesives with limited ductility [2, 3]. In highly ductile adhesive layers, such as polyethylene, flaw size has been found to reduce joint strength in proportion to the size of the defect [4].

Our recent interests have included the effect of relatively small, localized defects on the locus of failure and the associated fracture resistance of structural adhesive bonds. Several methods are used to induce such localized defects, including contamination with graphite and brief contact with silicone-based pressure sensitive adhesives [5]. Chemical vapor

Y.L. Guan • D.C. Ohanehi • J.G. Dillard • D.A. Dillard (✉)
Engineering Science & Mechanics Department, Virginia Tech, Blacksburg, VA 24061, USA
e-mail: dillard@vt.edu

S. Ranade
Macromolecular Science & Engineering Program, Virginia Tech, Blacksburg, VA 24061, USA

I. Vu
Mechanical Engineering Department, Virginia Tech, Blacksburg, VA 24061, USA

R.C. Batra
Chemistry Department, Virginia Tech, Blacksburg, VA 24061, USA

H.J. Qi et al. (eds.), *Challenges in Mechanics of Time-Dependent Materials, Volume 2: Proceedings of the 2014 Annual Conference on Experimental and Applied Mechanics*, Conference Proceedings of the Society for Experimental Mechanics Series,
DOI 10.1007/978-3-319-06980-7_7, © The Society for Experimental Mechanics, Inc. 2015

deposition of copper through special masks onto surface pretreated aluminum adherends is also being studied; this will be conveyed in the presentation but not herein. The results were obtained as part of a larger research effort that involved the development of experimental evidence and numerical procedures [6–8] to characterize the effects of mode mixity and weakened interfaces on fracture propagation in adhesive bonds.

7.2 Initial Experimental Observations

The double cantilever beam (DCB) specimens were made of 6061-T6 aluminum alloy, and bonded with one of two, commercially available toughened epoxy adhesives, referred to herein as Epoxy A and Epoxy B. Both of these were nominally room temperature cure systems. All adherends were first surface treated by placing them in a 10 % (wt) NaOH solution for 10 min, and then in 50 % (v) HNO_3 for 3 min, with a distilled water rinse after each step. Some initial mode I DCB tests showed that bonding to the interfaces was strong, resulting exclusively in cohesive failures within the adhesive layer. The average value of the fracture energy was found to be about 500 J/m^2 and appeared to be independent of crack propagation rate over the range of the tests conducted.

Initially a lead pencil was manually used to locally contaminate strips on the adherend surface, depositing with graphite particles. Subsequently the joints were bonded with epoxy adhesive, and tested in pure mode I loading. This local contamination, resulting in areas of reduced adhesion, was found to affect the path followed by the growing crack in some situations but not others. There are at least four factors that may influence the crack path selection: the width of the weakened strips, the distance between them, the degree of contamination, and, as reported here, the effective crack propagation rate at the crack tip. In order to obtain insights to guide the plans for further study, several patterns were applied on a single specimen. Out of the five approximately 5 mm wide weak zones, as shown in Fig. 7.1, four were clearly discovered by the growing crack, resulting in what are apparently failures within the weakened interfacial areas. But the fifth one was only partially detected by the growing crack tip. This test was repeated on another specimen with the same pattern and the same result was obtained. Because this DCB specimen was loaded at a constant crosshead displacement rate, the crack propagation rate was expected to be higher at the beginning but decrease as the crack length increased, at least in an average sense, assuming that the fracture energy is constant. The results suggested possible rate dependence for the detection of the weakened regions, as borne out further in the tests described below.

To verify this idea, other specimens with the same pattern of contaminated interfaces were tested at a constant crack propagation rate that was applied based on simply beam theory through custom function implemented in the controller used

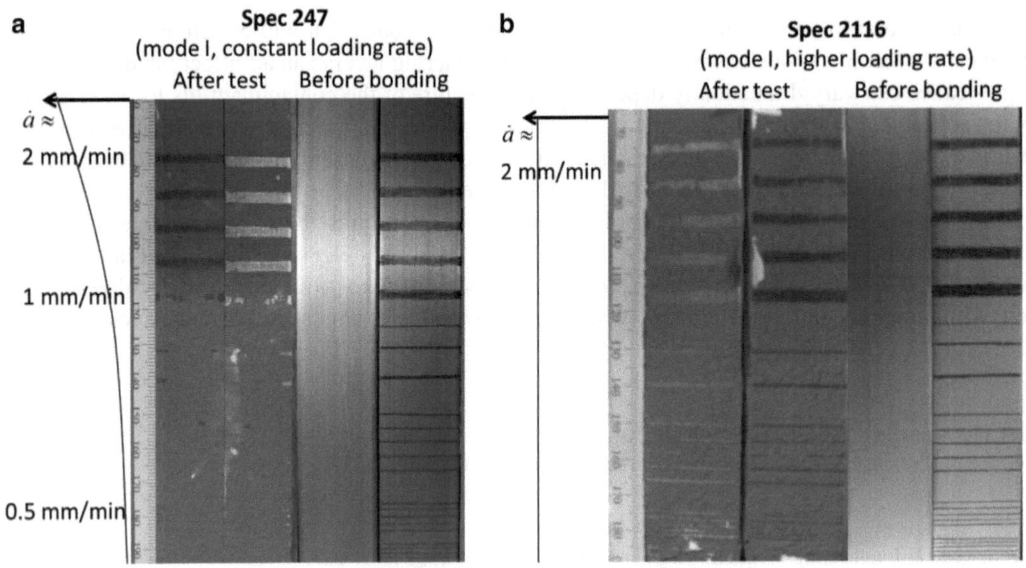

Fig. 7.1 Crack path selection at different displacement rates in Epoxy A adhesive bonded DCB joints: (**a**) constant crosshead loading rate: 0.1 mm/min, (**b**) constant crack propagation rate. (On the surfaces before bonding, the black bands were weakened areas, theoretically predicted crack propagation speed curves are shown on the left side of each picture.) The results appear very sensitive to propagation speed, which near the lower end of the two specimens differed by only a factor of 4

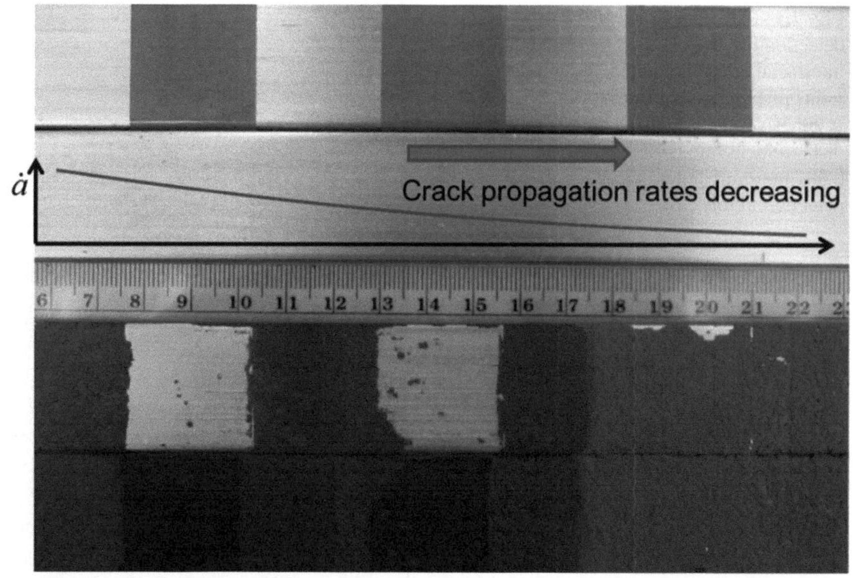

Fig. 7.2 Rate dependent detection of weakened interfacial areas in bonded joints prepared with Epoxy A adhesive, the crosshead displacement rate is 0.1 mm/min, and the crack propagation rate is decreasing

with the dual actuator load frame (DALF) [9, 10]. With the goal of inducing a constant rate of crack propagation, a constant fracture energy ($G_{Ic} = 500$ J/m^2) is anticipated for a DCB specimen. If this is the case, the crack propagation rate can be shown to be proportional to the reciprocal of the crack length a, as well as the crosshead displacement rate, as shown in the following relationship:

$$\dot{a} = \frac{\dot{\Delta}}{2a}\sqrt{\frac{EI}{G_{Ic}B}} \tag{7.1}$$

where $\dot{a}$ is the crack propagation rate, $\dot{\Delta}$ is the crosshead displacement rate, EI is the stiffness of the aluminum adherend, B is the width of the adherends and bondline, as well as the bond width. Therefore, by using this equation, an approximately constant $\dot{a}$ was attained by adjusting $\dot{\Delta}$ after every 1 mm increment in a manually. (Up to now, the crosshead displacement is calculated automatically and suggested, but applied by the user manually in order to avoid potential safety issue resulted from uncertainties during the tests.) The initial crosshead displacement rate of 0.12 mm/min in these tests was gradually increased to 2 mm/min for $a = 60$ mm to maintain a nearly constant crack propagation rate along the entire specimen length.

In summary, it was found that higher crack propagation rates appear much more likely to detect the weakened interfacial zones for these bonded joints. However, these weak zones were detected at low crack propagation rates. Similarly, most of the narrow weak zones ($\approx$1 mm) were not detected by the main crack when the propagation rate was low. But they were clearly identified when the growth rate was high in Fig. 7.1b.

A similar rate dependency was also found in specimens contaminated during brief contact by a commercially available silicone tape. Silicone tape segments were applied to the treated (by base and acid solution) and dried aluminum (the same aluminum alloy as used in the previous tests) surfaces manually, and then removed after a 1 min exposure. The size of the silicone contaminated areas was approximately 25 mm by 25 mm, and the spaces between these areas were also 25 mm. During the brief contact, some low molecular weight silicone contamination was potentially transferred to the covered areas of the adherends.

Following exposure, the joints were bonded using the same method described above. Quasi-static DCB tests were conducted at a crosshead displacement rate of 0.1 mm/min. After the test, it was found that the first and second weakened areas were detected by the main crack in the adhesive layer, where the crack propagation rate is relatively fast. However, the third weakened area, where the crack propagation rate is slower, was not detected by the main crack, as seen in Fig. 7.2. This test was also repeated and similar results were found. When similar tests were tested at a higher test rate (0.002 mm/min), detection of the contaminated areas was less pronounced, as shown in Fig. 7.3.

All of these tests showed that the detection of locally weakened areas by a propagating crack appears to be a rate dependent phenomena in adhesive bonds. Interestingly, similar rate dependence was found for two entirely different methods of contamination: graphite flakes and unreacted, low molecular weight silicone. Stimulated by the unexpected results of these initial and rather crude experiments, addition tests with greater control were performed to further elucidate this behavior.

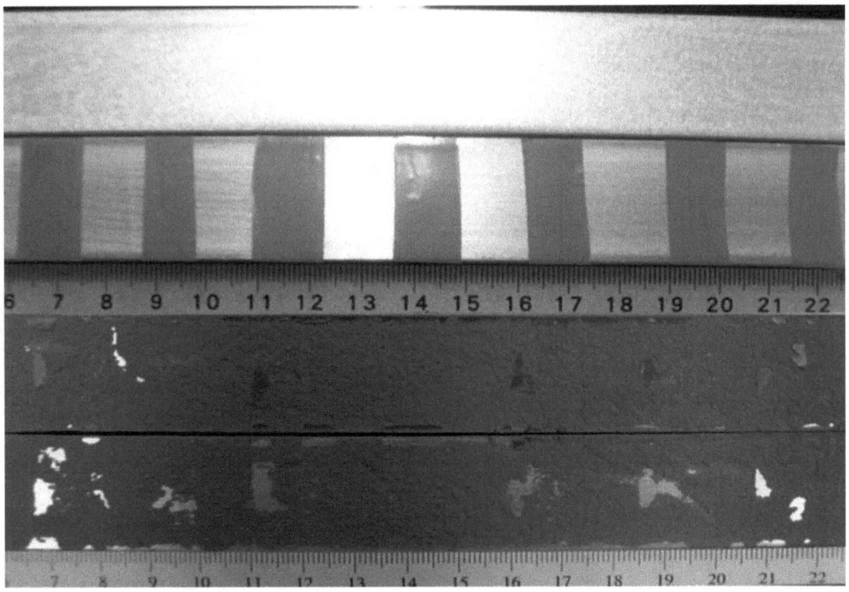

Fig. 7.3 Rate dependent detection of weakened interfacial areas in bonded joints prepared using Epoxy A adhesive, the crosshead displacement rate is 0.002 mm/min and no clear detection in this case

7.3 Subsequent Experimental Results Using Graphite Contamination

Graphite was used again to prepare contaminated areas with better spacing control and repeatable applications of the contaminant. Some half millimeter graphite lines with 8.5 mm spacing were applied to the treated and dried adherends, as discussed above. Six of these DCB specimens were bonded with the Epoxy A, and six more were bonded with the Epoxy B. Again, these specimens were tested at different crosshead displacement rates: some of them at 0.01 mm/min, some of them at 100 mm/min. After the DCB tests, specimens bonded with different epoxies showed different results. For the specimens bonded with the Epoxy A, high propagation rate tests detected almost all the weakened interfaces clearly, but low propagation rate tests didn't find any of them, as shown in Fig. 7.4. The fracture energy curve obtained by using a corrected beam theory analysis [11] of the resulting data is shown in Fig. 7.5, which resulted in slightly higher values than previously reported for DCB tests conducted at a crosshead displacement rate of 0.1 mm/min.

For the specimens bonded with the Epoxy B, neither the high propagation rate tests nor the low propagation rate tests detected the weakened interfaces clearly. This is shown in Fig. 7.6, where no clear interfacial failures are apparent visually, even though the crack appeared to be traveling quite close to the contaminated interface, especially at the higher loading rate. In summary, for a common general-purpose epoxy such as the Epoxy A adhesive, the crack path is quite sensitive to the graphite contamination on the interface when tested at higher rates. For a more highly formulated commercial adhesive such as Epoxy B, however, the crack path in the adhesive layer appears less sensitive to the graphite contamination on the interface, and at least visually, does not appear to be detect and debond the weakened interfaces under either slow or more rapid crosshead rates and crack propagation rates.

The creation of locally weakened regions is challenging, but these results suggest an interesting rate dependence associated with the detection and resulting propagation through weakened interfacial regions. Though relatively large non-bonding inserts or inclusions have been used [2–4], making small versions of these and locating them in controlled patterns is difficult. More refined methods for creating local defects are currently being developed in which copper is physical vapor deposited (PVD) onto pretreated aluminum adherends, resulting in very weak, carefully controlled, and precisely patterned defects [12]. Size, spacing, and fracture mode mixity are found to play roles in the resulting detection of and crack propagation paths for such specimens.

7.4 Experimental Results Using Silane Weakened Interfaces

Silane coupling agents are widely used as surface treatments to develop robust and durable bonds to a wide range of adherends. These products are available in a range of different chemical forms, each offering different abilities to enhance bond performance. Some silanes may even reduce adhesion, as will be described here. These were used to weaken one

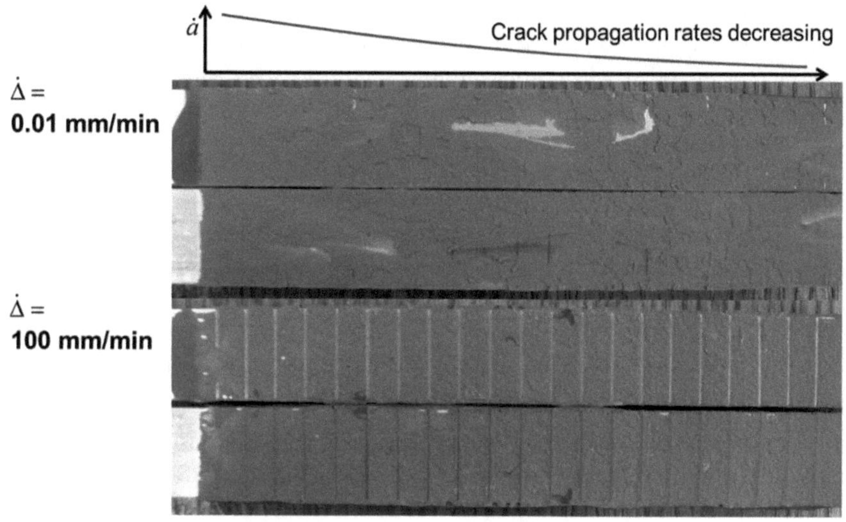

Fig. 7.4 Rate dependent detection of weakened interfacial areas in bonded joints prepared by the Epoxy A adhesive, the crosshead displacement rates are 0.01 mm/min, and 100 mm/min, respectively. High propagation rate tests resulted in the clear detection of almost all the weakened interfaces, but low propagation rate tests resulted in almost no detection, as judged visually

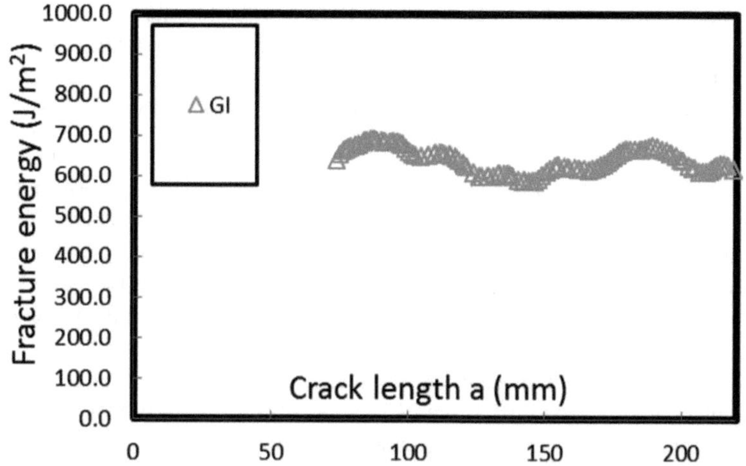

Fig. 7.5 Fracture energy curve for one high crosshead rate test (100 mm/min) of specimens bonded with the Epoxy A

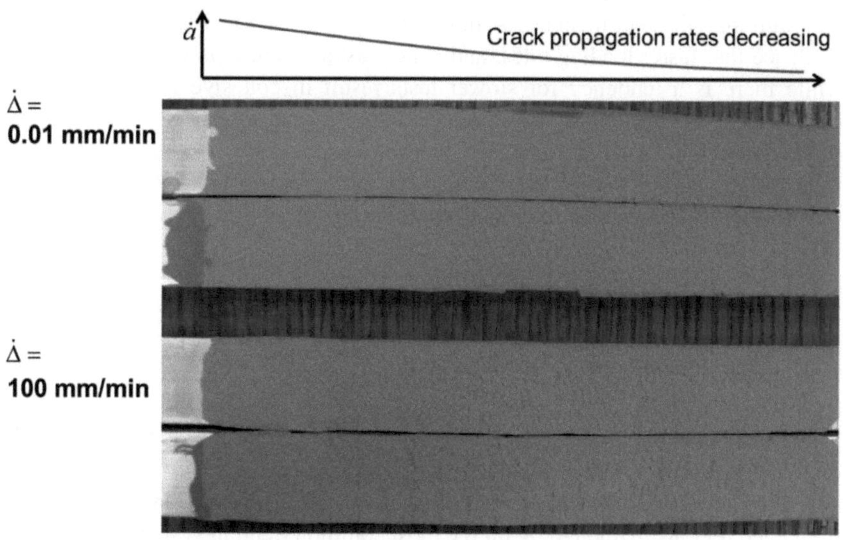

Fig. 7.6 Rate dependent detection of weakened interfacial areas in bonded joints prepared by the Epoxy B adhesive, the crosshead displacement rates are 0.01 mm/min, and 100 mm/min, respectively. Neither the high propagation rate tests nor the low propagation rate tests resulted in clear interfacial failures, though there are hints of the graphite covered strips showing through in the high rate test

aluminum interface with respect to the corresponding bond surface on the other adherend. These are thus not locally weakened areas, but rather globally weakened interfaces that allow one to discriminate between failures at one adherend interface or the other when propagating cracks grow interfacially rather than cohesively. For these tests, one of the two adherends was weakened by immersion in a silane solution chosen to provide lower bonding strength and affect the resulting crack path selection. Several different silane solutions were used to provide a unique and uniform treatments on a large area, thereby reducing inconsistency of the contamination process. In principle, if the surface was very weak, the crack in the adhesive should turn into the interface. If the surface is not so weak, the crack in adhesive could, sometimes, ignore the weak surfaces and remain cohesive within the adhesive, at least when loaded in pure opening mode fracture (mode I). The contact angle between the water drop and aluminum beam was used as an indication of the bonding strength: smaller angle meant stronger bonding and higher surface energy.

For some of the tests, propyltrimethoxysilane (PS) was used to weaken the adhesion to one adherend, while 3-aminopropyltrimethoxysilane (APS) was used to strengthen it. The application methods are the same for both silane coupling agents and can be applied through the same process:

1. Complete abrading and base–acid treatments of adherends as above, but without drying.
2. In a beaker prepare a solution of approximately 5 % (v/v) PS in 100 % (200 proof) ethanol.
3. To 100 mL of the solution prepared above, add about 5 mL of 0.1 M acetic acid. Stir this solution for about 10 min (Solution A).
4. In a separate beaker, add 5 mL of 0.1 M acetic acid to 100 mL of ethanol (Solution B). Immerse all or one end of the adherends in Solution B for about 10 min with stirring.
5. Add 100 mL of Solution A to Solution B, and continue stirring the combined solutions. Allow this derivatization reaction to continue at room temperature for at least 30 min.
6. Remove the adherends from the derivatization reaction vessel, rinse the adherends with deionized (DI) water and dry the adherends in an oven at 110 °C for at least 30 min.

In order to adjust the resulting surface energy and level of adhesion, mixtures of propyltrimethoxysilane (PS) and 3-amino propyltrimethoxysilane (APS) could be used, effectively varying the concentrations of the beneficial and detrimental coupling agents from one treatment to another. The same treatment sequence as described above was used for these mixed solutions as well, allowing us to effectively titrate the desired level of adhesion. Some of the pictures from the tests are shown in Fig. 7.7. It is worth noting that, when the contact angle changed from 67 to 83°, for the specimens tested, different crack paths were found in pure mode I fracture tests.

Therefore, to obtain a partially weakened interfacial area (neither very strong nor very weak) for this study, the water drop contact angle needed is between 67 and 83° for the Epoxy A/aluminum oxide interface. However, repeated tests indicated that this range of contact angle does not work for the Epoxy B/aluminum oxide interface. Instead, a new mixture with different silanes was developed to identify the transition between visually interfacial and cohesive failures for the Epoxy B/ aluminum oxide interface. The mixture consisted of 1 ml APS and 5 ml hexyltrimethoxysilane (HS). The contact angle is $\alpha \approx 90°$. Again, six specimens prepared the Epoxy B were tested at different crosshead displacement rates: three of them at 0.01 mm/min, three of them at 100 mm/min. Some of the DCB test results is shown in Fig. 7.8. Clearly, there is again strong evidence that faster loading rates, and hence faster fracture propagation rates, is more likely to detect weakened interfaces while there is a tendency for slower tests result in cohesive failures. Interestingly, the resulting fracture energies are dramatically different, with slow propagation resulting in respectable resistance of about 750 J/m^2, while fast propagation showed negligible fracture resistance over most of the debond length. Also of interest is that Epoxy B was shown to be more robust in avoiding locally weakened interfacial regions in Fig. 7.6, and required a lower adherend surface energy (higher contact area) to transition from cohesive to interfacial failures.

7.5 Discussion and Conclusions

The results conveyed here raise interesting questions about what appears to be a rather strong tendency for crosshead displacement and resulting crack propagation rate on the locus of failure in adhesively bonded joints. For the systems described herein, faster rates were more likely to cause growing cracks to detect and proceed along either locally or globally weakened regions. Although by far the most common trend in systems we studied, this is not believed to be universal, and in fact one system was found to do the opposite (e.g. epoxy bonds to aluminum adherends wiped with highlighter fluid). The rate dependence of adhesive bond fracture has typically looked to the viscoelastic properties of the bulk adhesive used

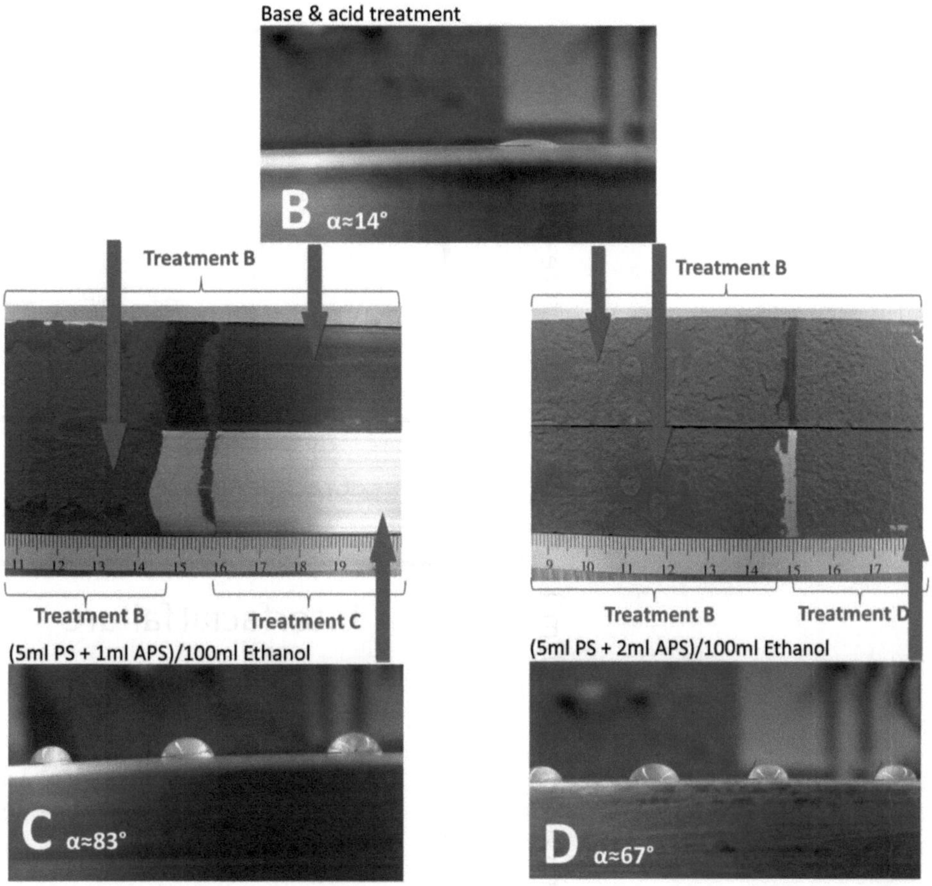

Fig. 7.7 Relationships between the water drop contact angle and the failure modes for Epoxy A bonds: cohesive failure when the contact angle is smaller than 67°; interfacial failure when the contact angle is larger than 83°. All adherends were pretreated with a base/acid treatment, then portions (right ends of lower adherends shown in central figures) were immersed in either the C or D silane treatment solution, resulting in the indicated contact angles. For mode I testing, a transition from visually interfacial to predominantly cohesive failures was obtained in going from 5:1 to 5:2 PS/APS solutions

[13–18], but clearly this dependence can result from interfacial regions as well. Further experimental investigations of this phenomena are recommended.

One might speculate that the size of the plastic zone ahead of the growing crack tip would play a role in the distribution of stresses, including their magnitudes at the interfaces. In this sense, a rate-dependent cohesive zone model, as illustrated schematically in Fig. 7.9, can be suggested. For simplicity, the interfacial region is assumed to be rate independent to emphasize that bulk viscoelastic behavior of the adhesive could still result in rate dependent locus of failure. Since yielding of the viscoelastic adhesives are known to be rate dependent, it is suggested that large yielded regions associated with slower crack propagation would transmit smaller peak tractions to the interface, potentially avoiding interfacial failures. At higher loading rates, however, smaller yield regions transmitting higher peak tractions could induce failures at the weakened areas [5], thereby inducing what appear to be more reliable detection of weakened interfacial regions.

Finally, it is important to recognize that changing the mode mixity by adding components of mode II loading can have a dramatic effect on crack propagation direction [19] or altering material properties [20] or residual stress state [21, 22] can all affect the locus of failure and sometimes the resulting fracture energies. Depending on the sign of the shearing imposed by mode II components of loading, growing cracks may either be steered towards or away from a given interface. CZM modeling has been used to confirm this behavior [7] and experiments in this area continue, including for local weakened interfacial zones [12]. Opportunities remain for characterizing, understanding, and accurately modeling crack path selection processes in adhesively bonded joints under mixed mode testing at different loading rates, including with locally weakened interfacial regions. The methods suggested herein may prove useful in such studies, as clear differences have been demonstrated for different adhesives, different loading rates, and different surface treatments.

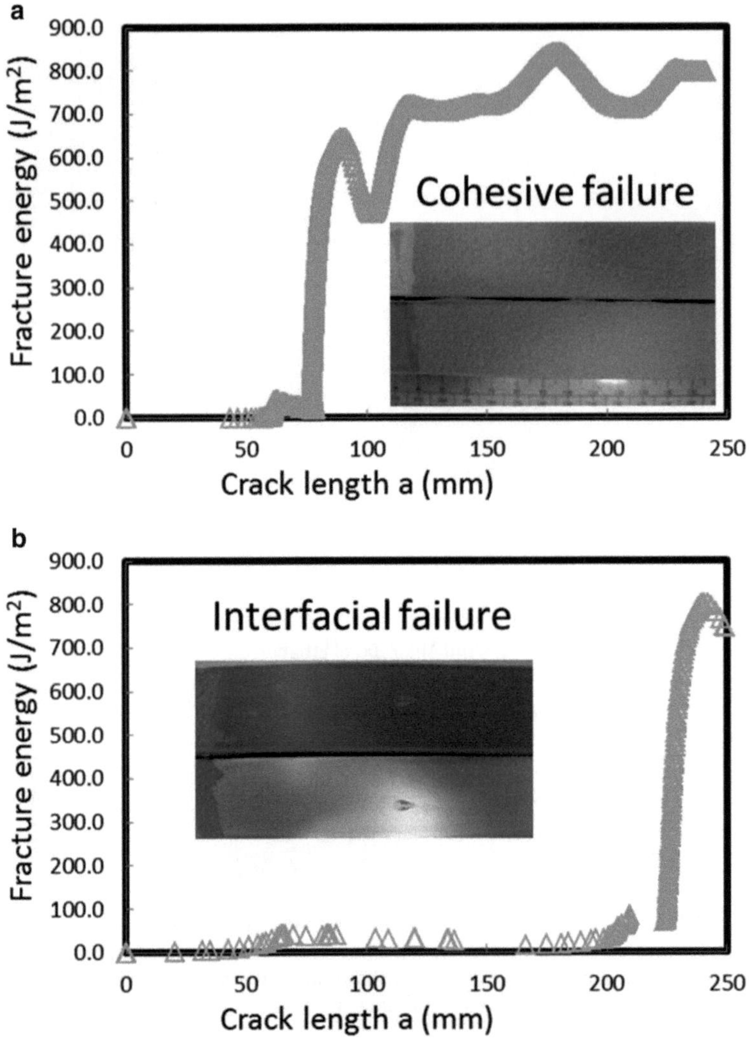

Fig. 7.8 Representative fracture energy curves of DCB specimens prepared using Epoxy B tested at: (**a**) 0.01 mm/min (cohesive failure); (**b**) 100 mm/min (interfacial failure)

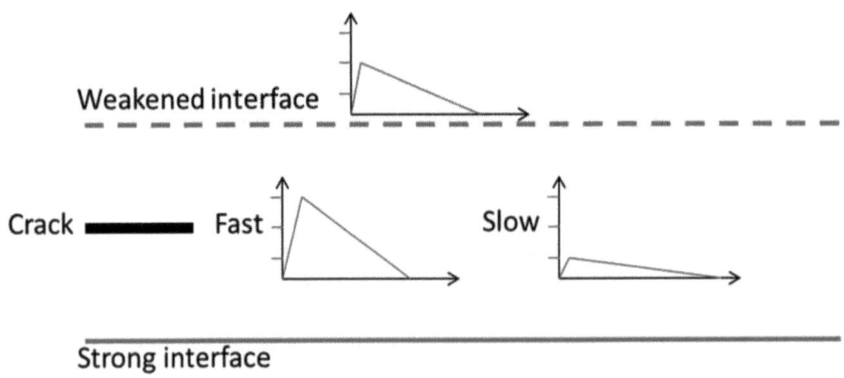

Fig. 7.9 Schematic of the explanation to the rate dependency of the crack path selection using the CZM for adhesive layer, and weakened interface. The small graphs of suggested traction vs separation laws illustrate how rate dependent yielding of the adhesive layer might transmit different stresses to the interface, here assumed to be time independent

Acknowledgments The authors gratefully acknowledge the financial support of the National Science Foundation through NSF/CMMI Award No. 0826143, as well as their support under DMR-0415840 for the construction of the dual actuator load frame used in this research. In addition, we acknowledge use of facilities in the Engineering Science and Mechanics and Chemistry Departments, as well as the Macromolecules and Interfaces Institute for fostering interdisciplinary research.

References

1. Goland M, Reissner E (1944) The stresses in cemented joints. J Appl Mech 11:A17–A27
2. Schonhorn H, Ryan FW, Wang TT (1972) Effect of bonding defects on shear strength in tension of lap joints having brittle adhesives. J Appl Polym Sci 16:1901–1909
3. Karachalios EF, Adams RD, da Silva LFM (2013) Strength of single lap joints with artificial defects. Int J Adhes Adhes 45:69–76
4. Schonhorn H, Ryan FW, Wang TT (1971) Effects of symmetrical bonding defects on tensile shear strength of lap joints having ductile adhesives. J Appl Polym Sci 15(5):1069–1078
5. Guan YL (2014) Crack path selection and shear toughening effects due to mixed mode loading and varied surface properties in beam-like adhesively bonded joints. In: Engineering mechanics 2014. Virginia Tech, Blacksburg
6. Tsai CL et al (2014) Analysis of cohesive failure in adhesively bonded joints with the SSPH meshless method. Int J Adhes Adhes 51:67–80
7. Guan Y et al (2014) Using cohesive elements and XFEM to study crack path selection in the presence of locally weakened interfaces in DCB specimens. In: Long TE (ed) The 37th annual meeting of the adhesion society. The Adhesion Society, San Diego
8. Tsai CL et al (2013) Comparison of the performance of SSPH and MLS basis functions for two-dimensional linear elastostatics problems including quasistatic crack propagation. Comput Mech 51(1):19–34
9. Dillard DA et al (2006) A dual-actuator load frame for mixed-mode fracture of laminated or adhesively bonded specimens. In: Society for experimental mechanics. Society for Experimental Mechanics, St. Louis
10. Singh HK et al (2008) Mixed mode fracture testing of adhesively-bonded wood specimens using a dual actuator load frame. In: 31st annual meeting of the adhesion society. The Adhesion Society, Austin
11. Blackman B et al (1991) The calculation of adhesive fracture energies from double-cantilever beam test specimens. J Mater Sci Lett 10 (5):253–256
12. Ranade SR et al (2014) Studying the interaction of a propagating crack with locally weakened interfaces in an adhesive joint. In: The 37th annual meeting of the adhesion society. The Adhesion Society, San Diego
13. Blackman BRK et al (1995) The failure of fibre composites and adhesively bonded fibre composites under high rates of test.1: Mode I loading: experimental studies. J Mater Sci 30(23):5885–5900
14. Jacob GC et al (2004) Strain rate effects on the mechanical properties of polymer composite materials. J Appl Polym Sci 94(1):296–301
15. Michael Starbuck J, Jacob GC, Dillard DA (2005) Dynamic testing for quantifying rate sensitivities in bonded composite structures. In: 2005 SEM annual conference and exposition on experimental and applied mechanics, 7 June 2005-9 June 2005. Society for Experimental Mechanics Inc., Portland
16. Pohlit DJ et al (2008) Evaluating the rate-dependent fracture toughness of an automotive adhesive. J Adhes 84(2):143–163
17. Dillard DA et al (2009) Observations of decreased fracture toughness for mixed mode fracture testing of adhesively bonded joints. J Adhes Sci Technol 23(10–11):1515–1530
18. Dillard DA et al (2011) On the use of a driven wedge test to acquire dynamic fracture energies of bonded beam specimens. J Adhes 87 (4):395–423
19. Chen B et al (2002) Crack path selection in adhesively bonded joints: the roles of external loads and specimen geometry. Int J Fract 114 (2):167–190
20. Chen B et al (2001) Crack path selection in adhesively-bonded joints: the role of material properties. J Adhes 75(4):405–434
21. Chen B, Dillard DA (2001) The effect of the T-stress on crack path selection in adhesively bonded joints. Int J Adhes Adhes 21(5):357–368
22. Fleck NA, Hutchinson JW, Suo ZG (1991) Crack path selection in a brittle adhesive layer. Int J Solids Struct 27(13):1683–1703

Chapter 8
Time Dependent Response of Composite Materials to Mechanical and Electrical Fields

K.L. Reifsnider

Abstract The object of this paper is to broaden the familiar concepts of viscoelasticity and viscoelastic behavior of polymeric materials to the discussion of the (highly) nonlinear mechanical behavior of composite materials and to introduce relationships between that time-dependent mechanical response and the dielectric character of those materials. The foundation of the paper is a discussion of damage accumulation as the introduction of a 'second phase' into heterogeneous material morphologies and the interpretation of the influence of that accumulation process on the time dependent mechanical and electrical properties and behavior of those materials. It is the principal objective of the paper to establish science based relationships between the mechanical behavior of heterogeneous/composite materials under high-strain conditions and the response of those materials to alternating current or voltage inputs. Applications of the concepts will be discussed, including "design-defected composite materials" such as batteries, fuel cells, and separation membranes, and "service-defected composite materials" such as the structural composites used in airplanes, vehicles, and bridges. The purpose of the present work is to provide some coherence to that subject and to provide a foundation for the design of composite material systems.

Keywords Composite materials • Time dependent response

8.1 Composite Materials: Defected Heterogeneous Materials by Design

Figure 8.1 shows a few examples of the composite materials under consideration in the present paper. Their general defining features are essential parts of the technical arguments we hope to construct. First, they are by definition heterogeneous, i.e., made up of a 'mixture' of different phases of material, at different scales. Those phases may be fiber 'reinforcement' in a polymer matrix such as the glass-epoxy system seen in Fig. 8.1a, a mixture of ceramic electronic and ionic conductive materials in a solid oxide fuel cell electrode (b), or separation membrane (d), or laminates of polymer materials in a proton exchange fuel cell (c), as examples. Clearly, these composites are defined not only by the bulk properties of the constituent materials, but (especially) by the morphology (shape, size, surfaces, voids, and interfaces) and arrangement of the constituents. These latter factors are extrinsic features, features that we can change by design and control by processing. They have a profound effect on the function of the composite material system. In the solid oxide fuel cell electrode, Fig. 8.1b, the voids provide conduits for the transport of fuels and reactant gases of the chemical materials that are used to generate electricity; the interfaces of the ionic and electronic conductor phases that are exposed to those conduits create a "triple point boundary" which is the only locality where the electrochemical reaction that makes the fuel cell work can occur. So controlling the extrinsic design features is the key to making a successful functional material system like a fuel cell or a separation membrane.

A first examination of structural composites such as the continuous fiber reinforced polymer matrix material shown in Fig. 8.1a may suggest that they are inherently "simpler" systems. When we design an airplane wing or a bridge structure we are primarily concerned with mechanical behavior and we teach our students to select materials and to design the global shape and size to achieve suitable stiffness and strength for the mechanical loads. Generally, that analysis uses "homogeneous" composite properties which are typically obtained from extensive testing and characterization. The "design" of

K.L. Reifsnider (✉)
University of South Carolina, 300 Main St., Columbia, SC 29209, USA
e-mail: Reifsnider@sc.edu

H.J. Qi et al. (eds.), *Challenges in Mechanics of Time-Dependent Materials, Volume 2: Proceedings of the 2014 Annual Conference on Experimental and Applied Mechanics*, Conference Proceedings of the Society for Experimental Mechanics Series,
DOI 10.1007/978-3-319-06980-7_8,

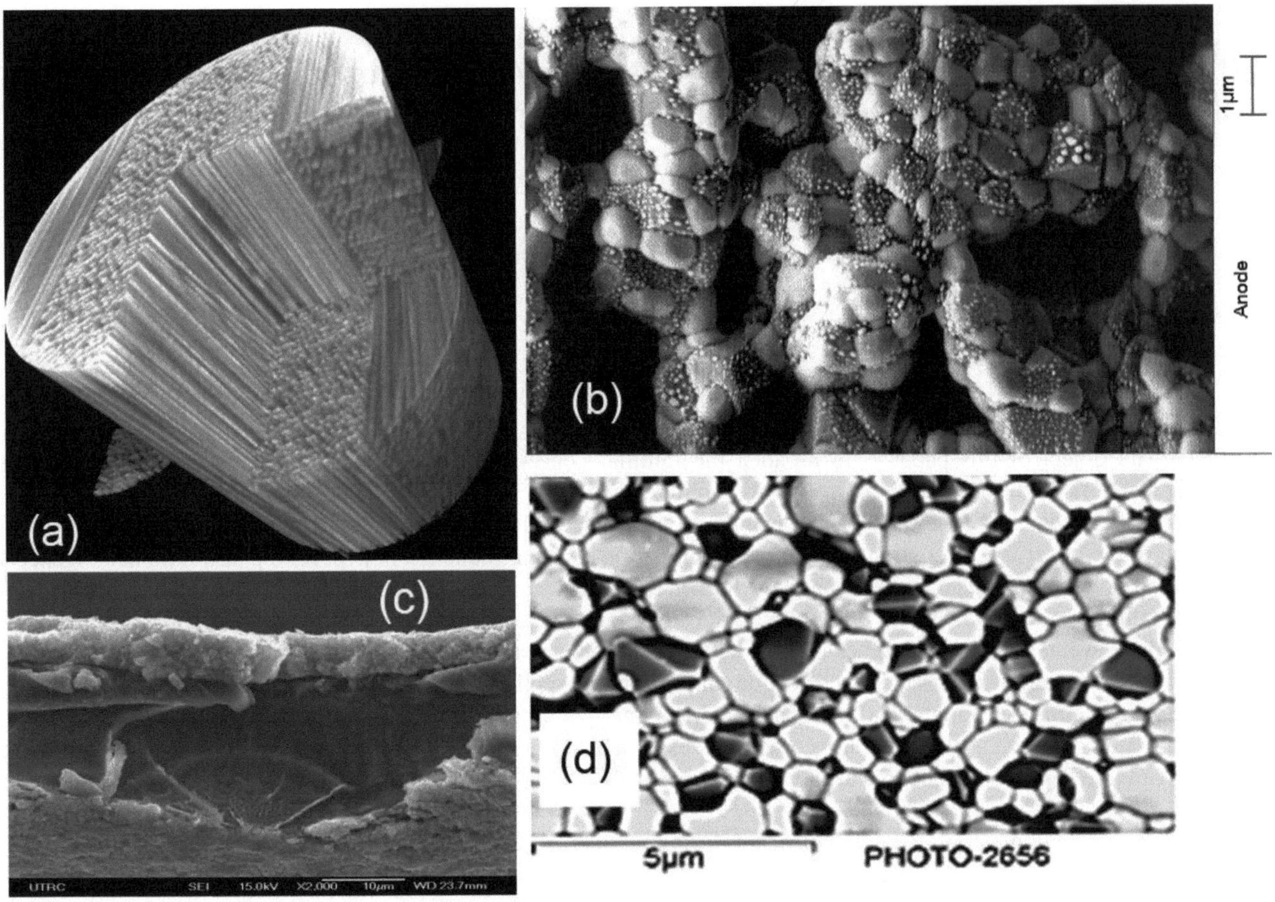

Fig. 8.1 Composite materials used in aerospace structures (**a**), solid oxide fuel cells (**b**), polymer membrane fuel cells and (**c**) ceramic separation membranes for oxygen transport (**d**)

microstructure of those materials is most often conceptual with the help of "micro-mechanics" or "effective property" analysis. With some important exceptions, those approaches are most often generalized mixture theories wherein the global resultant property is some sort of "sum" of the constituent properties and characteristics. These approaches are useful and often convenient, but they neglect an essential truth. Heterogeneous material systems are generally interactive material systems across multiple scales, wherein the constituents not only act together (and constrain one another), but also interact to create emergent composite response that is not any simple combination of the constituent characteristics. In short, all composite materials are composite material systems, and their response to mechanical, electrical, and thermal (or other) applied fields is greatly affected and controlled by their systems response, which is a rightful objective of design.

8.2 Composite Material System Response to Mechanical and Thermal Applied Fields

Figure 8.2 shows the stress-strain response of several continuous fiber plain weave glass-epoxy coupons subjected to uniaxial tensile mechanical loading at various angles to the principal material directions (details can be found in [1]). The "ductility" observed is the result of extensive microcracking and the subsequent relative motion of the constituents as micro-damage accumulates. This highly nonlinear behavior is the heart of the design of composite materials to sustain and mitigate impact events, such as 'catching' a jet engine blade in a "blade-out" event, or stopping military threats, etc. The representation of such behavior with constitutive laws is well established [2, 3]. An incremental plastic strain relationship can be derived in the usual way. Introducing an effective stress, $\sigma^{*}=\sqrt{3f}$, one can write

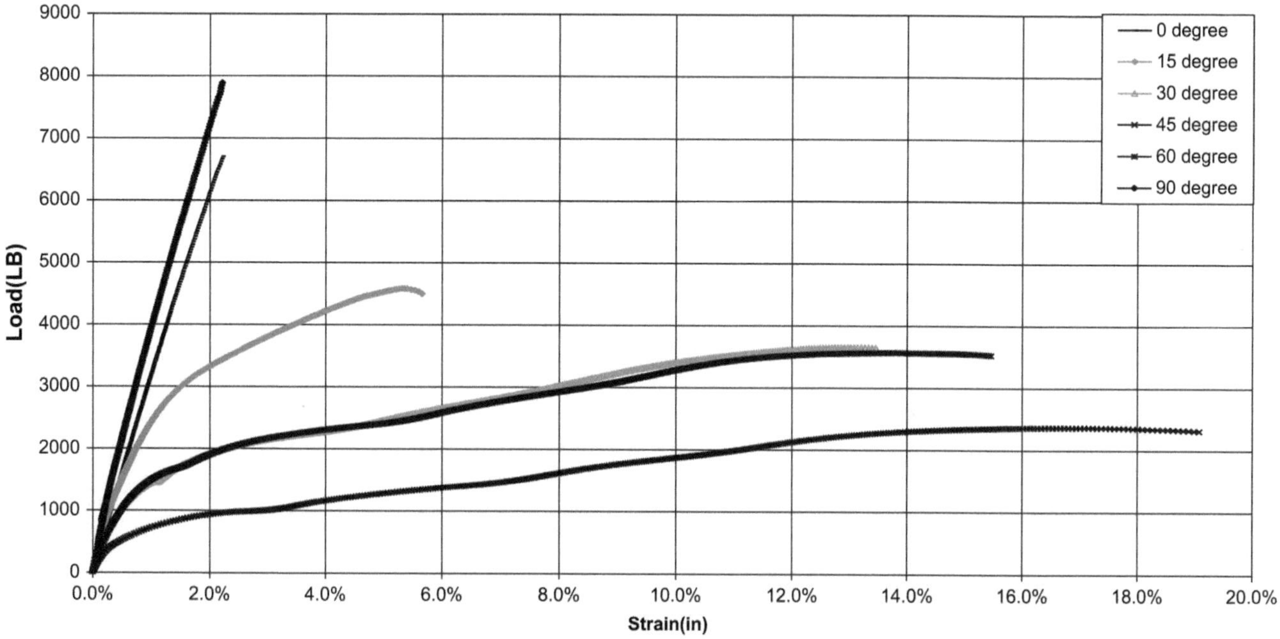

Fig. 8.2 Stress-strain response of continuous fiber plain weave glass-epoxy coupons subjected to uniaxial loading at various angles to the principal material directions

$$dW^p = \sigma_{ij} d\varepsilon_{ij}^p = 2fd\lambda = \sigma^* d\varepsilon^{*p} \Rightarrow d\varepsilon^{*p} = \frac{2}{3}\sigma^* d\lambda \text{ and } d\lambda = \frac{3}{2}\left(\frac{d\varepsilon^{*p}}{d\sigma^*}\right)\left(\frac{d\sigma^*}{\sigma^*}\right) \tag{8.1}$$

where ε^{*p} is the effective plastic strain and W^p is the plastic work per unit volume. The plastic slope, $\left(\frac{d\varepsilon^{*p}}{d\sigma^*}\right)$ is generally rate dependent. For plane stress equation (8.1) reduces to four terms. For unidirectional loading, σ_x in a direction that forms a positive angle θ with the fiber direction, x_1, the stress components in the material system are

$$\sigma_{11} = \sigma_x \cos^2(\theta) \qquad \sigma_{22} = \sigma_x \sin^2(\theta) \qquad \sigma_{12} = -\sigma_x \sin(\theta)\cos(\theta) \tag{8.2}$$

where upon $\sigma^* = \sigma_x h(\theta)$ and

$$h(\theta) = \sqrt{\frac{3}{2}(a_{11}\cos^4\theta + a_{22}\sin^4\theta + 2(a_{12} + a_{66})\sin^2\theta\cos^2\theta)} \tag{8.3}$$

For woven composites, Ogihara proposed that $a_{11} = 1$, resulting in the most general three parameter plasticity model for planar problems and uniaxial loading [4]. Then the plastic potential function reduces to

$$2f = \sigma_x^2 h^2(\theta); h(\theta) = \sqrt{\frac{3}{2}(\cos^4\theta + a_{22}\sin^4\theta + 2(a_{12} + a_{66})\sin^2\theta\cos^2\theta)} \tag{8.4}$$

Since the effective stress-strain relation should be a material property under monotonic loading, the material parameters in (8.7) must be chosen so that the $\sigma^*(\varepsilon^{*p})$ relations are independent of loading angle. If (8.4) is cast in the following form, it is possible to determine c_1 and c_2 in the following manner.

$$h(\theta) = \sqrt{\frac{3}{2}\cos^4\theta + c_1\sin^4\theta + 2c_2\sin^2\theta\cos^2\theta} \tag{8.5}$$

We can determine the parameters of the plastic potential function so that the effective-stress, $\sigma_x \cdot h(\theta)$ vs the nonlinear effective-strain, $\varepsilon^n/h(\theta)$, forms a single master curve. It is seen that for $\theta = 90^0$, $h(90^0)$ depends only on the potential

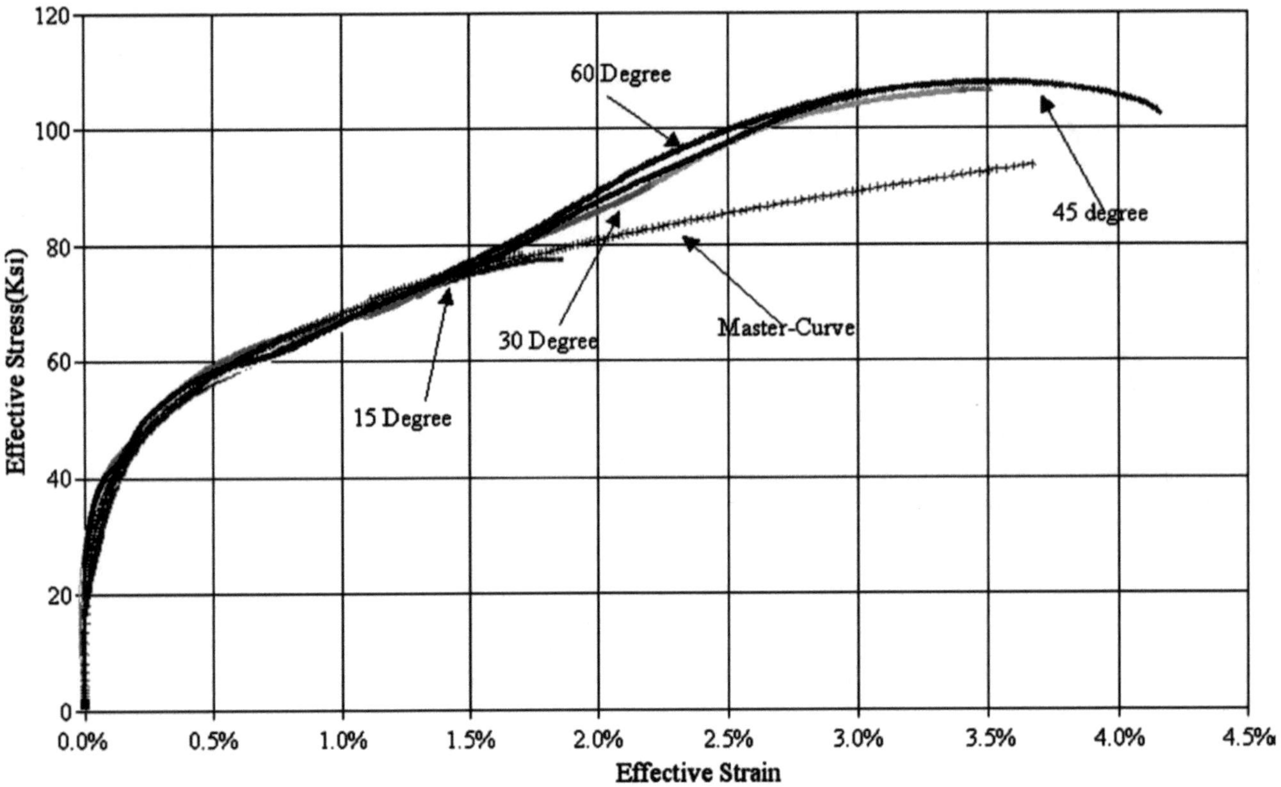

Fig. 8.3 Master curve representing nonlinear response for all angles of mechanical loading, from [1]

function parameter c_1. The result is shown in Fig. 8.3 when a simple power law is used to represent the data. However, this response is dependent on both the rate of deformation and the specimen temperature. For this polymer-based data, we formulated the master effective stress-effective strain curve by fitting the data by a power law as

$$\overline{\varepsilon}^p = \varepsilon^{*p} = A(\overline{\sigma})^n = A(\sigma^*) \tag{8.6}$$

in which amplitude A is a function of strain rate as discussed in our earlier publications [1–4].

In our work, we have used the form, $A = \kappa(\varepsilon^P)^m$ where κ and m are material constants and ε^P is the plastic strain, for all angles of loading. Finally, we construct a failure criterion which states that the "time to break," t_b, is related to the strain rate and angle of loading relative to the fiber directions by the simple relationship

$$t_b(d\varepsilon/dt)^m h(\theta) = \text{material strength constant} = X \tag{8.7}$$

where "m" is the same material constant as before, and $h(\theta)$ is the function used to construct the Master curve. With this representation, for this highly nonlinear response over large strain values (for what is thought of as a "brittle" material), the result predicts a single material strength, X, for uniaxial loading, for all angles of loading and all strain rates. The effectiveness of this concept can be demonstrated in Table 8.1 which shows comparisons of predicted stress at break values for a variety of angles of loading and strain rates compared to observed values. Clearly, strain rate and mechanical loading have a profound (and coupled) effect on the response of this material system. By accounting for this coupling in the systems response, we are able to construct a single material strength constant for all uniaxial loading conditions and angles of loading for this anisotropic laminate, for highly nonlinear behavior and strains of 14 % or more. We have also shown in our earlier publications that one can construct a time-temperature equivalence concept that uses this foundation to trade off temperature for time in laboratory tests to enable robust characterization of behavior for conditions otherwise difficult to accommodate in the laboratory, e.g., very high strain rates simulated by very low temperatures, etc. Of course, this observation needs to come with the usual reminder that we are dependent on material (especially molecular level) mechanisms being undisturbed by such equivalence concepts for the success of such equivalence methods.

Table 8.1 Predicted and observed strength for several strain rates and angles of loading

Strain rate	Failure stress (Psi)	15°	30°	45°	60°
0.0001/s	Experiment	29,849	25,261	24,160	25,109
	Model fit	29,107	23,102	22,894	23,102
	Percent difference	2.49 %	8.55 %	5.24 %	7.99 %
0.002/s	Experiment	34,366	27,138	25,815	27,806
	Model fit	33,218	26,169	23,636	26,169
	Percent difference	3.35 %	3.57 %	8.44 %	5.89 %
0.01/s	Experiment	36,489	28,515	28,144	29,089
	Model fit	35,225	27,431	25,896	27,431
	Percent difference	3.46 %	3.81 %	7.99 %	5.70 %

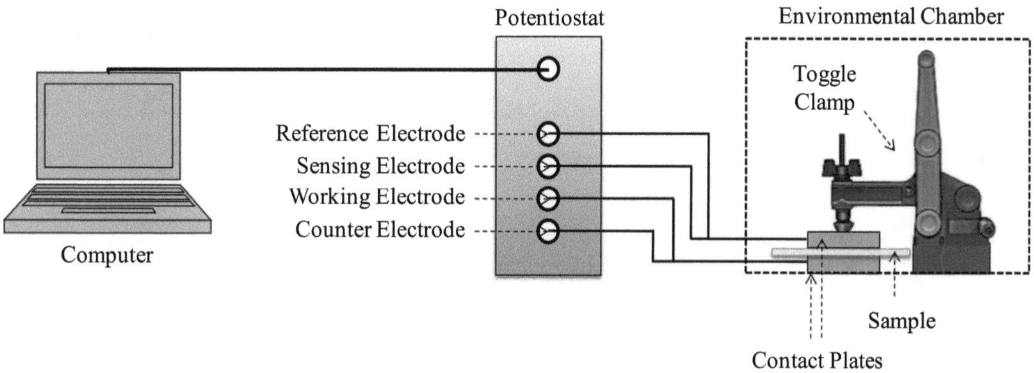

Fig. 8.4 Experimental configuration for recovering Broadband Dielectric Spectroscopy (BbDS) data from coupon specimens. Similar methods are also mounted on specimens during mechanical loading

8.3 Composite Materials Response to Combined Mechanical and Electrical Applied Fields

We have seen that the mechanical response of a heterogeneous material system is time and rate dependent for the materials considered. We have also seen in prior publications that this time and rate dependence is due not only to the basic constituent characteristics but also due to the interaction of the constituents with each other. For the electrical response of composite materials, the dependence of response on time and rate is even more general, and strikingly more emergent or anomalous. We start with the surprising statement that essentially all heterogeneous materials are dielectric in their response to an applied electrical field. One might expect this to be true when the constituents have dielectric responses, as is the case for the polymer-based composite materials in Fig. 8.1a, but it is also true for the mixture of conductive materials in the ceramic electrodes in Fig. 8.1b, d. From an experimental point of view, this result is useful, since an understanding of this dependence can provide a valuable method of determining material state, including some of the details of the internal microstructure and morphology from such observations. Even the nature and amount of void or micro-crack development is easily followed with dielectric measurements. The foundation of those details is the time dependent response of the heterogeneous materials to time-variable vector electric applied fields.

We illustrate these facts with data from glass-epoxy plain-weave composite materials, of the type we have discussed in earlier publications [1–4].

Figure 8.4 shows an example of the general test configuration used to recover data that represent the through-thickness specimen response to the application of an AC electrical signal over frequencies of about 0.01 to the megahertz range. Figure 8.5 shows two examples of the results of such observations.

Even at 25 % of the life of the specimens, the dielectric response is greatly influenced by the micro-defects that initiate and accumulate as the cyclic loading continues. The BbDS spectrum is remarkably sensitive to the presence of defects in the microstructure, and to the accumulation of those defects [5]. The physical foundation for these data is the time dependent

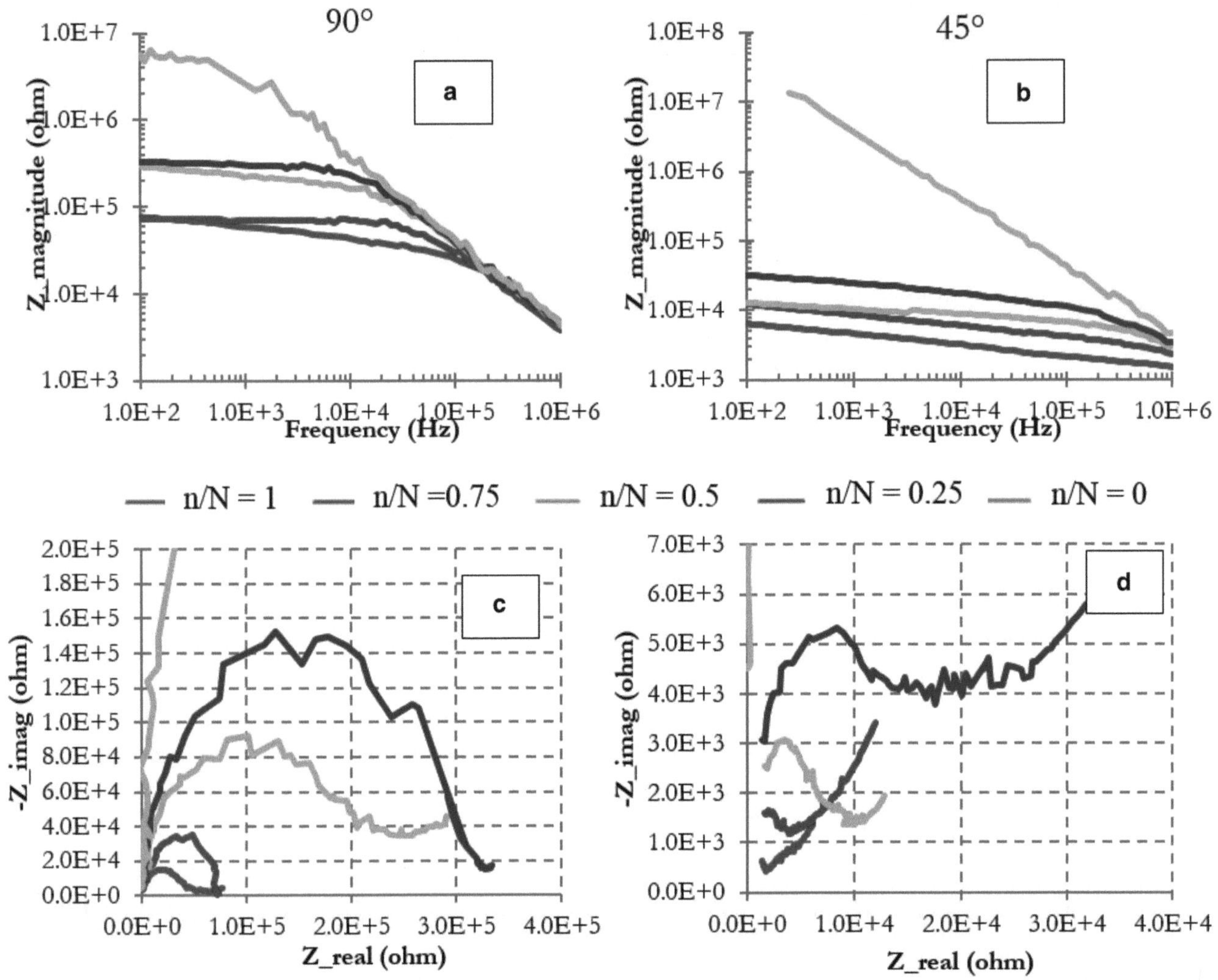

Fig. 8.5 Impedance magnitude vs. frequency of oscillation of the electric field for cyclic end-loaded bending at 90 and 45 °. to the material axes for woven glass epoxy laminates (**a** and **b**), and plots of the imaginary part vs. the real part of the complex Impedance for those tests, (**c** and **d**), at initial condition, 25, 50, 75, and 100 % of the life of the specimens

displacement of charge in materials or material mixtures in response to an alternating applied field. Fig. 8.6a shows an example of such charge displacement calculated using the commercial code COMSOL, which solves the equation

$$-\nabla d[(\sigma + j\omega \cdot \varepsilon_o)\nabla V - (J^c + j\omega P)] = \mathrm{dQ}_j \tag{8.8}$$

The direction of the applied field is shown relative to the inclusion in Fig. 8.6a. The time dependent voltage distribution of the voltage along a centerline, in the direction of the applied field, is shown in Fig. 8.6b. Two features are especially important to the current discussion. First, the simulation was done for a metal conductive insert in an epoxy matrix surrounding it. (The purpose of that material selection was to enable experimental validation of the results.)

The conductivity of the epoxy matrix material was small, but not zero; the relative dielectric permittivity of the matrix was of the order of 6, while that of the metallic insert was several orders of magnitude smaller. Hence, the metallic insert had very little voltage drop across it, while the 'insulating' epoxy showed a large voltage change due to it's dielectric charge displacement. But the arrangement of those charge polarizations is not instantaneous in this heterogeneous mixture of materials with different constituent responses to the applied electric fields. The inherent time dependence of the response is evident in the simulations shown in Fig. 8.6b. In some cases the charge displacement necessary to establish an equilibrium distribution of voltage takes tens of seconds for this material combination.

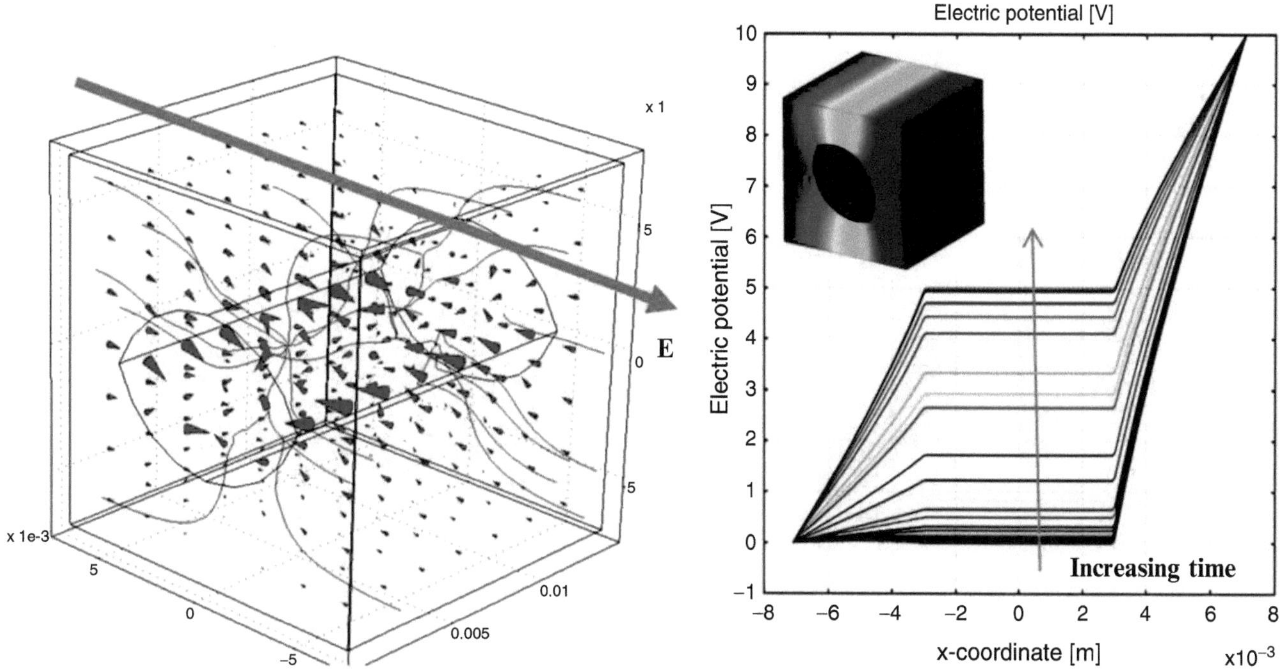

Fig. 8.6 Time dependent charge displacement in a two phase cube with a single inclusion that is more conductive than the surrounding matrix material

8.4 Summary and Conclusion

Time dependent response of heterogeneous materials to input fields (mechanical, electrical, thermal, etc.) is the rule rather than the exception. Such materials often demonstrate non-equilibrium, non-conservative response that results from the interactions of the constituents across their shared boundaries and volumes. The results of these interactions can enable engineering responses like the electrochemistry and charge storage that makes batteries and fuel cells possible in our society, or manage and control the flow of charge in a bridge or airplane during a lightning strike to create a safer environment, or they can be used to follow the development of defect states in composite structures to monitor physical integrity and support condition based maintenance. Our understanding of this behavior at the fundamental level is the foundation for applications of the concepts for "design-defected composite materials" such as batteries, fuel cells, and separation membranes, and for control of the development of damage in "service-defected composite materials" such as the structural composites used in airplanes, vehicles, and bridges. The engineering community has not yet developed a comprehensive foundation of science and understanding to support the design of heterogeneous materials to exploit these possibilities in a systematic and efficient way. That is an opportunity for continuing work.

References

1. Xing L, Reifsnider K, Huang X (2009) Progressive damage modeling for large deformation loading of composite structures. Compos Sci Technol 69:780–784
2. Tamuzs V, Dzelzitis K, Reifsnider KL (2004) Fatigue of woven composite laminates in off-axis loading. I. The mastercurves. Appl Compos Mater 11(5):259–79
3. Ogihara S, Reifsnider KL (2002) Characterization of nonlinear behavior in woven composite laminates. Appl Compos Mater 9:249–63
4. Reifsnider KL, Tamuzs V, Ogihara S (2006) Compos Sci Technol 66:2473–2478
5. Reifsnider K, Majumdar P, Fazzino P (2009) Material state changes as a basis for prognosis in aeronautical structures. J Aeronaut Soc 113:1150

Chapter 9
Characterizing the Temperature Dependent Spring-Back Behavior of Poly(Methyl Methacrylate) (PMMA) for Hot Embossing

Danielle Mathiesen and Rebecca Dupaix

Abstract Characterization of the temperature dependent spring-back behavior of poly(methyl methacrylate) (PMMA) is essential to model hot embossing. The constitutive model must capture several deformation modes including uniaxial compression and stress relaxation with cooling in order to predict spring-back. In this work, the thermo-mechanical coupling of spring-back is investigated through finite element simulations utilizing a constitutive model that captures stress relaxation. It was found that the material model successfully predicts spring-back trends under a variety of heat transfer conditions. At the cooling times used experimentally, spring-back decreased with held strain and increased with embossing temperature. Initial simulations utilizing the experimentally obtained platen temperature under predicted spring-back. After performing a simple heat transfer simulation, spring-back predictions were improved by altering the temperature profile according to the heat transfer simulation and matched experimentally obtained values. To investigate the effect of a thermal gradient, a fully coupled thermo-mechanical simulation was performed. From this, it was found that the thermal gradient had a minimal effect on spring-back. Rather, the rapid cooling upon release of PMMA was found to cease spring-back and can be modeled without a fully coupled simulation. These results indicate that temperature, along with strain level, and cooling time are important to the process of spring-back.

Keywords PMMA • Glass transition • Spring-back • Stress relaxation • Large strain

9.1 Introduction

Hot embossing is a commonly used polymer processing technique that allows for quick, accurate replication of surface patterns on the micron scale. The unique thermal mechanical properties of polymers are well-suited to the hot embossing process. At temperatures greater than glass transition, T_g, polymers behave as highly viscous fluids while below T_g they behave as viscoelastic solids. By applying the stamp at temperatures greater than T_g, the polymer can quickly fill the mold at low stress levels due to its highly viscous nature. After, by reducing the temperature to below T_g, the polymer hardens into the relief shape of the stamp. Using this process reduces the pressure requirements while applying the stamp, and reduces hold time. However, the process of cooling the necessary 30–100 °C is not well understood and is an area where optimization through simulations could potentially reduce cycle time and cost.

While several polymers are used in hot embossing, one of particular interest is poly(methyl methacrylate) (PMMA). PMMA is of particular interest because it is a highly biocompatible transparent thermoplastic that is used in several different biomedical devices, such as biochips and intraocular lenses [1, 2]. PMMA has an ideal low glass transition temperature of about 105–110 °C making it an ideal polymer for several processing techniques such as hot embossing. In addition, high aspect ratio features such as microchannels and wells have been successfully embossed in PMMA [3, 4]. With this and PMMA's high biocompatibility, it has been shown to be an ideal substrate for directed cell growth [5–8]. Once the hot embossing process has been optimized for PMMA, its use within the medical industry can increase.

Hot embossing involves several deformation processes that need to be accurately described in order to predict the final shape of the polymer. Several investigators have considered die-filling of the embossing process and have successfully simulated the process [9–12]. However, several key components of hot embossing are ignored in these simulations such as

D. Mathiesen • R. Dupaix (✉)
Scott Laboratory, The Ohio State University, 201 West 19th Ave, Columbus, OH 43210, USA
e-mail: dupaix.1@osu.edu

H.J. Qi et al. (eds.), *Challenges in Mechanics of Time-Dependent Materials, Volume 2: Proceedings of the 2014 Annual Conference on Experimental and Applied Mechanics*, Conference Proceedings of the Society for Experimental Mechanics Series,
DOI 10.1007/978-3-319-06980-7_9,

cooling, stress relaxation, and spring-back after stamp removal. First, by ignoring the concurrent processes of cooling and stress relaxation, there is no way to know how stress in the polymer is going to change during the hold and cool portion of the hot embossing process. It has been shown that energy stored in the polymer will be released upon load removal causing some recovery of its initial shape known as spring-back [13–16]. The amount of recovery is related to the strain level, amount of stress relaxation, and temperature making it highly difficult to predict [14, 17]. By not knowing how the stress changes during the relaxation period, there is no way to predict the amount of time required to hold and cool the sample to reduce spring-back.

To develop a material model capable of predicting the entire hot embossing process, multiple deformation processes such as compression, stress relaxation, creep, and spring-back must be considered. The bulk of material during hot embossing undergoes uniaxial compression, however, the portion in the vicinity of the stamp experiences three-dimensional deformation. Previously, uniaxial compression in PMMA has been heavily characterized [18–24], however, this has not yielded enough information to produce accurate embossing simulations including spring-back. Stress relaxation under isothermal conditions has also been recently considered to characterize the short-term relaxation behavior of PMMA that is seen in hot embossing [25]. A new model using these experiments has been developed that can capture this relaxation behavior that previous models have been unable to capture [25]. After this, modified unconstrained recovery tests to incorporate cooling and quantify spring-back under conditions commonly experienced during hot embossing were performed [26]. The model was then modified to fit the cooling data and was shown to be successful at capturing the stress profile with cooling incorporated [26]. However, from these experiments the temperature dependence of PMMA during spring-back was found to be highly difficult to capture through experiments alone. Here we attempt to quantify the temperature dependence of PMMA through the use of finite element simulations.

9.2 Experimental Methods

Temperature assignment of PMMA for spring-back simulations has been shown to be extremely important for accurate spring-back prediction. Previous simulations using the model developed by Mathiesen et al. have been shown to be capable of producing a range of spring-back values that span the measured spring-back values [26]. Here temperature prediction and application is further investigated to attempt to yield the most accurate simulation possible.

The material model used here is the same used in Mathiesen et al. and is described fully elsewhere but is briefly described here [25, 26]. Shown in Fig. 9.1 is a one dimensional schematic but the model itself is fully three-dimensional. First, it is broken down into two sets of interactions, intermolecular and network. Intermolecular interactions are captured through one branch consisting of a linear spring and thermally activated dashpot in series. These interactions capture the initial elastic behavior and subsequent strain softening at temperatures less than T_g. As temperature increases the material constants change such that the initial elastic region decreases in magnitude and strain softening ceases at temperatures greater than T_g. Network interactions consist of two highly temperature dependent branches in parallel. At temperatures less than T_g, network interactions are dominated by the highly non-linear 8-chain hyperelastic spring to capture the significant strain hardening at these temperatures [18, 27–29]. Meanwhile, the second branch of network interactions, consisting of a Rolie-Poly element with finite extensibility, contributes minimally [30, 31]. As temperature increases, the hyperelastic spring reduces its contribution by increasing the number of rigid links between entanglements. This allows the stress at temperatures greater than T_g to be dominated by the Rolie-Poly component to capture the melt behavior seen at these temperatures. A previous

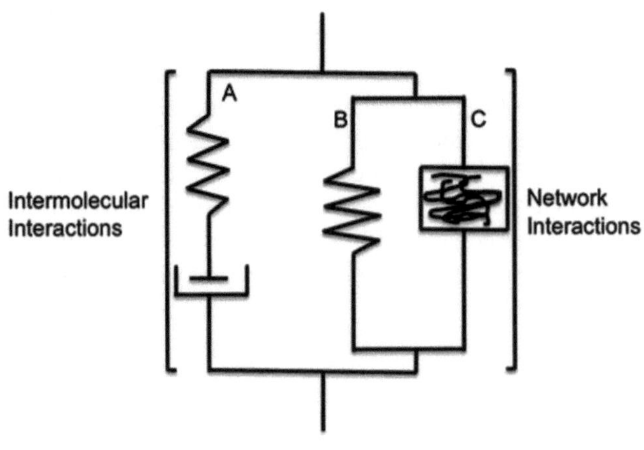

Fig. 9.1 Schematic of material model used in finite element simulations

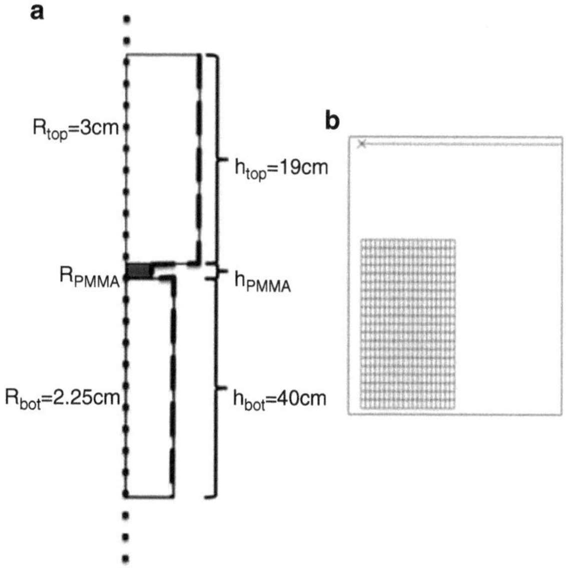

Fig. 9.2 (**a**) Shows the coupled thermal-mechanical simulation with boundary conditions and (**b**) shows the mesh of the PMMA in the decoupled simulations

model by De Focatiis uses a similar structure to this model, but uses significantly more material constants and is unable to capture strain hardening at temperatures less than T_g and is therefore not investigated here [32]. By combining the Rolie-Poly model with the hyperelastic spring in parallel and making the spring highly temperature dependent, the model by Mathiesen et al. successfully describes the vastly different behavior exhibited by PMMA across the glass transition temperature [25].

To investigate the temperature dependent behavior of spring-back, finite element simulations are used to apply different temperature definitions to attempt to achieve the most accurate spring-back prediction possible. Four sets of simulations are performed, labeled as (1) Exp Cool, (2) New Cool, (3) Coupled, and (4) Quench. The first set of simulations, (1), use a uniform temperature definition of PMMA that is fit to the experimentally obtained platen temperature [26]. These simulations investigate spring-back using information that can easily be measured during a hot embossing process. Next, a heat transfer simulation is performed separate from the spring-back simulations to quantify the actual temperature of the PMMA during embossing. The second set of mechanical simulations, (2), use a curve fit to the temperature of the center of the PMMA from the separate heat transfer simulation [26] to produce a second set of simulations. A third set of fully coupled thermo-mechanical spring-back simulations, (3), is performed to investigate the temperature gradient dependence. Finally, a fourth set of simple mechanical simulations (4) is performed using a uniform temperature definition that incorporates the quenching effect after release. By investigating these four sets of simulations, the amount of accuracy needed in the temperature assignment of PMMA can be quantified for future micro-channel hot embossing simulations.

The first two sets of simulations, labeled as (1) 'Exp Cool' and (2) 'New Cool' in the figures that follow, have been reported previously [26]. Exp cool applies a curve fit temperature to the measured experimental platen temperature uniformly to the PMMA. New cool utilizes a separate heat transfer simulation to assign a uniform temperature to the PMMA. Both of these simulations do not account for any increase in cooling that may occur during release when the top of the PMMA is exposed to the air. Rather, they assume that cooling continues at the same rate prior to release.

Next, a set of fully coupled thermal-mechanical simulation is performed to determine the effects of a temperature gradient on spring-back referred to as (3) 'Coupled'. First, conductivity, specific heat, and density of the stainless steel and PMMA are assigned to each material [26]. Convective heat transfer coefficients were then assigned to the outer surfaces shown by the dashed line in Fig. 9.2. Using the same process as New Cool, the coefficients were optimized such that the measured platen temperature from the experiment matched that of the simulation in the same location [26]. Doing this found that there is a short period of forced convection when the chamber door is opened and it quickly reduces to free convection after that. PMMA's mechanical properties were described through the user material and stainless steel was assumed to be linearly elastic. This set of simulations account for the thermal gradient and rapid cooling at the top of PMMA upon release that the previous two simulations did not account for. While temperature is more accurately modeled in this set of simulations, the goal here is to determine whether the improvement in spring-back accuracy is enough to warrant the increased complexity in future finite element models.

In the final set of simulations, called (4) 'Quench', a decoupled simulation is run that also accounts for the rapid cooling of the top surface of PMMA. Temperature is estimated in the PMMA by fitting the fully coupled simulation temperature after release. Prior to release, it follows the same temperature curves as New Cool, however, once the PMMA is released from

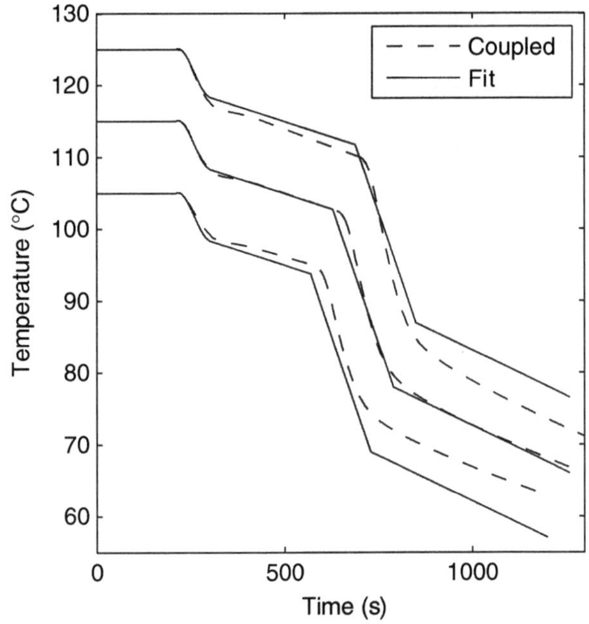

Fig. 9.3 Temperature assignment as given by Eq. (9.1) for PMMA in 'Quench' simulations. 'Coupled' refers to the coupled heat transfer simulations that Eq. (9.1) is curve fit to, 'Fit' is curve given by Eq. (9.1)

hold there is a period of rapid cooling followed by a slower steadier rate as given by Eq. (9.1) and shown in Fig. 9.3. Equation (9.1) applies the temperature uniformly to the PMMA.

All simulations were performed in ABAQUS/Standard. In the decoupled simulations, PMMA was represented with axisymmetric CAX4R elements. PMMA was modeled using a user material that describes the constitutive model. The PMMA had an initial radius and height of 5 units and 8.8 units respectively. The part was meshed with 400 elements having an element aspect ratio of 1:1.76 to prevent severe mesh deformation as shown in Fig. 9.2. A mesh refinement study was performed to verify mesh independence. The die was 50 units wide and considered rigid. Contact was assumed to be well lubricated and modeled as frictionless. To improve convergence, contact was softened using the ABAQUS exponential pressure-overclosure. Die motion was restricted in the horizontal direction and was prescribed in the vertical direction. Boundary conditions of the PMMA prevented motion in the vertical direction at the bottom of the sample. Motion was prevented in the horizontal direction on the line of axisymmetry and horizontal motion was uniform on the free edge.

In the fully coupled thermo-mechanical simulations, axisymmetry was also used and the PMMA and steel modeled with CAX4RT elements. PMMA was again described through a user material and steel was considered linearly elastic. Convection coefficients were applied as in Fig. 9.2 along the dashed lines. Pure conduction was assumed between the PMMA and the steel. PMMA used the same element aspect ratio as in the decoupled simulations and the mesh in the steel was refined to an appropriate level. A mesh refinement study was performed to ensure mesh independence. Contact was assumed frictionless and was softened as in the decoupled simulations. The motion of the bottom steel platen was restricted in the horizontal and vertical directions. The top platen was restricted in the horizontal and prescribed in the vertical. PMMA motion was prevented in the horizontal on the line of axisymmetry and vertical motion of the bottom surface was tied to the top of the bottom platen and therefore restricted.

Both simulations used the same deformation scheme. First, contact was initiated over 1 s followed by the die or platen motion downward at a rate such that the sample is strained at a true strain rate of $-1.0\ \text{min}^{-1}$ under an isochoric assumption. Once the desired held strain is reached, deformation is held for a period of 180 s prior to initiation of cooling. The amount of time spent in cooling depends on the held strain and embossing temperature being investigated and equals that of previous experiments [26]. After completion of cooling at time t_{rel}, release of the sample is achieved by raising the die or upper platen to 3 units about the original sample over 30 s. Spring-back is allowed to occur for 600 s after deformation release.

Equation 9.1 Temperature curve for PMMA—where $t_a = t - t_{begin}$, t_{begin} = time at which cooling begins, t_{rel} = time at which PMMA is released from hold period.

For $t_a < 90$ s

$$\theta = \theta_{emb} + .00002t_a^3 - .003t_a^2 + .035$$

for 90 s < t_a < t_{rel}

$$\theta = \theta_{emb} - 5.122 - 0.017t_a$$

for t_{rel} < t_a < t_{rel} + 160 s

$$\theta = \theta_{160} - .155(t_a - t_{rel}) \text{ where } \theta_{160} = \theta(t_{rel} + 160)$$

for t_{rel} + 160 s < t_a

$$\theta = \theta_{160} - 20.736 - .0254(t_a - t_{rel})$$

9.3 Results

Displacement, temperature, and stress data were collected from each simulation. From the displacement data the percent spring-back can be calculated from Eq. (9.2), and is shown for each simulation in Fig. 9.4. From this, it is clear that the material model predicts the trends of spring-back but appears to under predict the total amount for most simulations. A promising feature of the model is its ability to capture the decrease in spring-back at large held strains no matter what cooling process is used. Additionally, the model exhibits the increase in spring-back found at high embossing temperatures. By capturing these trends, it indicates that the model is qualitatively describing the held strain and embossing temperature dependence, but quantitatively underestimates the amount of spring-back.

First, by comparing the different thermal loading conditions in Fig. 9.4 the 'New Cool' condition is found to be the most predictive of spring-back. The 'New Cool' simulation applied temperature uniformly to the PMMA from a separate heat transfer experiment without increasing the amount of cooling during release from exposure of the top of the PMMA. At high embossing temperatures, it does over predict the amount of spring-back present, however, at lower temperatures it is highly predictive of spring-back. The decoupled simulation that used platen temperature directly, 'Exp Cool' is also promising

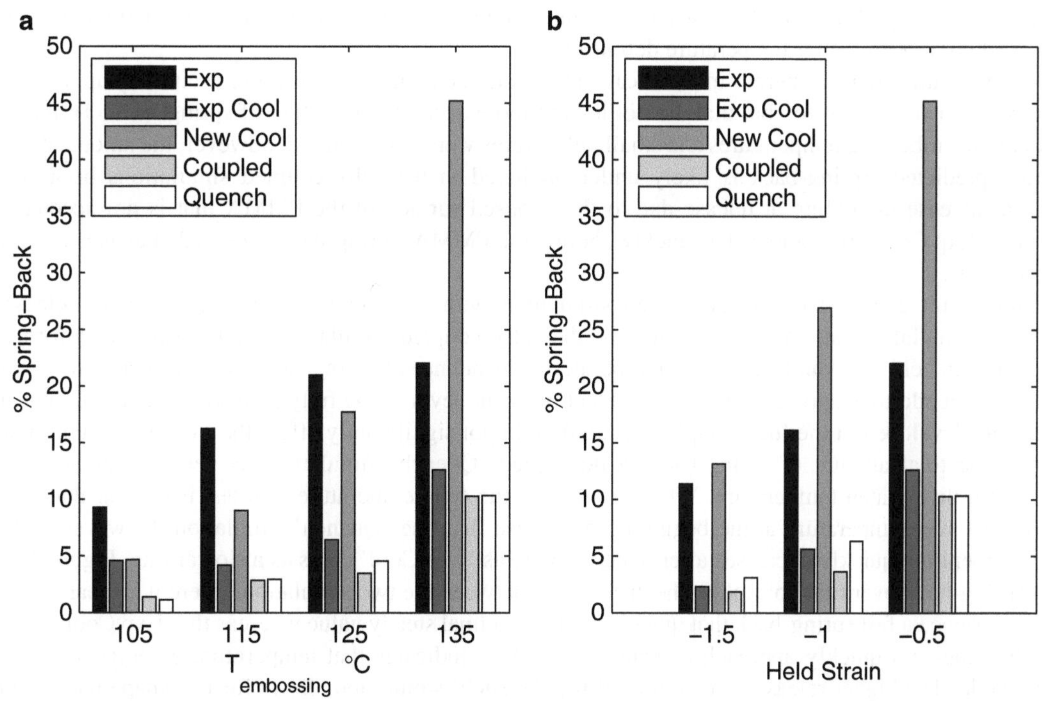

Fig. 9.4 Spring-Back of samples at (**a**) held strain of −0.5 and (**b**) 135 °C at various held strains. In these graphs, 'Exp' is the experimentally obtained spring-back, 'Exp Cool' is a decoupled simulation with temperature assigned according to the experimental platen temperature, 'New Cool' is a decoupled simulation with temperature assigned according to a separate heat transfer experiment without quenching at release, 'Coupled' is a fully coupled thermal-mechanical simulation, and 'Quench' is a decoupled simulation that uses the temperature profile from 'New Cool' until release, after which it uses the quenching profile found from 'Coupled'

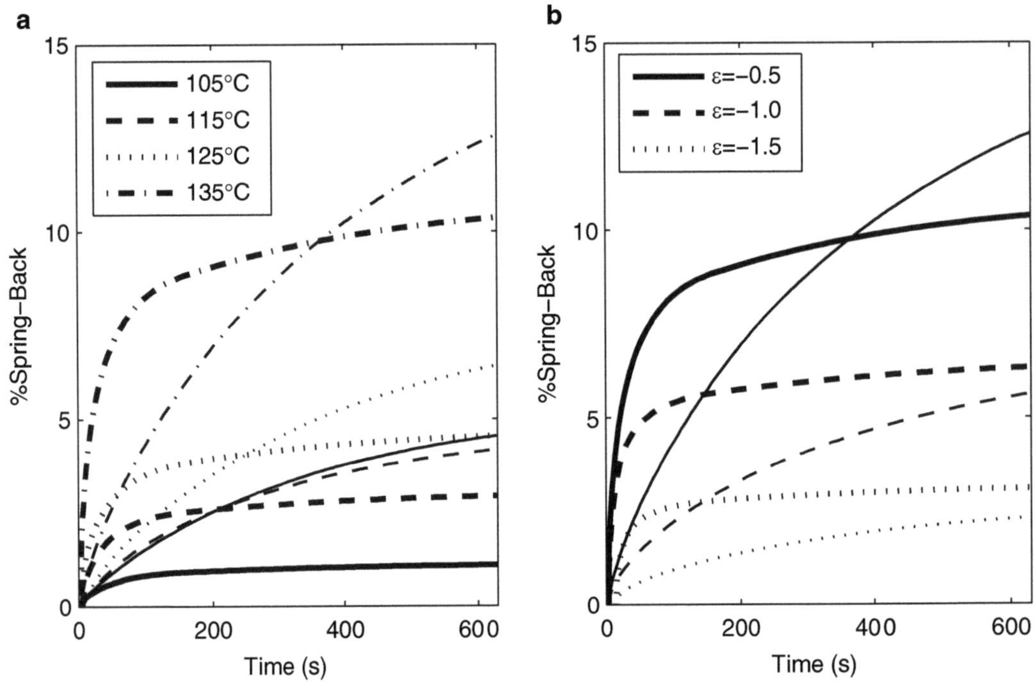

Fig. 9.5 Spring-back v. time of samples at (**a**) held strain of −0.5 and (**b**) 135 °C at various held strains. In these graphs the *thin lines* are the original simulations with temperature assigned from experimental platen temperature and the *thick lines* are the quenched simulations

given its simplicity. Without considering any heat transfer and assuming PMMA is the same temperature as the platen, the spring-back trends are also fully captured despite under predicting at all embossing temperatures. The temperature of the PMMA is more accurately predicted in the 'New Cool' simulations, which leads to the expectation of increasing accuracy of spring-back predictions with better temperature definition.

By performing a fully coupled thermo-mechanical simulation, the temperature is more accurately applied to the PMMA and a gradient is allowed to develop but that did not lead to more accurate prediction of spring-back. Trends of increasing spring-back at high embossing temperatures and small held strains were still captured, however, the amount of spring-back is severely under predicted. Spring-back is likely under predicted in the fully coupled simulations because the PMMA experiences an increase in cooling at release due to the exposed surface of the PMMA that is not accounted for in the 'New Cool' and 'Exp Cool' simulations. By quickly cooling the PMMA, spring-back is ceased at an earlier time than in the original simulations.

A promising result comes from the decoupled simulation with the increased cooling rate after release, known as 'Quench'. These simulations attempted to reproduce the fully coupled simulation results using a uniform temperature profile. The trends in held strain and embossing temperature dependence of spring back were predicted in 'Quench' as in all other simulations. In addition, spring-back is predicted at the same level as the fully coupled simulation, indicating that the thermal gradients developed in the fully coupled simulation do not significantly affect the overall amount of spring-back.

In addition, the total amount of spring-back in decoupled 'Quench' simulations is very similar to the 'Exp Cool' simulations that utilize platen temperature. When comparing the two temperature profiles, it is clear that the 'Exp Cool' simulation has a lower temperature at the beginning of release than the 'Quench' simulation. However, in the 'Quench' scenario the temperature quickly decreases after release where as 'Exp Cool' cools as a slower rate. Figure 9.5 presents the evolution of spring-back over time to explore the differences between the two simulations. Here it is clear that the 'Quench' simulations experience a fast spring-back that quickly settles to a final steady value whereas the 'Exp Cool' springs back at a slower rate that does not quickly approach a steady value. This indicates that temperature essentially is controlling the spring-back profile. The higher release temperature of the 'Quench' simulation allows for fast shape recovery that quickly ceases as the temperature decreases enough below T_g. In 'Exp Cool' the temperature is initially lower after release causing slower spring-back, but it takes longer to reach the cooler temperature required to stop spring-back. Therefore, although the total amount of spring-back of the two simulations is similar, the actual path of recovery is different.

Equation 9.2 The percent spring-back where h_f is the height after release, h_{held} is the held height, and h_0 is the initial height

$$\%\ Spring - Back = \frac{h_f - h_{held}}{h_0 - h_{held}} \times 100\ \%$$

9.4 Discussion

Several important features of the material model's prediction of spring-back have been investigated through the use of finite element simulations. First, it is clear that the model underestimates the total amount of spring-back. When considering the model's formulation, this is not wholly unexpected. The model was developed using stress relaxation data under isothermal conditions. Stress relaxation accurately describes the hold period of the embossing process prior to release but not the actual process of spring-back. Spring-back is likely better described by a creep phenomenon. Therefore, the model could likely be improved by considering creep experiments in the initial fitting of material constants. However, even without considering creep, the model is capable of predicting spring-back trends with careful consideration of temperature.

Temperature definition of PMMA plays a dominant role in the simulation of spring-back. It is apparent that spring-back will not cease until it is sufficiently below T_g. This means that by cooling the polymer specific amounts prior to release, it is likely that the amount of spring-back can be controlled. Surprisingly, despite the large thermal gradients that occur in PMMA during cooling, they do not appear to affect the total amount of spring-back overall. This is highly promising for future simulations of the micro hot-embossing process. As opposed to running a large fully coupled thermal-mechanical model, a simple heat transfer simulation can be performed first to quantify temperature. Then temperature can be uniformly assigned to a mechanical simulation, thereby reducing complexity in future simulations.

9.5 Conclusion

The prediction of spring-back in PMMA requires accurate characterization of several deformation processes, especially temperature. It was found that using experimentally obtained platen temperature and an isothermal definition was sufficient to capture the increase in spring-back at high embossing temperatures and small held strains although under predicting the overall amount of spring-back. A simple heat transfer simulation was used to quantify the temperature of the PMMA in relation to the experimental platen temperature. After applying this temperature to a decoupled mechanical simulation, the most accurate spring-back predictions were achieved. Also, it was found that including the increased rate of cooling of PMMA after release is necessary to quickly cease spring-back. Thermal gradients were found to not have a large effect on the total amount of spring-back. This indicates that decoupled simulations using an isothermal temperature definition are sufficient to predict spring-back using this material model. From these simulations important spring-back dependencies on temperature have been defined that will be highly useful in future micro-channel hot embossing simulations.

Acknowledgements The authors thank National Science Foundation for funding this work under the NSF CMMI Grant No. 0747252 and Plaskolite, Inc. for supplying the material.

References

1. Chien RD (2006) Micromolding of biochip devices designed with microchannels. Sens Actuators A Phys 128(2):238–47
2. Hollick EJ, Spalton DJ, Ursell PG, Pande MV, Barman SA, Boyce JF, Tilling K (1999) The effect of polymethylmethacrylate, silicone, and polyacrylic intraocular lenses on posterior capsular opacification 3 years after cataract surgery. Ophthalmology 106(1):49–54
3. Lu C, Cheng MM, Benatar A, Lee LJ (2007) Embossing of high-aspect-ratio-microstructures using sacrificial templates and fast surface heating. Polym Eng Sci 47(6):830–840
4. Narasimhan J, Papautsky I (2004) Polymer embossing tools for rapid prototyping of plastic microfluidic devices. J Micromech Microeng 14(1):96–103
5. Schneider P, Steitz C, Schafer KH, Ziegler C (2009) Hot embossing of pyramidal micro-structures in PMMA for cell culture. Phys Status Solidi A Appl Mater Sci 206(3):501–7
6. Mattotti M, Alvarez Z, Ortega JA, Planell JA, Engel E, Alcántara S (2012) Inducing functional radial glia-like progenitors from cortical astrocyte cultures using micropatterned PMMA. Biomaterials 33(6):1759–70
7. Gorelick S, Rahkila P, Sagari AAR, Sajavaara T, Cheng S, Karlsson LB, van Kan JA, Whitlow HJ (2007) Growth of osteoblasts on lithographically modified surfaces. Nucl Instrum Methods Phys Res B 260(1):130–5

8. Gamboa JR, Mohandes S, Tran PL, Slepian MJ, Yoon J-Y (2013) Linear fibroblast alignment on sinusoidal wave micropatterns. Colloids Surf B Biointerfaces 104:318–25
9. Srivastava V, Chester SA, Ames NM, Anand L (2010) A thermo-mechanically-coupled large-deformation theory for amorphous polymers in a temperature range which spans their glass transition. Int J Plast 26(8):1138–82
10. Kim NW, Kim KW, Sin HC (2008) Finite element analysis of low temperature thermal nanoimprint lithography using a viscoelastic model. Microelectron Eng 85(9):1858–65
11. Kim SM, Kang JH, Lee WI (2011) Analysis of polymer flow in embossing stage during thermal nanoimprint lithography. Polym Eng Sci 51 (2):209–17
12. Worgull M, Kabanemi K, Marcotte J-P, Hétu J-F, Heckele M (2008) Modeling of large area hot embossing. Microsyst Technol 14(7):1061–6
13. Hasan OA, Boyce MC (1993) Energy storage during inelastic deformation of glassy polymers. Polymer 34(24):5085–92
14. Quinson R, Perez J, Rink M, Pavan A (1996) Components of non-elastic deformation in amorphous glassy polymers. J Mater Sci 31(16):4387–94
15. Jones RL, Hu T, Soles CL, Lin EK, Reano RM, Pang SW, Casa DM (2006) Real-time shape evolution of nanoimprinted polymer structures during thermal annealing. Nano Lett 6(8):1723–8
16. Martin C, Ressier L, Peyrade JP (2003) Study of PMMA recoveries on micrometric patterns replicated by nano-imprint lithography. Physica E Low Dimens Syst Nanostruct 17(1):523
17. Arzhakov MS, Arzhakov SA (1995) A new approach to the description of mechanical behavior of polymer glasses. Int J Polym Mater 29(3–4):249
18. Arruda EM, Boyce MC, Jayachandran R (1995) Effects of strain rate, temperature and thermomechanical coupling on the finite strain deformation of glassy polymers. Mech Mater Int J 19(2/3):193
19. Ghatak A, Dupaix RB (2010) Material characterization and continuum modeling of poly(methyl methacrylate) (PMMA) above the glass transition. Int J Struct Changes Sol 2(1):53–63
20. Hasan OA, Boyce MC (1995) A constitutive model for the nonlinear viscoelastic viscoplastic behavior of glassy polymers. Polym Eng Sci 35 (4):331
21. Hasan OA, Boyce MC (1993) Energy storage during inelastic deformation of glassy polymers. Polymer 34(24):5085–92
22. Palm G, Dupaix RB, Castro J (2006) Large strain mechanical behavior of Poly(methyl methacrylate) (PMMA) near the glass transition temperature. J Eng Mater Technol 128(4):559–563
23. Ames NM, Srivastava V, Chester SA, Anand L (2009) A thermo-mechanically coupled theory for large deformations of amorphous polymers. part II: applications. Int J Plast 25(8):1495–539
24. Dooling PJ, Buckley CP, Rostami S, Zahlan N (2002) Hot-drawing of poly(methyl methacrylate) and simulation using a glass-rubber constitutive model. Polymer 43(8):2451–65
25. Mathiesen D, Vogtmann D, Dupaix RD (2014) Characterization and constitutive modeling of stress-relaxation behavior of poly(methyl methacrylate) (PMMA) across the glass transition temperature. Mech Mater 71:74–84
26. Mathiesen D, Kakumani A, Dupaix RB (2014) Experimental characterization and finite element modeling of temperature and strain dependent spring-back of poly(methyl methacrylate) (PMMA) near the glass transition (submitted)
27. Dupaix RB, Boyce MC (2007) Constitutive modeling of the finite strain behavior of amorphous polymers in and above the glass transition. Mech Mater 39(1):39–52
28. Arruda EM, Boyce MC (1993) A three-dimensional constitutive model for the large stretch behavior of rubber elastic materials. J Mech Phys Solids 41(2):389–412
29. Arruda EM, Boyce MC (1993) Evolution of plastic anisotropy in amorphous polymers during finite straining. Int J Plast 9(6):697–720
30. Likhtman AE, Graham RS (2003) Simple constitutive equation for linear polymer melts derived from molecular theory: Rolie-Poly equation. J Nonnewton Fluid Mech 114(1):1–12
31. Kabanemi KK, Hétu J (2009) Nonequilibrium stretching dynamics of dilute and entangled linear polymers in extensional flow. J Nonnewton Fluid Mech 160(2–3):113–21
32. De Focatiis DSA, Buckley CP, Embery J (2010) Large deformations in oriented polymer glasses: experimental study and a new glass-melt constitutive model. J Polym Sci B 48(13):1449–63

Chapter 10
Thermomechanical Fatigue Evaluation of Haynes® 230® for Solar Receiver Applications

Bonnie R. Antoun, Kevin J. Connelly, Steven H. Goods, and George B. Sartor

Abstract Haynes® 230® is a Ni–Cr–W–Mo alloy commonly used in aerospace and chemical process industries because of its excellent oxidation resistance, fatigue and creep performance at very high temperatures. In this study, the alloy was evaluated as a candidate for its use as tubing material in solar receivers, where coupled thermal-mechanical cycling is imposed in-use by heating imposed during diurnal cycles. The effect of temperature, 425 and 677 °C, and hold times on isothermal fatigue was evaluated and fatigue-life curves were developed for the alloy in its as-received condition and after aging at 677 °C for 3 months. Experimental apparatus and techniques were developed to apply thermomechanical cycles, between 425 and 677 °C, in an expedited manner to determine fatigue life at low strain ranges, again for both material conditions. The influence of stress ratio, R = −1 and R = ∞, was also assessed. The experimental techniques developed and resulting data and findings will be presented.

Keywords Fatigue • Thermal-mechanical • Haynes 230 • Solar • Temperature

10.1 Material

The material used in this study was a 5/8 in. (15.88 mm) diameter bar of Haynes® 230® steel. The chemical composition is shown in Table 10.1 and the average grain size was ASTM 4 (100 μm). The material was tested in two states, as-received and aged. Aging was induced by holding machined specimens in air, in a furnace at 650 °C for a period of 3 months. Two specimen geometries were used, tensile specimens and fatigue specimens. The tensile specimens, 4.25 in. (107.95 mm) in total length, were machined to have an overall straight gage section of 1.5 in. (38.1 mm) and gage diameter of 0.35 in. (8.89 mm), with 5/8 in. (15.875 mm) threaded ends that couple to the tensile test fixtures. The fatigue specimens were designed for optimal performance in the low cycle fatigue regime. Fatigue specimens were 4.45 in. (113.0 mm) in total length, with a straight gage section of 0.5 in. (12.7 mm) and gage diameter of 0.25 in. (6.35 mm) and a generous radius to transition to 5/8 in. (15.875 mm) threaded ends. The fatigue specimens were machined with no remaining tooling marks and finished by vertical lapping of the gage section, to eliminate any fatigue initiation sites for accurate measurement of fatigue life.

10.2 Experimental Equipment

All experiments were conducted on a 50 Kip (178 kN) MTS axial-torsional test frame, although only the axial channel was used. The experimental setup for tensile testing is shown in Fig. 10.1. The MTS test system controller and software, Flextest GT and Flexware, were used to program the tension tests to run in displacement (stroke) control to give approximate strain rates of 5.6E−03, 5.6E−04 and 6E−05 s^{-1}. Strains were measured directly on the specimen using various extensometers, depending on the displacement range needed, fitted with alumina rods and secured to the specimen with nichrome wire.

B.R. Antoun (✉) • K.J. Connelly • S.H. Goods • G.B. Sartor
Sandia National Laboratories, Livermore, CA 94551, USA
e-mail: brantou@sandia.gov

H.J. Qi et al. (eds.), *Challenges in Mechanics of Time-Dependent Materials, Volume 2: Proceedings of the 2014 Annual Conference on Experimental and Applied Mechanics*, Conference Proceedings of the Society for Experimental Mechanics Series,
DOI 10.1007/978-3-319-06980-7_10,

Table 10.1 Chemical composition (%) of Haynes® 230® material

Al	B	C	Co	Cr
0.32	<0.002	0.110	0.17	21.35
Cu	Fe	Mn	Mo	N
0.04	0.83	0.47	1.32	
Ni	O	P	S	W
BAL		0.005	<0.002	13.51
Si	Ti	V	La	
0.33	<0.01		0.012	

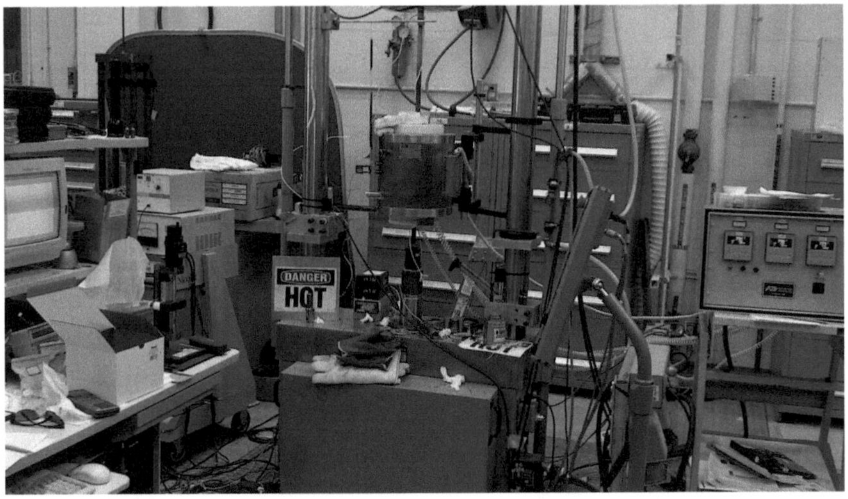

Fig. 10.1 Experimental setup for tension testing

A three zone furnace and controller were used to heat to the elevated temperatures, with excellent temperature uniformity monitored by two type K thermocouples attached directly to each specimen.

The fatigue experiments required a substantially different and much more complicated setup, although the same test frame was used. The setup is shown in Fig. 10.2. Fixtures were designed to be robust to withstand reversed (tension and compression) fatigue loading and elevated temperatures for extended periods. Since the fatigue specimen design had threaded ends, the fixtures also were designed to prevent any backlash at any point during loading. The entire system was precision aligned and locked using a series of spiral locking washers because rigid fixtures were required for compressive ($R = -1$) loading. All fatigue tests were conducted in extensometer (strain) control, the extensometer and attachment mechanisms should not cause damage or initiate failure so the alumina rods were custom machined to the specimen geometry and ceramic cords were used for attachment. Induction heating was used for all fatigue tests. An induction coil was designed through an iterative process to provide a uniform temperature along the gage length while allowing spaces for the extensometer rods. The coil is shown in Fig. 10.3, note the specimen failed in the gage section and within the extensometer rods. Induction heating was necessitated by the thermomechanical fatigue tests, to minimize the length of each applied cycle, and then also used during isothermal fatigue tests. Considerable time was spent optimizing methods to synchronize the induction heating with mechanical straining, leading eventually to a very short thermomechanical cycle period of 78 s.

10.3 Experimental Results: Tension

Tensile characterization experiments were conducted on as-received and aged material at the three temperatures, 425, 565, and 677 °C, and three strain rates, 6E−05 s^{-1} for comparison with historic Haynes International data [1] and 5.6E−03 s^{-1} which is close to the fatigue cycle straining rates. The tensile results are shown in Fig. 10.4. At the cyclic strain levels of interest, 1.2 % or less, no strain rate effect was noted, although there is an effect at higher strain levels. The results of the aged

Fig. 10.2 Experimental setup for isothermal and thermomechanical testing

Fig. 10.3 Photograph showing six-wind induction coil with custom alumina extensometer rods and ceramic attachment cords

material are important because they can give insight into the expected fatigue response of the material, such as the different hardening behavior observed. Aging had a noticeable effect on the tensile behavior, increased yield and lower ductility. Some strain ratcheting (aging) was observed at each temperature and was similar for the aged and as-received conditions. Tests were also conducted at room temperature, not included here, and the same comments apply except no strain ratcheting was observed.

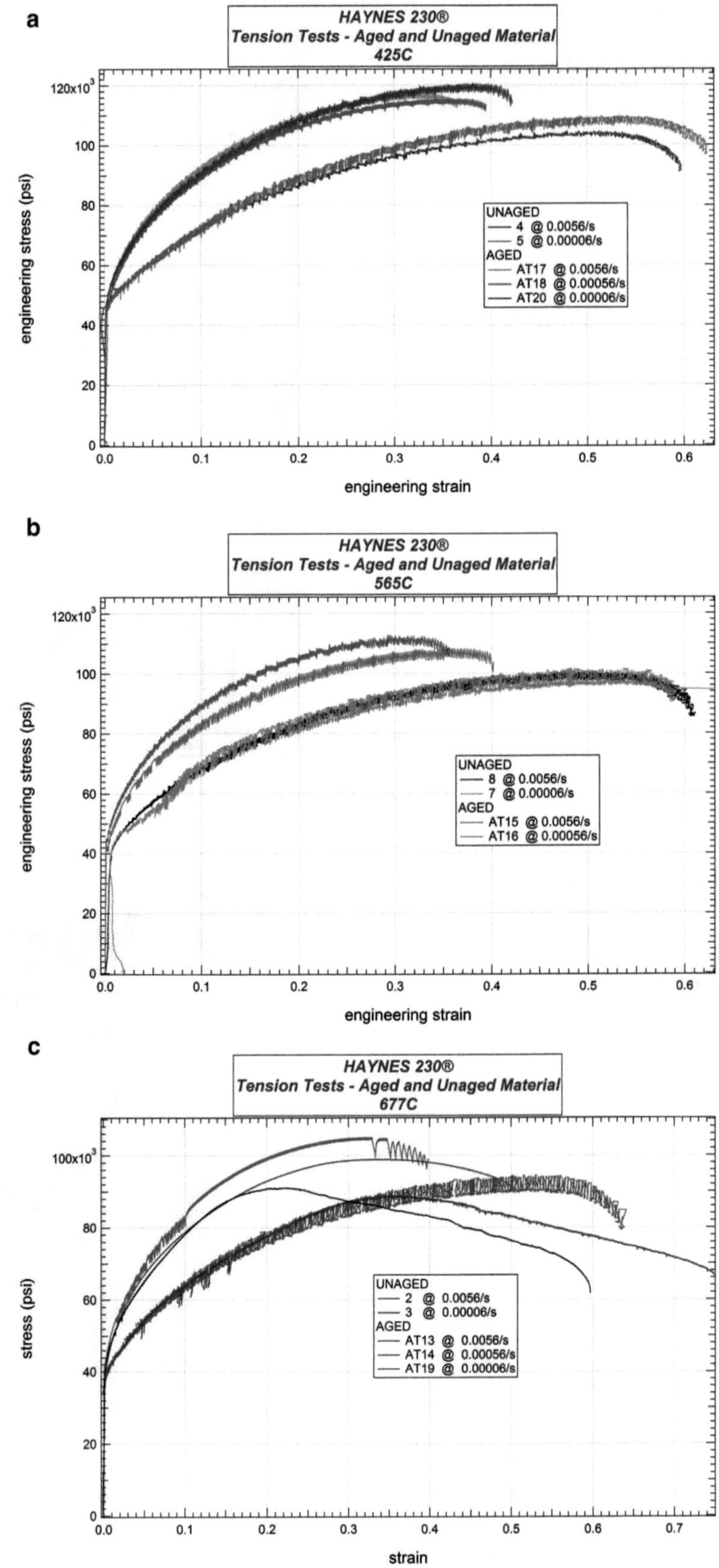

Fig. 10.4 Tensile results for as-received and aged Haynes® 230® steel at (**a**) 425 °C, (**b**) 565 °C and (**c**) 677 °C

10.4 Experimental Results: Isothermal and Thermomechanical Fatigue

An extensive set of experiments were completed to evaluate the Haynes® 230® alloy in fatigue. A set of example results will be included herein, there are many ways to observe the data acquired from S–N curves to stress versus strain cycles. Documentation of the complete results is currently in progress [2]. The effect of temperature, 425 and 677 °C, and hold times on isothermal fatigue was evaluated and fatigue-life curves were developed for the alloy in its as-received condition and after aging at 677 °C for 3 months. Figure 10.5 shows the S–N curve comparing results from this study with historic Haynes International data [1]. As expected, increasing temperature results in a decrease in the fatigue life.

The effect of hold times in compression was studied, an example of the results is shown in Fig. 10.6. Both tests HAY29 and HAY32 were conducted at 677 °C, R $= \infty$ and a strain range of 1.2 %, but the hold times were zero and 120 s, respectively.

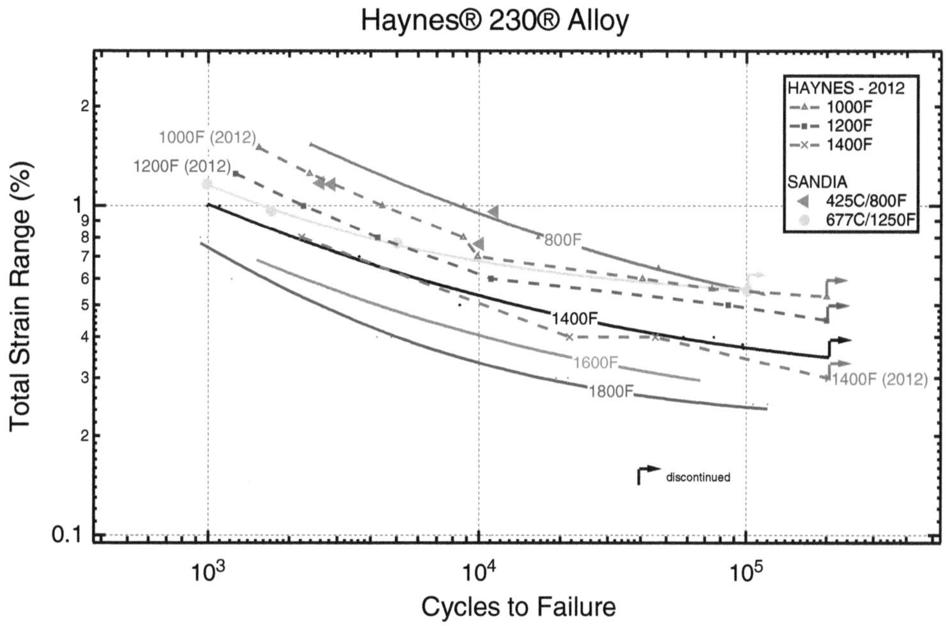

Fig. 10.5 S–N curve for isothermal fatigue of Haynes® 230® steel

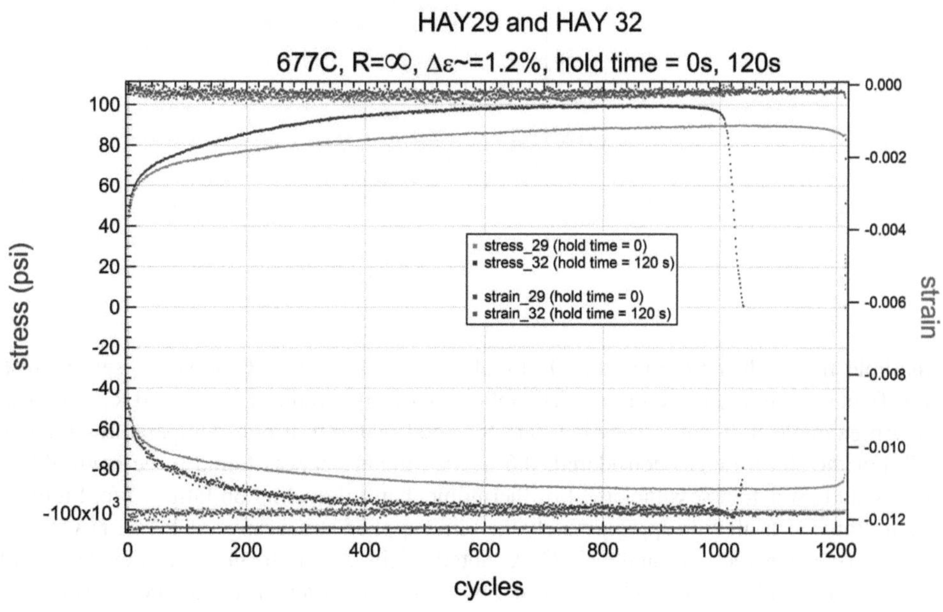

Fig. 10.6 Effect of 2 min. compressive hold time on fatigue life Haynes® 230® steel at 677 °C

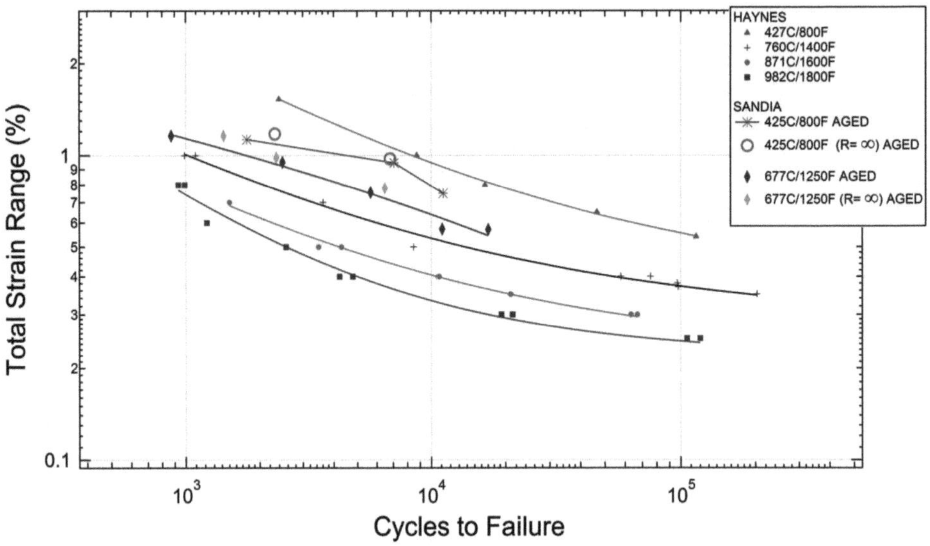

Fig. 10.7 Effect of stress ratio on fatigue life of aged Haynes® 230® steel, R = −1 unless noted

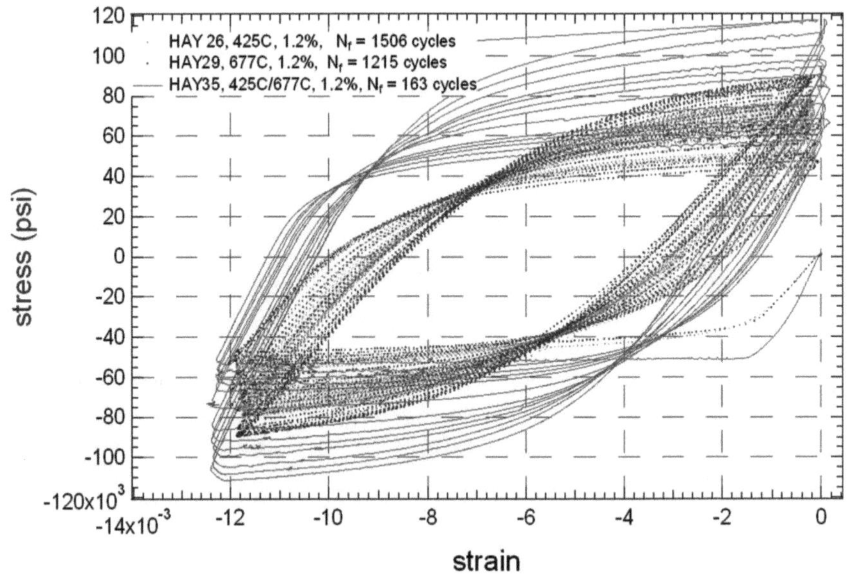

Fig. 10.8 Effect of thermomechanical cycling on fatigue life of Haynes® 230® steel, R = −1

The addition of the hold time reduced the fatigue life by almost 20 %, at 425 °C however, several tests throughout the S–N curve show no effect of compressive hold time. The influence of stress ratio, R = −1 and R = ∞, was also assessed at both temperatures. An example S–N curve showing data for aged material at both temperatures for both stress ratios is shown in Fig. 10.7. For the strain ranges considered, 0.6–1.2 %, the effect was found to be negligible. One of the most important parts of this study was to assess the effect of thermomechanical cycling on fatigue life. Figure 10.8 shows stress versus strain cycles for three tests all with approximately the same applied strain range of 1.2 %: HAY26 at 425 °C, HAY29 at 677 °C and HAY35 cycled between 425 and 677 °C. As noted on the graph, the number of cycles to failure was reduced about 20 % between the 425 and 677 °C isothermal tests, by almost an order of magnitude between the isothermal tests and the thermomechanical test. The difference between thermomechanical and isothermal cycling on fatigue life is also

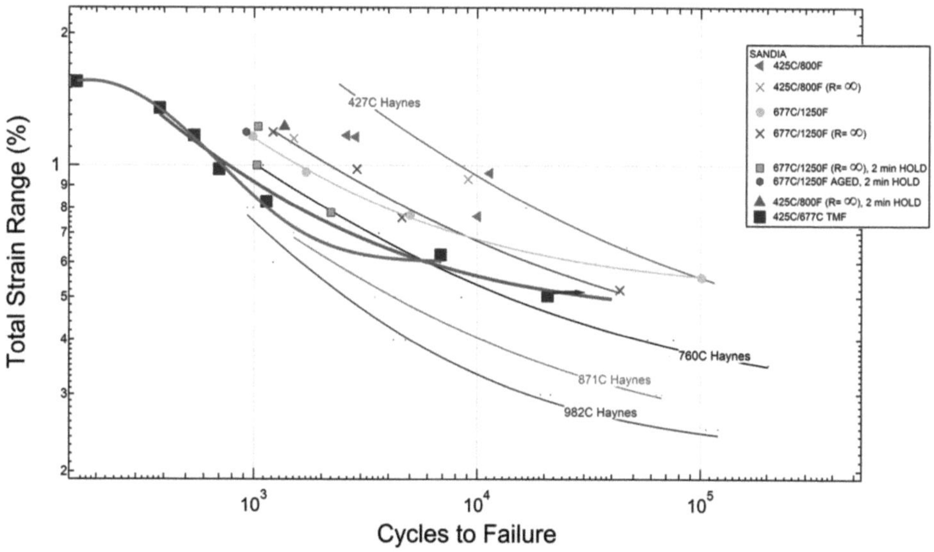

Fig. 10.9 S–N curve comparing thermomechanical and isothermal fatigue life of Haynes® 230® steel

shown in Fig. 10.9. A simple comparison between cycling continuously at 677 °C (light blue circles and line) to applying alternating cycles between 425 and 677 °C (purple squares and line) shows that thermomechanical cycling is much more severe and is critical to account for in design.

10.5 Summary

Haynes® 230® was found to have excellent fatigue performance at solar receiver application temperatures and be relatively insensitive to stress ratio effects, aging and compressive hold times. Experimental apparatus and techniques were developed to apply thermomechanical cycles in an expedited manner. This allowed completion of critical experiments for design of solar receiver tube material since the most severe loading is coupled thermal-mechanical cycling by heating imposed during diurnal cycles. A limited set of the complete data was presented to illustrate the key findings.

Acknowledgements Sandia National Laboratories is a multi-program laboratory managed and operated by Sandia Corporation, a wholly owned subsidiary of Lockheed Martin Corporation, for the U.S. Department of Energy's National Nuclear Security Administration under contract DE-AC04-94AL85000.

References

1. High-Temperature Alloys (2007) Haynes® 230® ALLOY, H3000.pdf. Haynes International, Inc
2. Antoun BR, Connelly K, Sartor GS (2014) Characterization of Haynes® 230® alloy in fatigue for solar receiver applications. SAND Report in Preparation, Sandia National Laboratories (2014)

Chapter 11
Viscoelastic Characterization of Fusion Processing in Bimodal Polyethylene Blends

Aaron M. Forster, Wei-Lun Ho, Kar Tean Tan, and Don Hunston

Abstract Polyethylene is an advantageous material for the construction of buried pipelines. It is corrosion resistant, seismic tolerant, and utilizes low cost installation methods. Pipe sections are often joined using thermal fusion processes. The strength of the joint is related to the ability of the polyethylene chains to inter-diffuse and form inter-crystalline tie-chains across the two polyethylene surfaces. Testing the strength of the fusion bond is difficult and a number of different destructive and non-destructive tests have been developed, but it is not possible to understand the impact of the fusion process on the local microstructure from these tests. In this work, instrumented indentation with a flat punch is used to measure the local viscoelastic behavior of five different polyethylene resins used for pipe manufacturing at short times. High strain behavior related to slow crack growth is measured using a strain hardening measurement under tension. The impact of thermal processing is investigated by imposing three different thermal cooling histories (0.4, 9, and 100 °C/min) on the polyethylenes. The goal is to determine if short-term creep under indentation is capable of accurately measuring; (a) bulk creep behavior, (b) impact of resin architecture, and (c) the impact of thermal processing. The results show that indentation using a flat punch is capable of measuring creep within the range of bulk creep behavior, but not sensitive to the slow crack growth resistance of the resin.

Keywords Indentation • Polyethylene • Slow crack growth • Creep • Fusion bond

11.1 Introduction

Polyethylene (PE) is important for critical infrastructure applications such as drinking water, waste water, and fossil fuels transportation. The advantages of polyethylene are competitive cost compared to metallic pipe, lower corrosion rates, and toughness. Central to the durability of polyethylene is not only its low reactivity with the environment, but also the microstructure of the material. Slow crack growth (SCG) is the predominant long-time failure mode for these materials. The process of SCG failure is understood as a stress concentrator, such as a contamination, catalyst particle, or damage in the pipe causes small voids to form a small crack in the pipe. These voids grow and form craze fibrils ahead of the stress concentration. The craze fibrils grow in length as the polyethylene chains disentangle from each other and the crystalline lamellae break up locally. The crack advances when the craze fibril breaks and stress shifts to the remaining fibrils, which causes them to break. Eventually the pipe fails in a brittle, rather than ductile, manner. This process is a creep driven failure and largely controlled by the tie-chains present in the pipe material [1].

Early pipe technologies utilized a single molecular weight distribution with short chain branching (SCB) to reduce SCG failures. Short chain branching along the backbone of the main polymer encourages the formation of tie-chains and increases entanglements within the amorphous region of the polyethylene. High performance resins used for piping applications are

A.M. Forster (✉) • W.-L. Ho • D. Hunston
Engineering Laboratory, National Institute of Standards and Technology (NIST), 100 Bureau Dr, Gaithersburg, MD 20899, USA
e-mail: aaron.forster@nist.gov

K.T. Tan
Engineering Laboratory, National Institute of Standards and Technology (NIST), 100 Bureau Dr, Gaithersburg, MD 20899, USA

Department of Materials Science and Engineering, University of Maryland, College Park, MD 20742, USA

H.J. Qi et al. (eds.), *Challenges in Mechanics of Time-Dependent Materials, Volume 2: Proceedings of the 2014 Annual Conference on Experimental and Applied Mechanics*, Conference Proceedings of the Society for Experimental Mechanics Series,
DOI 10.1007/978-3-319-06980-7_11,

composed of a bimodal blend of two different polyethylene molecular masses: a medium to high density linear chain and a medium density branched chain (<10 branches per 1,000 monomers). The advantage of the bimodal distribution is that the linear chains crystallize to provide the pipe its strength, while the branched chains resist chain pull-out giving the material exceptional slow crack growth resistance [1]. The SCG failure of polyethylene resins is sensitive to the microstructure of the resin, which depends on the polymer as well as the flow and thermal history during processing. For example, Ting and coworkers have shown that bimodal formulations require higher force to initiate failure and have higher ultimate failure strains than unimodal resins [2]. In the case of a fusion bond, which is often used to join pipe segments, the polymer is the same, but the thermal and flow history in the joint region can change local properties so they differ from those of the parent pipe. The heat affected zone (HAZ) may have altered crystallinity, molecular orientation and, in the case of the bimodal system, changes in composition.

Recently, the nuclear industry has been interested in the application of greater than 1 m diameter pipe for water transportation. While the structure–property relationships for pipe are generally understood, less is known about long term performance of thick walled pipe, particularly the fusion joints used to join these pipe sections. There has been concern that standard test methods for fusion joints do not adequately predict failure time for SCG [3]. In the past few years, the Welding Institute (TWI) in the U.K. has developed a whole pipe creep rupture test that is uniquely suited for long term failure measurements of fusion joints [4]. A challenge for any full pipe based test method is the cost required for evaluating large diameter pipe fusion joints. It would be advantageous to develop a coupon level metrology for conducting design of experiments and sensitivity analysis prior to testing large diameter pipes. The current research program at the National Institute of Standards and Technology (NIST) aims to develop an understanding of the viscoelastic and viscoplastic behaviors for PE resin materials as a function of temperature, pressure, and cooling profiles observed during the formation of butt fusion joints in polyethylene. The initial focus is the heat affected zone (HAZ) around the fusion. Instrumented indentation is investigated as a metrology to quantify the local viscoelasticity across the HAZ zone and into the fusion bond. If instrumented indentation is capable of measuring the local viscoelasticity of the polyethylene and this behavior connected to the local microstructure of the polyethylene; this would provide an experimental route to maximize fusion bond conditions and provide the mechanical framework for modeling these thermal joints.

11.2 Experimental Details

Five different PE resins (DOW) were measured for this work [5]. One was a unimodal resin which has a Pennsylvania Edge Notch Test (PENT) (ASTM F1473) value of ~200 h; while the other four were bimodal resins with PENT times from greater than 5,000 h to greater than 15,000 h as shown in Table 11.1. Each resin was compression molded into a 15 cm × 15 cm × 500 μm plaque at 160 °C according to a procedure described in [7]. After pressing, the samples were removed from the press and placed in between two brass platens and manually clamped at all edges using C-clamps. The samples were reheated to 160 °C in an oven and cooled using either controlled cooling (0.4 and 9 °C/min) or quenched in an ice water bath (>400 °C/min). The effect of thermal processing on the crystallinity, melting temperature, and lamellae thickness was characterized using Differential Scanning Calorimetry (DSC) scanning from room temperature to 160 °C at 10 °C/min. The lamellar thickness was calculated using the Gibbs-Thomson equation [6].

$$T_m = T_m^o\left(1 - \frac{2\sigma_e}{\Delta h_m l}\right) \tag{11.1}$$

Table 11.1 Properties of the five different polyethylene resins

		PE-U	PEb-1	PEb-2	PEb-3	PEb-4
		Unimodal	Bimodal	Bimodal	Bimodal	Bimodal
Density (g/cm^3)		0.946	0.949	0.949	0.949	0.941
Melt index (g/10 min)	190 °C/2.16 kg	0.1	0.08	0.08	0.06	0.16
	190 °C/21.6 kg	8.3	8.5	7.0	5.5	9.5
Tensile elongation (%)		850	>500	>500	>500	>600
Slow crack growth (h)		200	>5 k	>10 k	>10 k	>15 k
Flexural modulus (psi)		120 k	150 k	150 k	150 k	>90 k
Tensile strength (psi)		3,200	>3,500	>3,500	>3,500	>2,600

where T_m is the measured melting temperature, T_m^0 is the equilibrium melting point of an infinite crystal (415 K), σ_e is the surface free energy of the basal plane (60.9×10^3 J/m^2), and Δh_m is the enthalpy of fusion per unit volume (2.88×10^8 J/m^3), l is the lamellar crystal thickness.

Bulk creep of selected resins was measured in tension using 10×10 mm rectangular bars. An initial stress of 2.75 MPa was applied to the bar at time equal to zero. The resulting change in strain from 0.17 to 0.51 % was measured over the 10,000 s. The creep compliance was calculated as a function of change in displacement with time.

In order to determine whether the thermal treatment impacted the SCG resistance of the resins, the SCG resistance was measured using a strain hardening (SH) technique [7, 8] conducted at 80 °C. In order to correlate with the indentation measurements, the strain hardening modulus was measured at room temperature. The strain hardening modulus was measured as the slope of the stress vs. strain curve, ($\langle G_p \rangle$), after the polyethylene natural draw ratio between an extension ratio of 8–12. The current study uses SH measurements of dog bone samples (ASTM D-638). The tests were performed at a constant displacement rate of 10 mm/min [7, 8] using an electromechanical test machine (Instron) and laser extensometer to measure strain. Prior to SH testing, the specimens were allowed to equilibrate at 80 °C in the oven for 30 min. A minimum of three tests were run for each type of specimen.

The creep behavior of the resins was measured using instrumented indentation with a 50 μm flat punch indenter over 600 s following the method of Herbert [9]. In this method a large step load ($P_1 = 50$ mN) is applied to the sample to achieve conformal contact with the flat punch. An iterative control loop holds the load constant while measuring the displacement over time. The loop calculates whether the slope of displacement over time is below a drift threshold of 0.3 nm/s, in this case. If the tip has significantly drifted, then the controller moves the tip back to the original displacement in the sample and holds the new load, P_2, for another iterative round. Once the drift threshold has been met the indenter applies a smaller load and the displacement as a function of time is measured under this constant P_2. The creep under three different constant loads of 3, 6, and 9 mN was investigated. In the time domain for instrumented indentation experiments the creep compliance using a flat punch may be calculated from Eq. (11.2) [10].

$$J_c(t) = \frac{2Rh(t)}{P(1 - \upsilon^2)} \tag{11.2}$$

where $J_c(t)$ is the creep compliance, R is the radius of the flat punch, $h(t)$ is the displacement time response, P is the applied constant load, and ν is the time independent Poisson's ratio (=0.35).

11.3 Results and Discussions

The morphological characteristics within polyethylene are important for the ultimate properties of the material. The SCG behavior of polyethylene is a function of the molecular weight distribution, the amount of short chain branching and the distribution of the short chain branches. The crystalline lamellae provide the strength for the material, but the chains that cross the amorphous regions to form tie chains increase the yield strength. The harder it is to disentangle those chains in the amorphous region, through the presence of short chain branching, the better the SCG resistance of the material. Slowing the speed of the material through the crystallization transition can increase the amount of order within the system, which may allow for the formation of larger crystals and less amorphous region. Quenching the material will decrease the order within the system, which will decrease the size of the crystalline lamellae and increase the amount of amorphous material. The question of whether cooling rate will change SCG resistance is difficult to establish for a broad range of molecular weight distributions since the formation of tie chains is not directly correlated to % crystallinity (%X) or lamellar thickness [11]. Sardashanti [12] has shown that accelerating the cooling rate decreases the % crystallinity and decreased the lamellar thickness. The authors also reported that decreasing the cooling rate slightly decreased the corrected strain hardening modulus for several polyethylenes. This was indicative of worse SCG performance for the resins the slower the cooling rate of the material.

The thermal measurements from the DSC show that decreasing the cooling rate does induce a significant change in the % crystallinity within the sample. It does increase the melting temperature. Figure 11.1 shows the lamellar thickness change for the polyethylene resins as a function of cooling rate. Decreasing the cooling rate shows an increase in the lamellar thickness. The increase in lamellar thickness is greatest for the bimodal materials compared to the unimodal polyethylene. While there is a difference between the lamellar thickness of the unimodal and the bimodal materials for all cooling rates, there is no significant difference between the different bimodal polyethylenes (PEb-1 to PEb-4). The increase in lamellar thickness

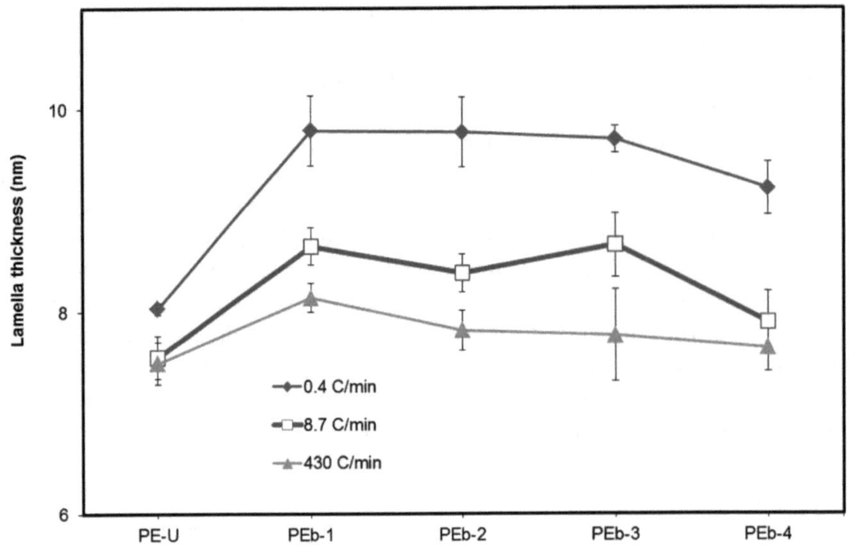

Fig. 11.1 Lamellar thickness of the polyethylene resins, calculated from Eq. (11.2), as a function of different cooling rates

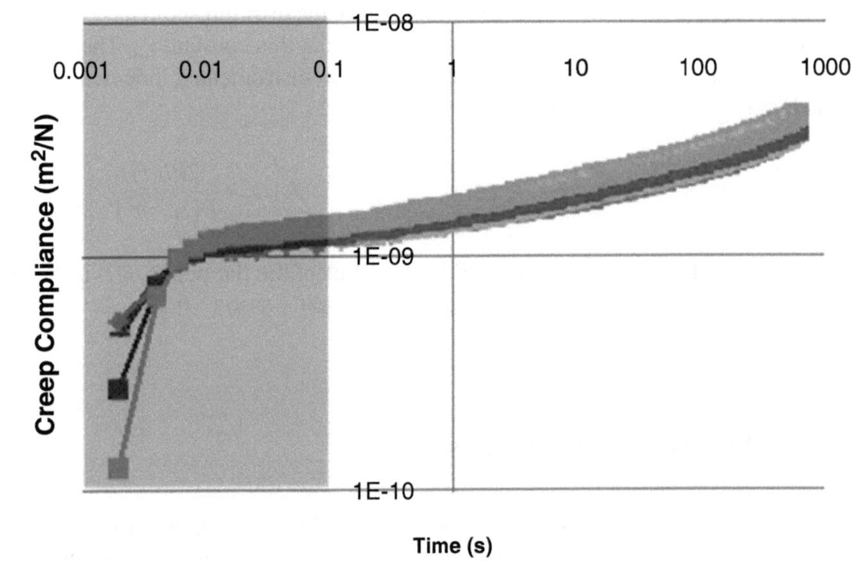

Fig. 11.2 J(t) measured for the PE-U sample using a 50 μm flat punch indenter at creep loads of 6 and 9 mN. The highest curve is 0.4°C /min cooling and the lowest is 430 °C /min

suggests that increasing the cooling rate might decrease the SCG resistance. The strain hardening modulus was measured in order to estimate the changes in the SCG resistance for these materials.

The strain hardening modulus (<G_p>) at 80 °C was 33 ± 3 for the unimodal material and 48 ± 4 for the bimodal materials. These values agree with those found in the literature [7, 8, 11, 12] and show that this technique is able to measure differences between materials with large differences in <G_p> and PENT values. The strain hardening method was not able to discern differences between the bimodal materials. These materials only exhibit a difference in PENT failure time of a factor of 3 compared to a difference factor of ~30 for the unimodal. The strain hardening modulus did not change significantly with thermal treatment for any of the polyethylene materials. This suggests that the strain hardening modulus is not significantly impacted by cooling rate for these polyethylene materials. If the strain hardening modulus is a corollary for SCG resistance, then this suggests the SCG resistance of the polyethylene is not affected by cooling rate. Fracture tests are currently under way at NIST to verify that conclusion, but it may be that this technique is not sensitive in the strain hardening range of these high performance resins. The cooling rate does have an effect on the lamellar thickness and melting temperature of the polyethylenes, but it is not a large enough effect to alter the tie-chain structure within the material. This suggests that creep measurements of the materials should not be affected by the cooling rate either.

Figure 11.2 shows the creep compliance measured with instrumented indentation on the unimodal polyethylene, PE-U, under a 6 mN load. This figure shows the creep compliance for the complete load-hold segment. A transparent red rectangle has been placed on the figure to represent compliance data that would be affected by the load step. The size of this box is based on the convention that 10x the loading time should be given for the material to relax the P_2 load step. In this figure the different cooling rates lay close to of each other. The curve has a slight curvature on the log-log scale with curvature increasing at long times. The magnitude of the creep compliance measured from indentation is slightly higher than bulk measurements for this material. Bulk tension creep measurements (not shown) indicate that creep compliance starts at 6.0×10^{-10} and increases to 1.0×10^{-9} over 600 s. This is lower than the instrumented indentation measurements and may reflect the differences in the loading methods.

11.4 Conclusion

The impact of cooling rate on the microstructure, strain hardening, and creep behavior of five different polyethylene pipe materials was investigated. It was found that increasing the cooling rate does not significantly change the % crystallinity, but does reduce the order in the system. This was characterized through the melting temperature and lamella crystal thickness. The high strain behavior of the materials was characterized using strain hardening, which has been shown to correlate to SCG resistance. Strain hardening measurements showed a difference between the unimodal and bimodal materials, but could not discern the differences in the SCG resistance of the four bimodal resins. This technique did not measure a difference in the strain hardening modulus as a function of cooling rate either. The viscoelasticity of the materials was characterized using both small-strain bulk creep and instrumented indentation creep. The bulk and indentation measured creep differed, but were within a decade of each other for compliance. The source of error between these two techniques is under investigation. Both the instrumented indentation and the strain hardening measurements show for PE-U that the cooling rate didn't significantly affect the viscoelastic behavior of the polyethylene resin.

Acknowledgements Dr. Jimmy Zhou kindly supplied the polyethylene resins for this study.

References

1. Davis P, Burn S, Gould S, Cardy M, Tjandraatmadja G, Sadler P (2007) Long-term performance prediction for PE pipes. AWWA Research Foundation, Denver
2. Ting SKM, Williams JG, Ivankovic A (2006) Characterization of the fracture behavior of polyethylene using measured cohesive curves. II variation of cohesive parameters with rate and constraint. Poly Eng Sci 46:792–798
3. Focht E (2010) NRC issues regarding the use of HDPE piping in safety-related nuclear applications. In: Use of HDPE for power plant piping systems workshop, Charlotte, 7–10 June 2010
4. Troughton M, Brown C, Hessel J, Piovano M (2006) Proceedings of the plastics pipes XIII conference, Washington, DC, October 2006
5. Any mention of commercial products within this paper is for information only; it does not imply recommendation or endorsement by NIST
6. Wlochowicz A, Eder M (1984) Distribution of lamellar thickness in isothermally crystallized polypropylene and polyethylene by differential scanning calorimetry. Polymer 9:1268–1270
7. Kurelec L, Teeuwen M, Schoffeleers H, Deblieck R (2005) Strain hardening modulus as a measure of environmental stress crack resistance of high density polyethylene. Polymer 46:6369–6379
8. McCarthy M, Deblieck R, Mindermann P, Kloth R, Kurelec L, Martens H (2008) Plastic pipes XIV, Budapest, 22–24 September 2008
9. Herbert EG, Oliver WC, Lumsdaine A, Pharr GM (2009) Measuring the constitutive behavior of viscoelastic solids in the time and frequency domain using flat punch nanoindentation. J Mater Res 24:626–637
10. Lu H, Wang B, Ma J, Huang G, Viswanathan H (2003) Measurement of creep compliance of solid polymers by nanoindentation. Mech Time Depend Mater 7:189–207
11. Cheng J, Polak M, Penlidis A (2009) A tensile strain hardening test indicator of environmental stress cracking resistance. J Macromol Sci Part A Pure Appl Chem 45:599–611
12. Sardashti P, Scott AJ, Tzoganakis C, Polak M, Penlidis A (2014) Effect of temperature on environmental stress cracking resistance and crystal structure of polyethylene. J Macromol Sci Part A Pure Appl Chem 51:189–202

Chapter 12
Viscoelastic Properties for PMMA Bar over a Wide Range of Frequencies

T. Tamaogi and Y. Sogabe

Abstract Viscoelastic properties for a polymethyl methacrylate (PMMA) bar were examined using both ultrasonic wave propagation experiments in the higher frequency range of 25–200 kHz and longitudinal wave propagation experiments in the lower frequency range of up to 15 kHz. Since the geometrical dispersion due to three-dimensional deformation was caused by higher frequency components involved in the ultrasonic waves, the three-dimensional wave theory was employed to analyze experimental data of wave propagation. It was found that the 5-element model based on the three-dimensional theory could evaluate the viscoelastic properties of a wide range of frequencies. The peak value of the attenuation coefficient moved to the higher frequency as the diameter of the bar became small. Moreover, the viscoelastic properties could be evaluated only by the solution of the first mode based on the three-dimensional wave theory when the diameter of the bar was thin.

Keywords Dynamic properties • Propagation • Viscoelastic • Ultrasonic • Transducer

12.1 Introduction

The use of viscoelastic materials has been increasing in various industries such as automotive and electronic industries because of their excellent impact resistance or damping capability. Precise knowledge of their dynamic properties is needed to further widen the applicability of viscoelastic materials. The deformation of the materials remarkably depends on the time or strain rate. The viscoelastic theory is generally applied to their impact behavior. The characteristics of viscoelastic medium are controlled by the different mechanical properties according to the various frequencies.

Much research has been reported the dynamic properties for polymethyl methacrylate (PMMA), which were generally used in viscoelastic impact tests for viscoelastic materials, could be identified as a 3-element solid model based on the one-dimensional theory by the longitudinal wave propagation experiments [1–3]. However, the geometrical dispersion due to three-dimensional deformation will be caused by the high frequency component involved in a wave, the three-dimensional theory should be employed to process the experimental data of wave propagation [4–6].

This paper deals with longitudinal viscoelastic wave propagating in solid PMMA bars of two kinds of thickness (8, 15 mm). Attenuative and dispersive properties of viscoelastic materials are discussed three dimensionally.

T. Tamaogi (✉)
Department of Mechanical Engineering, Niihama National College of Technology, 7-1 Yakumo-cho, Niihama, Ehime 792-8580, Japan
e-mail: takatamajp@yahoo.co.jp

Y. Sogabe
Department of Mechanical Engineering, Ehime University, 3 Bunkyo-cho, Matsuyama, Ehime 790-8577, Japan

H.J. Qi et al. (eds.), *Challenges in Mechanics of Time-Dependent Materials, Volume 2: Proceedings of the 2014 Annual Conference on Experimental and Applied Mechanics*, Conference Proceedings of the Society for Experimental Mechanics Series,
DOI 10.1007/978-3-319-06980-7_12,

12.2 Attenuation and Dispersion for Viscoelastic Bar

12.2.1 Elementary Theory

In case of a thin and uniform viscoelastic bar, let $\bar{\varepsilon}(x,\omega)$ be the Fourier transform of a strain-time relation $\varepsilon(x,t)$.

When the material is linear viscoelastic, the following equation can be obtained [2]:

$$\bar{\varepsilon}(x,\omega) = \bar{\varepsilon}(0,\omega)\cdot \exp\{-(\alpha + ik)\}x, \tag{12.1}$$

where x, ω and i are the coordinate along the rod axis, angular frequency and imaginary unit, respectively.

The attenuation coefficient α and wave number k are the functions of ω, and are related to the complex compliance as

$$\alpha^2(\omega) = \frac{1}{2}\rho\omega^2\{-J_1^*(\omega) + |J^*(\omega)|\}, \tag{12.2}$$

$$k^2(\omega) = \frac{1}{2}\rho\omega^2\{J_1^*(\omega) + |J^*(\omega)|\}, \tag{12.3}$$

where $|J^*(\omega)| = \sqrt{J_1^{*2}(\omega) + J_2^{*2}(\omega)}$, ρ is the material density. The complex compliance $J^*(\omega)$, which represents one of the viscoelastic properties of the material, is defined as

$$J^*(\omega) = J_1{}^*(\omega) - i\,J_2{}^*(\omega). \tag{12.4}$$

The phase velocity $C(\omega)$, which corresponds to dispersive properties, is given by

$$C(\omega) = \frac{\omega}{k(\omega)} \tag{12.5}$$

12.2.2 Pochhammer-Chree Theory

Consider a stress wave propagating in an infinite cylindrical elastic bar. The equation of motion is written in the following vector form:

$$\rho\frac{\partial^2 \mathbf{u}}{\partial t^2} = (\lambda + 2\mu)\text{grad}\Delta - 2\mu\text{rot}\ \Omega, \tag{12.6}$$

where $\mathbf{u}$ denotes the displacement vector, λ and μ are the Lame coefficients, $\Delta = \text{div}\mathbf{u}$, $2\Omega = \text{rot}\mathbf{u}$, ν is Poisson's ratio. Assuming axial symmetry, and applying the Fourier transformation and the correspondence principle [7] to Eq. (12.6), the following equations for a viscoelastic medium on the cylindrical coordinate plane are deduced:

$$\left.\begin{aligned} -\rho\omega^2 U_r &= (\lambda^* + 2\mu^*)\frac{\partial D}{\partial r} - 2i\xi\mu^* W \\ -\rho\omega^2 U_z &= (\lambda^* + 2\mu^*)(-i\xi)D - 2\mu^*\left(\frac{\partial W}{\partial r} + \frac{W}{r}\right) \end{aligned}\right\}, \tag{12.7}$$

where λ^* and μ^* are the complex Lame coefficients, the displacement $\bar{u}_r(r,z,\omega) = U_r(r,\omega)\cdot\exp(-i\xi z)$ and $\bar{u}_z(r,z,\omega) = U_z(r,\omega)\cdot\exp(-i\xi z)$, the volumetric strain $\bar{\Delta}(r,z,\omega) = D(r,\omega)\cdot\exp(-i\xi z)$, the rotation vector $\bar{\Omega}_\theta(r,z,\omega) = W(r,\omega)\cdot\exp(-i\xi z)$, $\xi(\omega) = k(\omega) - i\alpha(\omega)$ respectively. Solving the Eq. (12.7) in D and W, the Bessel's differential equations of order zero and one are obtained. The solutions can be expressed as follows:

$$\left.\begin{array}{l} D(r,\omega) = A_0 J_0(pr) \\ W(r,\omega) = A_1 J_1(qr) \end{array}\right\}, \tag{12.8}$$

where J_0, J_1 are Bessel functions of order zero and one, A_0 and A_1 are the arbitrary functions of ω, $p^2 = \rho\omega^2/(\lambda^* + 2\mu^*) - \xi^2$, $q^2 = \rho\omega^2/\mu^* - \xi^2$, respectively.

Considering stress free boundary conditions at the external surface of the bar, the following frequency equation results:

$$(q^2 - \xi^2)^2 J_0(pa)J_1(qa) - \frac{2p}{a}(q^2 + \xi^2)J_1(pa)J_1(qa) + 4\xi^2 pqJ_0(qa)J_1(pa) = 0, \tag{12.9}$$

where a is radius of the bar. Solving Eq. (12.9) for ξ numerically, the attenuation coefficient $\alpha(\omega)$, the wave number $k(\omega)$ and the phase velocity $C(\omega)$ are given by

$$\alpha(\omega) = \mathrm{Im}[-\xi(\omega)] \tag{12.10}$$

$$k(\omega) = \mathrm{Re}[\xi(\omega)] \tag{12.11}$$

$$C(\omega) = \frac{\omega}{k(\omega)} \tag{12.12}$$

12.3 Experiments

Polymethyl methacrylate (PMMA) materials are used for the specimen. Firstly, the longitudinal wave propagation experiments are carried out. The length and diameter of the PMMA bar are 2,000 and 15 mm, respectively. The striker bar is also made of PMMA, and has a length and diameter of 20 and 15 mm. Four strain gages are situated at positions separated by equal intervals (200 mm) at a distance from the impact face. The striker bar is launched by the air compressor, and impacts the front end of the PMMA bar [6].

Secondly, the ultrasonic wave propagation experiments are performed. Figure 12.1 indicates a schematic diagram of an ultrasonic wave propagation experiment. Two kinds of the diameters are used; $D = 8$ and 15 mm, and the length of the bars are 1,000 mm. Sixteen kinds of ultrasonic transducers having several characteristic frequencies from 20 to 200 kHz are prepared as shown in Table 12.1. The ultrasonic transducer types from (1) to (10) are used for PMMA bar with the diameter of 15 mm, from (11) to (16) are used for the diameter of 8 mm. The ultrasonic transducer is attached to the one side of the specimen. Four semiconductor strain gages are situated at positions separated by equal intervals (100 mm) at a distance from the transducer's side. The transducer is vibrated at the natural frequency by giving the voltage amplified with an AC amplifier.

The measured strain waves using the ultrasonic transducer type (11) (50 kHz) and the diameter of the specimen bar 8 mm are shown in Fig. 12.2a as typical examples. It is found that the attenuation and dispersion generate as the waves propagate. The frequency spectrum of each wave are represented in Fig. 12.2b. The frequency spectrums have a lot of frequency elements of in the frequency around 50 kHz. The values of 45–55 kHz, which are 70 % of the maximum values of the frequency spectrums, are used for evaluation of the attenuation and dispersion properties.

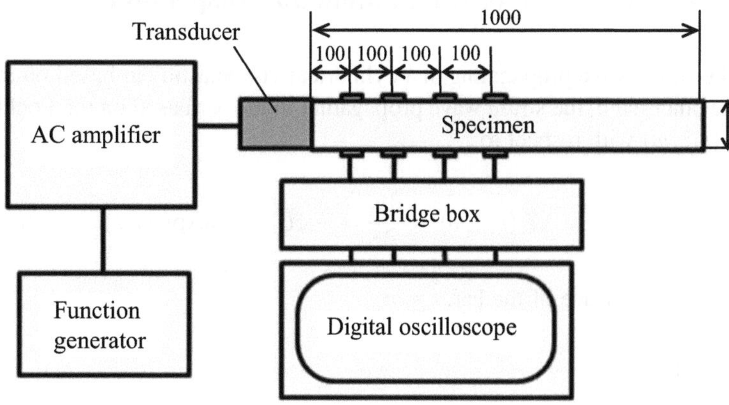

Fig. 12.1 Schematic diagram of propagation experiment using wave packets generated by ultrasonic transducer

Table 12.1 Properties of ultrasonic transducers

Type	Frequency (kHz)	Diameter of radial plane (mm)	Length (mm)	Capacitance (pF)
(a) Ultrasonic transducers for process machinery (NGK Spark Plug Co., Ltd.) (15 mm)				
(1) DA2228	28.17	20	92.4	1,250
(2) DA2240	38.94	20	64.4	1,090
(3) DA21560A	60.04	15	40.4	680
(4) DA2275A	74.79	20	30.4	1,740
(b) Piezoelectric ceramics transducers (Fuji Ceramics Corporation) (15 mm)				
Type	Frequency (kHz)	Element diameter (mm)	Length (mm)	Capacitance (pF)
(5) 0.05Z15D	50.00	15	26.20	136
(6) 0.075Z15D	74.60	15	16.40	220
(7) 0.1Z15D	99.20	15	10.50	339
(8) 0.13Z10D	131.40	10	8.40	192
(9) 0.15Z20D	148.20	20	8.40	788
(10) 0.2Z15D	199.30	15	6.20	564
(c) Piezoelectric ceramics transducers (Fuji Ceramics Corporation) (8 mm)				
Type	Frequency (kHz)	Element diameter (mm)	Length (mm)	Capacitance (pF)
(11) 0.05Z8D	50.00	8	26.68	41
(12) 0.075Z8D	73.30	8	17.85	61
(13) 0.1Z8D	98.60	8	13.05	82
(14) 0.125Z8D	121.50	8	10.33	104
(15) 0.15Z8D	147.60	8	8.09	133
(16) 0.2Z5D	193.50	5	6.51	66

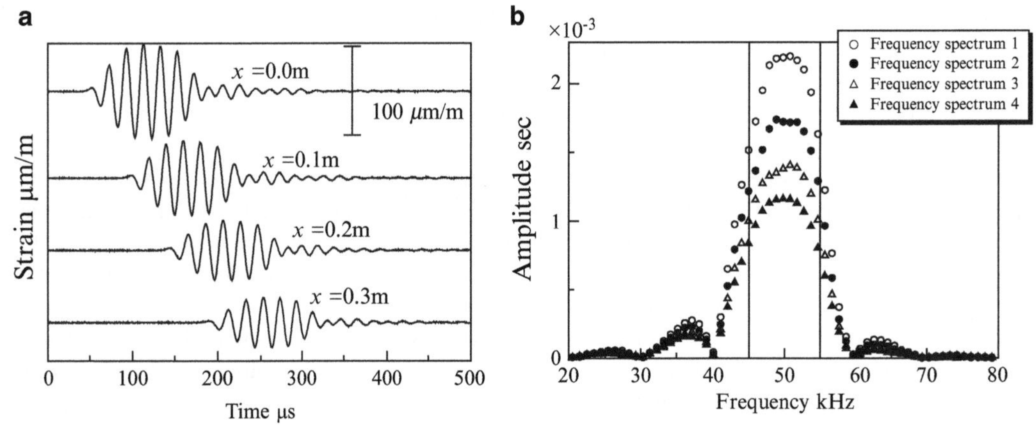

Fig. 12.2 Experimental results on propagation test using wave packets generated by ultrasonic transducer (50 kHz, 8 mm). (**a**) Measured strain waves, (**b**) frequency spectrums

12.4 Evaluation of Attenuation and Dispersion

The strain wave propagating in a cylindrical viscoelastic bar based on the elementary theory is expressed by Eq. (12.1). On the other hand, the strain wave propagating in the bar based on the Pochhammer-Chree theory is obtained by differentiating $\overline{u}_z(r, z, \omega)$ with respect to z.

$$\overline{\varepsilon}_z(r, z, \omega) = \frac{\partial \overline{u}_z}{\partial z} = -i\xi \overline{U}_z(r, \omega)\exp(-i\xi z) \quad = -(\alpha + ik)\overline{U}_z(r, \omega)\exp\{-(\alpha + ik)\}z. \tag{12.13}$$

On the surface of the bar, $r = a$:

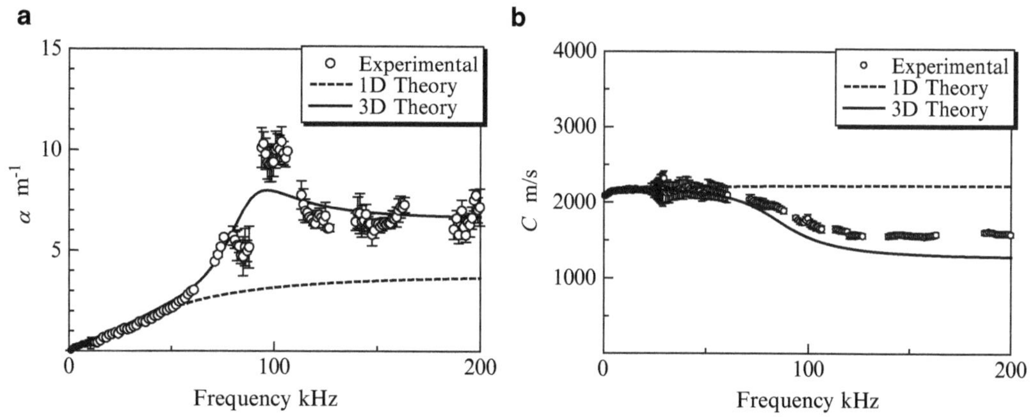

Fig. 12.3 Experimental and analytical values of $\alpha(\omega)$ and $C(\omega)$ obtained by connecting two kinds of experiments (diameter: 15 mm). (**a**) Attenuation coefficient (15 mm), (**b**) phase velocity (15 mm)

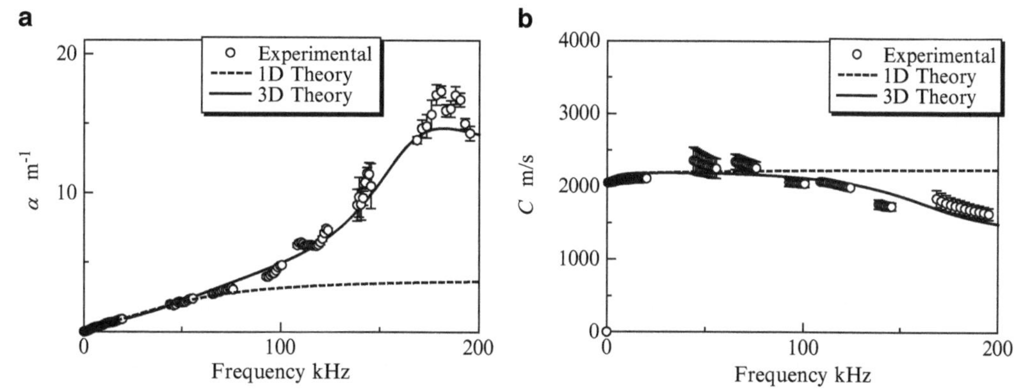

Fig. 12.4 Experimental and analytical values of $\alpha(\omega)$ and $C(\omega)$ obtained by connecting two kinds of experiments (diameter: 8 mm). (**a**) Attenuation coefficient (8 mm), (**b**) phase velocity (8 mm)

$$\bar{\varepsilon}_z = \bar{\varepsilon}_0(\omega)\exp\{-(\alpha + ik)\}z, \tag{12.14}$$

where $\overline{\varepsilon_0}(\omega) = -(\alpha + ik)\overline{U}_z(a,\omega)$. Therefore, the strain wave in a cylindrical viscoelastic bar is calculated by the same equation. Using the least square method, $\alpha(\omega)$ and $k(\omega)$ can be determined from the experimental data.

$$\left.\begin{aligned} \alpha(\omega) &= \frac{\sum z_m \sum \log|\bar{\varepsilon}_m| - 4\sum z_m \log|\bar{\varepsilon}_m|}{4\sum z_m^2 - \left(\sum z_m\right)^2} \\ k(\omega) &= \frac{\sum z_m \sum \theta_m - 4\sum z_m \theta_m}{4\sum z_m^2 - \left(\sum z_m\right)^2}, \end{aligned}\right\} \tag{12.15}$$

where m indicates gage position number (1 ~ 4), $|\bar{\varepsilon}_m|$ is absolute value of $\bar{\varepsilon}_m$, θ_m is phase angle $\arg(\bar{\varepsilon}_m)$, respectively.

Then, the phase velocity can be given by Eq. (12.5) or (12.12). Based on the Eqs. (12.15) and (12.5), the attenuation coefficient $\alpha(\omega)$ and phase velocity $C(\omega)$ are obtained by connecting the experimental values of the longitudinal impact experiments in the low frequency area (from 0 to the 15 kHz) and those of the ultrasonic propagation experiments in the high frequency area (from 20 to 200 kHz, measured by 16 kinds of ultrasonic transducers). Figures 12.3a, b and 12.4a, b indicate $\alpha(\omega)$ and $C(\omega)$ with 15 and 8 mm in diameters of the bars, respectively. The plots in the figures show average experimental values, and the vertical bars indicate the standard deviation. The dotted and the solid lines are the analytical values using 5-element model shown in Fig. 12.5 based on the one-dimensional theory (1D theory) and the three-dimensional theory

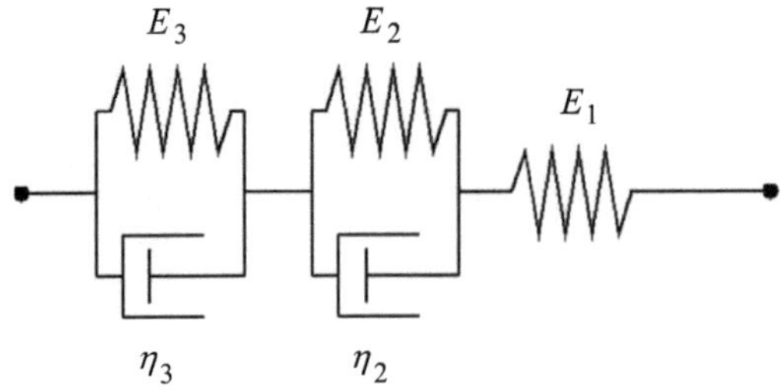

Fig. 12.5 Viscoelastic models for determining mechanical properties of PMMA material

(3D theory). The viscoelastic values E_1, E_2, η_2, E_3 and η_3 for 15 mm in diameters of the specimen bars are 5.89 GPa, 58.4 GPa, 2.80 MPa s, 122 GPa, 0.39 MPa s, and for 8 mm are 5.67 GPa, 68.4 GPa, 2.20 MPa s, 114 GPa, 0.21 MPa s, respectively.

It is found that the experimental and model's predicted values up to 15 kHz are almost identical. It is necessary to use a 5-element model based on the three-dimensional theory to evaluate the properties of a wide range of frequencies. It is found that the peak value of the attenuation coefficient move to the higher frequency as the diameter of the bar become small. The reason why the experimental and analytical values are different in the higher frequency domain than about 100 kHz for attenuation coefficient and phase velocity with the diameter of the bar 15 mm is that the wavelength becomes equal with the diameter of the bar, and the high-order mode is generated.

12.5 Conclusions

The conclusions obtained from the present study are summarized as follows:

- The attenuation and dispersion properties for viscoelastic material over the wide range of frequency were examined by two kinds of propagation experiments.
- It was found that the 5-element model based on the three-dimensional theory could evaluate the viscoelastic properties of a wide range of frequencies.
- The viscoelastic properties could be evaluated only by the solution of the 1st mode based on the three-dimensional wave theory when the diameter of the bar was thin.

References

1. Sackman JL, Kaya I (1968) On the determination of very early-time viscoelastic properties. J Mech Phys Solids 16(2):121–132
2. Sogabe Y, Tsuzuki M (1986) Identification of the dynamic properties of linear viscoelastic materials by the wave propagation testing. Bull Jpn Soc Mech Eng 29(254):2410–2417
3. Lundberg B, Blanc RH (1988) Determination of mechanical material properties from the two-point response of an impacted linearly viscoelastic rod specimen. J Sound Vib 126(1):97–108
4. Zhao H, Gary G (1995) A three dimensional analytical solution of the longitudinal wave propagation in an infinite linear viscoelastic cylindrical bar. Application to experimental techniques. J Mech Phys Solids 43(8):1335–1348
5. Bacon C (1999) Separation of waves propagating in an elastic or viscoelastic Hopkinson pressure bar with three-dimensional effects. Int J Impact Eng 22(1):55–69
6. Tamaogi T, Sogabe Y (2013) Attenuation and dispersion properties of longitudinal waves in PMMA bar over a wide range of frequencies. Jpn Soc Mech Eng 13(3):264–269 (in Japanese)
7. Flügge W (1975) Viscoelasticity. Springer-Verlag, Berlin-Heidelberg-New York. 159

Chapter 13
Implementation of Fractional Constitutive Equations into the Finite Element Method

L. Gaul and A. Schmidt

Abstract The damping properties of materials, joints, and assembled structures can be modeled efficiently using fractional derivatives in the respective constitutive equations. The respective models describe the damping behavior accurately over broad ranges of time or frequency where only few material parameters are needed. They assure causality and pure dissipative behavior. Due to the non-local character of fractional derivatives the whole deformation history of the structure under consideration has to be considered in time-domain computations. This leads to increasing storage requirements and high computational costs. A new concept for an effective numerical evaluation makes use of the equivalence between the Riemann–Liouville definition of fractional derivatives and the solution of a partial differential equation (PDE). The solution of the PDE is found by applying the method of weighted residuals where the domain is split into finite elements using appropriate shape functions. This approach leads to accurate results for the calculation of fractional derivatives where the numerical effort is significantly reduced compared with alternative approaches. Finally, this method is used in conjunction with a spatial discretization method and a simple structure is calculated. The results are compared to those obtained from alternative formulations by means of accuracy, storage requirements, and computational costs.

Keywords Fractional derivatives • Damping • Creep • Constitutive equations • Finite element method

13.1 Introduction

It is known that all structural materials show damping to some extent. When subjected to time periodic loads, a hysteresis can be observed and as a response to a Heaviside step in stress or strain, creep or stress relaxation occurs. The damping properties of some materials, such as rubbers or polymers, are quite pronounced and cannot be neglected when a structure containing these materials is modeled.

Damping models which show good adaptivity to measured material data can be obtained by introducing fractional derivatives in the respective constitutive equations. The application of fractional derivatives to viscoelasticity was studied substantially by Caputo and Mainardi [5] and is physically founded [3]. This concept results in fractional-order differential stress–strain relations, that provide good curve-fitting properties, require only few parameters, and lead to causal behavior [1, 2]. Bagley and Torvik [4] derived constraints for the material parameters of the 'fractional 3-parameter model' in order to ensure a non-negative internal work and rate of energy dissipation. Koeller [7] suggested to replace the viscous dashpots in rheological models by fractionally generalized elements which he called 'spring-pots'. The resulting constitutive relations then are consistent with thermodynamical principles [8]. An implementation of fractional constitutive equations into FE formulations is given by Padovan [10]. Parameter identifications in the time domain and in the frequency domain for the fractional 3-parameter model in conjunction with 3D FE calculations were presented by Schmidt and Gaul [11]. Enelund and Josefson [6] studied formulations of hereditary integral type in the FEM.

L. Gaul • A. Schmidt (✉)
Institute of Applied and Experimental Mechanics, University of Stuttgart, Pfaffenwaldring 9, 70550 Stuttgart, Germany
e-mail: schmidt@iam.uni-stuttgart.de

H.J. Qi et al. (eds.), *Challenges in Mechanics of Time-Dependent Materials, Volume 2: Proceedings of the 2014 Annual Conference on Experimental and Applied Mechanics*, Conference Proceedings of the Society for Experimental Mechanics Series,
DOI 10.1007/978-3-319-06980-7_13,

Since fractional derivatives are non-local operators, the actual behavior of 'fractional models' depends on the entire deformation history. Thus, in contrast to classical models, the numerical effort and the storage requirements increase with simulation time if the response of a system is computed in the time-domain. Consequently, different researchers [6, 12–14] made attempts to overcome this drawback. Two of these concepts will be discussed and compared in the following Sections.

13.2 Numerical Evaluation of Fractional Derivatives

In this Section, the approach made by Schmidt an Gaul [12] (concept A) will be introduced.

Using the Grünwald definition of fractional derivatives, see [9]

$$ {}_0\mathrm{D}_t^\alpha f(t) = \lim_{n\to\infty} \left(\frac{t}{n}\right)^{-\alpha} \sum_{j=0}^{n-1} A_{j+1} f_j \,, \tag{13.1} $$

where $f_j = f(t_\mathrm{a} - j\,t/n)$ and the Grünwald coefficients are given by

$$ A_{j+1} = \frac{\Gamma(j-\alpha)}{\Gamma(-\alpha)\Gamma(j+1)} = \frac{j-1-\alpha}{j} A_j \,. \tag{13.2} $$

A time-discrete approximation is obtained for a finite number n for which the expression $t/n = \Delta t$ (G1-algorithm, see [9])

$$ {}_0\mathrm{D}_t^\alpha f(t) = \Delta t^{-\alpha} \sum_{j=0}^{n-1} A_{j+1} f_j \tag{13.3} $$

can easily be obtained. As can be seen from Eq. (13.3), the numerical costs explode with the number n of time steps under consideration.

One basic idea for the reduction of the numerical effort is to reduce the computational costs by adapting the temporal resolution. Due to the fact that the Grünwald coefficients are converging towards zero for any order of derivative $\alpha > 0$ [12], events are faded out and possess a decreasing influence on the evaluation of the fractional derivative as time elapses. Using Eq. (13.3), the fractional derivative of a function $f(t)$ evaluated at the actual time t_a can be written as

$$ {}_0\mathrm{D}_{t_\mathrm{a}}^\alpha f(t) = {}_0\mathrm{D}_{t_\mathrm{l}}^\alpha(t_\mathrm{a}) f(t) + {}_{t_\mathrm{l}}\mathrm{D}_{t_\mathrm{a}}^\alpha(t_\mathrm{a}) f(t) \,, \tag{13.4} $$

where the notation ${}_a\mathrm{D}_b(t_\mathrm{a})$ shall denote the evaluation of the fractional derivative at the time t_a which may be different from the upper terminal b . By application of the G1 algorithm one obtains

$$ {}_0\mathrm{D}_{t_\mathrm{l}}^\alpha(t_\mathrm{a}) f(t) = \Delta t^{-\alpha} \sum_{j=i}^{i+k-1} A_{j+1} f_j, \qquad {}_{t_\mathrm{l}}\mathrm{D}_{t_\mathrm{a}}^\alpha(t_\mathrm{a}) f(t) = \Delta t^{-\alpha} \sum_{j=0}^{i-1} A_{j+1} f_j \,. \tag{13.5} $$

From Eq. (13.5) it is obvious that $i + k = n$ is the total number of time steps up to the actual time t_a. The first term ${}_0\mathrm{D}_{t_\mathrm{l}}^\alpha(t_\mathrm{a}) f(t)$ represents the contribution of the older part of the function's history to the fractional derivative. According to the basic idea, this interval, consisting of k time steps, will be kept fixed during further time integration and its influence on the fractional derivative will be treated separately. After n additional time steps one obtains

$$ {}_0\mathrm{D}_{t_\mathrm{l}}^\alpha(t_\mathrm{a} + n\Delta t) f(t) = \Delta t^{-\alpha} \sum_{j=i}^{i+k-1} A_{j+n+1} f_j \,. \tag{13.6} $$

Using the recursive relationship (13.2) yields

$$\begin{aligned} {}_0\mathrm{D}^{\alpha}_{t_1}(t_a+n\Delta t)f(t) &= \Delta t^{-\alpha}A_{i+n+1}\left[1f_i+\frac{i+n+\alpha}{i+n+1}f_{i+1}+\frac{(i+n+\alpha)(i+n+1-\alpha)}{(i+n+1)(i+n+2)}f_{i+2}+\cdots\right.\\ &\quad\left.+\frac{(i+n+\alpha)(i+n+1-\alpha)\cdots(i+n+k-2-\alpha)}{(i+n+1)(i+n+2)\cdots(i+n+k-1)}f_{i+k-1}\right]\\ &= \Delta t^{-\alpha}A_{i+n+1}T_n\,, \end{aligned} \tag{13.7}$$

where T_n will be called the 'transfer function'. From Eq. (13.7) it can be seen that all weighting factors in the square brackets are positive but smaller or equal to unity. In addition, as time elapses (n increases), they are monotonically increasing and tend to unity. In order to reduce the numerical costs, the factor T_n , which includes the contribution of k function values, will be approximated as *one* contribution to the fractional derivative. This is done by calculating the starting value T_0 according to Eq. (13.7) at the time t_a and the value

$$T_\infty = f_i + f_{i+1} + f_{i+2} + \cdots + f_{i+k-1} \tag{13.8}$$

for $t \to \infty$. The time-dependent transfer from T_0 to T_∞ is then approximated by a test function $f(t) = 1$ such that

$$T_n \approx T_0 + \frac{T_n^1 - T_0^1}{T_\infty^1 - T_0^1}(T_\infty - T_0) = T_0 + w_n(T_\infty - T_0)\,, \quad w_n \in [0,\ 1] \tag{13.9}$$

in which the upper right bold-face index 1 denotes that the respective transfer function is calculated using the test function $f(t) = 1$. The quality of the approximated transfer function is discussed in detail in [12].

For longer calculation times several intervals, each of length $k\,\Delta\,t$, are introduced as the number of time steps increases.

13.2.1 Implementation of Fractional Constitutive Equations into the Finite Element Method

Any general 3d fractional constitutive equation relating the stresses $\boldsymbol{\sigma}$ and strains ε can be written in the form

$$\sum_{i=1}^{n} a_i\, {}_0\mathrm{D}_t^{\alpha_i}\boldsymbol{\sigma} = \sum_{j=1}^{m} b_j\, {}_0\mathrm{D}_t^{\beta_j}\varepsilon. \tag{13.10}$$

Using the Grünwald approximation (13.3) for all fractional derivatives, Eq. (13.10) can explicitly be solved for the actual stresses $\boldsymbol{\sigma}(t)$ which then depends on the actual strains $\varepsilon(t)$, the stress history and the strain history (cf [10, 11]). From d'Alemberts principle, the equation of motion can be written in the form

$$\int_V \mathrm{B}^{\mathrm{T}}\,\boldsymbol{\sigma}\,\mathrm{d}V + \mathrm{M}\ddot{u} = f \tag{13.11}$$

where **B** contains the respective derivatives of the finite element shape functions, **M** is the mass matrix, $\boldsymbol{u}$ is the vector of the displacements, and $\boldsymbol{f}$ is the vector of the external forces. If one inserts the actual stresses from Eq. (13.10) into (13.11), the system can be solved by an explicit time integration scheme, such as the central difference method

$$\ddot{u}\,(t) = \frac{u(t+\Delta t) - 2u(t) + u(t-\Delta t)}{\Delta t^2} \tag{13.12}$$

For further details see e.g. [10].

13.3 Calculating a Fractional Derivate by Solving a Partial Differential Equation

In the following, the approach published by Singh and Chatterjee [13] (concept B) is briefly summarized and extended for a fractional 3-parameter model. Then, an adaption for finite element calculations is presented.

A special partial differential equation (PDE)

$$\frac{\partial}{\partial t}u(\xi,t) + \xi^{\frac{1}{q}}u(\xi,t) = \delta(t), \quad u(\xi,0^-) = 0 \tag{13.13}$$

with the solution

$$u(\xi,t) = e^{-\xi^{\frac{1}{q}}t} \tag{13.14}$$

can be used to derive a method to calculate a fractional derivative [13]. Integration over ξ yields the transfer function $g(t)$ of the system

$$g(t) = \int_0^\infty e^{-\xi^{\frac{1}{q}}t}\,\mathrm{d}\xi = \frac{\Gamma(1+q)}{t^q}. \tag{13.15}$$

In general, the output $r(t)$ of a linear time invariant dynamical system is obtained by convolution of the transfer function $g(t)$ and the system input. By replacing the input $\delta(t)$ in (13.13) by $\dot{x}(t)$, one obtains

$$r(t) = g(t) * \dot{x}(t) = \int_0^t g(t-\tau)\dot{x}(\tau)\mathrm{d}\tau = \int_0^t \frac{\Gamma(1+q)}{(t-\tau)^q}\dot{x}(\tau)\mathrm{d}\tau \tag{13.16}$$

where $*$ denotes the convolution operator. Comparison of (13.16) with the well-known Riemann–Liouville definition of fractional derivatives [9] with vanishing initial conditions

$$\frac{\mathrm{d}^q}{\mathrm{d}t^q}x(t) = \frac{1}{\Gamma(1-q)}\int_0^t \frac{\dot{x}(\tau)}{(t-\tau)^q}\mathrm{d}\tau, \quad 0 \le q < 1, \quad x(t \le 0) = 0 \tag{13.17}$$

yields an alternative description of a fractional derivative

$$\frac{\mathrm{d}^q}{\mathrm{d}t^q}x(t) = \frac{1}{\Gamma(1+q)\Gamma(1-q)}r(t) = \frac{1}{\Gamma(1+q)\Gamma(1-q)}\int_0^\infty u(\xi,t)\mathrm{d}\xi \tag{13.18}$$

which can be specified after solving the underlying PDE

$$\frac{\partial}{\partial t}u(\xi,t) + \xi^{\frac{1}{q}}u(\xi,t) = \dot{x}(t), \quad u(\xi,0^-) = 0. \tag{13.19}$$

13.3.1 Solving the PDE Using the Method of Weighted Residuals

The PDE (13.19) cannot be solved exactly for general q and $\dot{x}(t)$, therefore an approximation of $u(\xi, t)$

$$u(\xi,t) \approx \sum_{i=1}^{n} a_i(t)\phi_i(\xi), \tag{13.20}$$

consisting of weighting functions $a_i(t)$ and shape functions $\phi_i(\xi)$, is used. Inserting (13.20) into (13.19) yields a residual

$$R(\xi,t) = \sum_{i=1}^{n}\left(\dot{a}_i(t)\phi_i(\xi) + \xi^{\frac{1}{q}}a_i(t)\phi_i(\xi)\right) - \dot{x}(t), \tag{13.21}$$

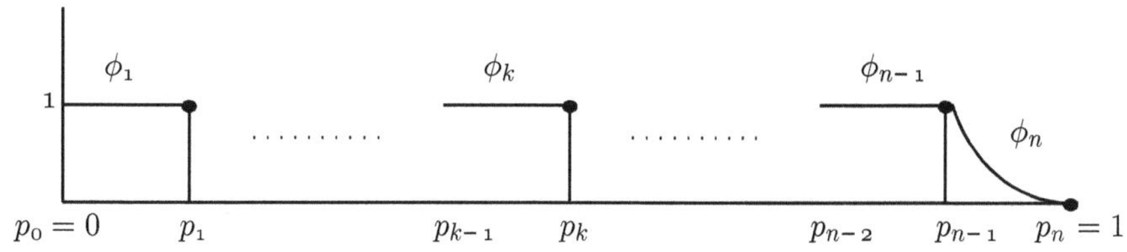

Fig. 13.1 Constant shape functions on unit interval $\eta \in [0, 1]$

which is minimized using the weak form and the shape function $\phi_m(\xi)$

$$\int_0^\infty \left(\sum_{i=1}^n \left(\dot{a}_i(t)\phi_i(\xi) + \xi^{\frac{1}{q}} a_i(t)\phi_i(\xi) \right) - \dot{x}(t) \right) \phi_m(\xi) \mathrm{d}\xi = 0. \tag{13.22}$$

In order to get rid of the improper integral in (13.22), a transformation of ξ on the unit interval [0, 1] is performed by

$$\eta(\xi) = \frac{\xi}{1 + \xi} \tag{13.23}$$

which yields

$$\int_0^1 \left(\sum_{i=1}^n \left(\dot{a}_i(t)\phi_i(\eta) + \left(\frac{\eta}{1-\eta} \right)^{\frac{1}{q}} a_i(t)\phi_i(\eta) \right) - \dot{x}(t) \right) \phi_m(\eta) \frac{1}{(1-\eta)^2} \mathrm{d}\eta = 0. \tag{13.24}$$

As we are interested to solve (13.24), we need to specify shape functions used to approximate $u(\eta, t)$. Choosing for example constant shape functions (see Fig. 13.1), one obtains

$$\phi_i(\eta) = \begin{cases} 1 & p_{i-1} < \eta \le p_i \\ 0 & \text{elsewhere} \end{cases} \quad \text{and} \quad \phi_n(\eta) = \begin{cases} \left(\dfrac{1-\eta}{1-p_{n-1}} \right)^{\frac{1+2q}{2q}} & p_{n-1} < \eta \le p_n \\ 0 & \text{elsewhere} \end{cases} \quad i = 1, 2, \ldots, n-1. \tag{13.25}$$

In order to allow analytical integration of the matrix entries presented in the next part, we use an adaptive shape function $\phi_n(\eta)$.

13.3.2 Formulation in Matrix Notation

Later on, we want to solve a fractional derivative in conjunction with a structural finite element discretization. Therefore, we switch to matrix notation noting that (13.24) can be expressed as a system of first order differential equations

$$\boldsymbol{A}\dot{\boldsymbol{a}} + \boldsymbol{B}\boldsymbol{a} = \boldsymbol{c}\dot{x}(t) \tag{13.26}$$

whereas

$$\boldsymbol{A}_{mi} = \int_0^1 \phi_m(\eta)\phi_i(\eta) \frac{1}{(1-\eta)^2} \mathrm{d}\eta, \tag{13.27}$$

$$\boldsymbol{B}_{mi} = \int_0^1 \left(\frac{\eta}{1-\eta} \right)^{\frac{1}{q}} \phi_m(\eta)\phi_i(\eta) \frac{1}{(1-\eta)^2} \mathrm{d}\eta \quad \text{and} \tag{13.28}$$

$$\boldsymbol{c}_m = \int_0^1 \phi_m(\eta) \frac{1}{(1-\eta)^2} \mathrm{d}\eta. \tag{13.29}$$

Using this notation, one can finally express a fractional derivative as

$$\frac{\mathrm{d}^q}{\mathrm{d}t^q} x(t) \approx \frac{1}{\Gamma(1+q)\Gamma(1-q)} \int_0^1 \sum_{i=1}^n a_i(t)\phi_i(\eta)\mathrm{d}\eta = \frac{1}{\Gamma(1+q)\Gamma(1-q)} \boldsymbol{c}^{\mathrm{T}} \boldsymbol{a} \tag{13.30}$$

keeping in mind from the Riemann–Liouville definition (13.17) that (13.30) is only valid if $q \in [0, 1)$.

13.4 A System Based on Fractional 3-Parameter Material Model

Damped structures can be modeled efficiently by the use of a fractional 3-parameter material model [11] (see Fig. 13.2) whose constitutive equation is given by

$$\sigma(t) + \frac{R}{E_1} \frac{\mathrm{d}^q}{\mathrm{d}t^q} \sigma(t) = E_0 \epsilon(t) + R \frac{E_0 + E_1}{E_1} \frac{\mathrm{d}^q}{\mathrm{d}t^q} \epsilon(t). \tag{13.31}$$

The appropriate equation of motion has the form

$$\tilde{\boldsymbol{B}} \frac{\mathrm{d}^q}{\mathrm{d}t^q} \ddot{\boldsymbol{x}} + \tilde{\boldsymbol{C}} \ddot{\boldsymbol{x}} + \tilde{\boldsymbol{D}} \dot{\boldsymbol{x}} + \tilde{\boldsymbol{E}} \frac{\mathrm{d}^q}{\mathrm{d}t^q} \boldsymbol{x} + \tilde{\boldsymbol{F}} \boldsymbol{x} = \tilde{\boldsymbol{G}} f(t) + \tilde{\boldsymbol{H}} \frac{\mathrm{d}^q}{\mathrm{d}t^q} f(t). \tag{13.32}$$

Since (13.30) can only be applied if $q \in [0, 1)$, the fractional derivative of $\ddot{x}$ in (13.32) cannot be evaluated directly by setting

$$\frac{\mathrm{d}^q}{\mathrm{d}t^q} \ddot{\boldsymbol{x}} = \frac{\mathrm{d}^{q+2}}{\mathrm{d}t^{q+2}} \boldsymbol{x}. \tag{13.33}$$

Instead, by introducing an additional dashpot $\tilde{R} \ll 1$ in parallel to the fractional dashpot (see Fig. 13.3), one gets an extended fractional 3-parameter model which finally leads to a third order equation of motion of the form

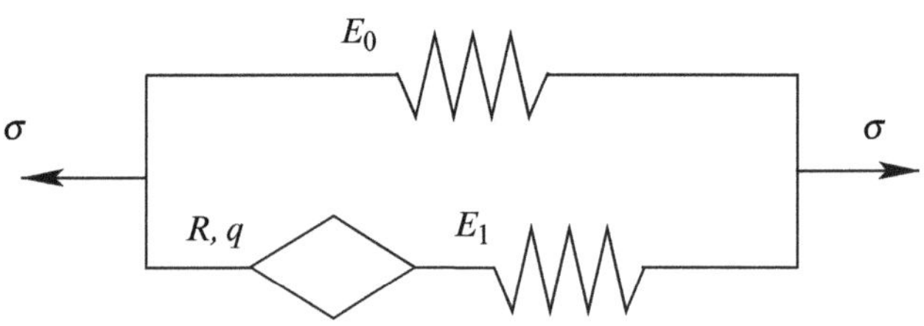

Fig. 13.2 Fractional 3-parameter model

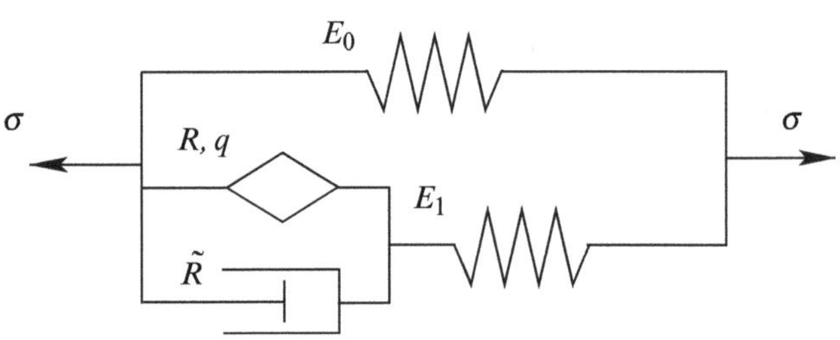

Fig. 13.3 Extended fractional 3-parameter model

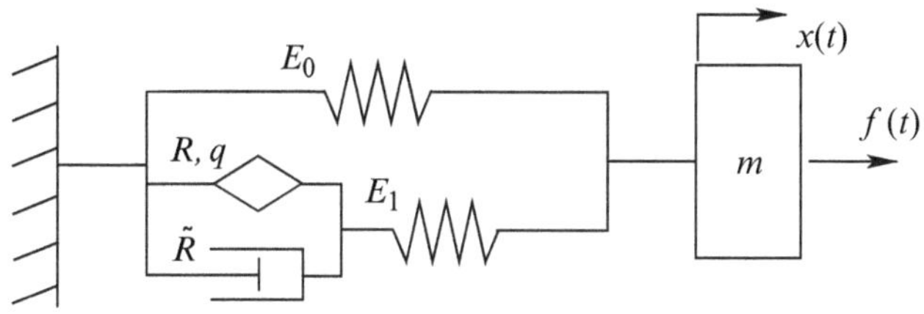

Fig. 13.4 Mass attached to extended 3-parameter material model

$$\tilde{\boldsymbol{A}}\,\dddot{x} + \tilde{\boldsymbol{B}}\frac{\mathrm{d}^q}{\mathrm{d}t^q}\ddot{x} + \tilde{\boldsymbol{C}}\ddot{x} + \tilde{\boldsymbol{D}}\dot{x} + \tilde{\boldsymbol{E}}\frac{\mathrm{d}^q}{\mathrm{d}t^q}x + \tilde{\boldsymbol{F}}x = \tilde{\boldsymbol{G}}f(t) + \tilde{\boldsymbol{H}}\frac{\mathrm{d}^q}{\mathrm{d}t^q}f(t) + \tilde{\boldsymbol{I}}\dot{f}(t) \tag{13.34}$$

which can easily be transformed to a set of first order differential equations.

13.4.1 Solving a 1-DOF System

After transforming (13.34) into a set of first order differential equations using

$$x_1 = x, \quad x_2 = \dot{x} \quad \text{and} \quad x_3 = \ddot{x}, \tag{13.35}$$

on can replace each fractional derivative by (13.30). In the following, this process is illustrated for a 1-DOF system as shown in Fig. 13.4. Introducing the abbreviations $E_{01} = E_0 + E_1$, the ratio of damping constants $\gamma = R/\tilde{R}$ and

$$\tilde{f} = \frac{E_1}{\tilde{R}}f(t) + \gamma\frac{\mathrm{d}^q}{\mathrm{d}t^q}f(t) + \dot{f}(t), \tag{13.36}$$

one finally gets $\dot{x}_1 = x_2$, $\dot{x}_2 = x_3$ and

$$\dot{x}_3 = -\gamma\frac{\mathrm{d}^q}{\mathrm{d}t^q}x_3 - \frac{E_1}{\tilde{R}}x_3 - \frac{E_{01}}{m}x_2 - \gamma\frac{E_{01}}{m}\frac{\mathrm{d}^q}{\mathrm{d}t^q}x_1 - \frac{E_0E_1}{m\tilde{R}}x_1 + \frac{1}{m}\tilde{f}. \tag{13.37}$$

If the external force $f(t)$ is known in advance, then $\tilde{f}$ can be calculated after evaluating

$$\dot{a}^{(f)} = -\boldsymbol{A}^{-1}\boldsymbol{B}\boldsymbol{a}^{(f)} + \boldsymbol{A}^{-1}\boldsymbol{c}\dot{f}(t) \tag{13.38}$$

whereas the index (f) in $\boldsymbol{a}^{(f)}$ denotes the application of (13.38) for the calculation of the external force $f(t)$. The remaining fractional derivatives in (13.37) can be replaced by

$$-\gamma\frac{\mathrm{d}^q}{\mathrm{d}t^q}x_3 \approx -\gamma\tilde{\Gamma}^{-1}\boldsymbol{c}^{\mathrm{T}}\boldsymbol{a}^{(x_3)} \quad \text{and} \quad -\gamma\frac{E_{01}}{m}\frac{\mathrm{d}^q}{\mathrm{d}t^q}x_1 \approx -\gamma\frac{E_{01}}{m}\tilde{\Gamma}^{-1}\boldsymbol{c}^{\mathrm{T}}\boldsymbol{a}^{(x_1)} \tag{13.39}$$

whereas the abbreviation $\tilde{\Gamma} = \Gamma(1+q)\Gamma(1-q)$ is used. (13.39) can be evaluated calculating two systems of differential equations

$$\dot{a}^{(x_3)} = -\boldsymbol{A}^{-1}\boldsymbol{B}\boldsymbol{a}^{(x_3)} + \boldsymbol{A}^{-1}\boldsymbol{c}\dot{x}_3(t) \quad \text{and} \quad \dot{a}^{(x_1)} = -\boldsymbol{A}^{-1}\boldsymbol{B}\boldsymbol{a}^{(x_1)} + \boldsymbol{A}^{-1}\boldsymbol{c}\dot{x}_1(t). \tag{13.40}$$

Since (13.37) and (13.40) have to be calculated together, one can introduce an extended system of differential equations containing both systems

$$\begin{bmatrix} \dot{x}_1 \\ \dot{x}_2 \\ \dot{x}_3 \\ \dot{a}_1^{(x_1)} \\ \vdots \\ \dot{a}_n^{(x_1)} \\ \dot{a}_1^{(x_3)} \\ \vdots \\ \dot{a}_n^{(x_3)} \end{bmatrix} = \underbrace{\begin{bmatrix} A_{\text{ext1}} & A_{\text{ext2}} & A_{\text{ext3}} \\ A_{\text{ext4}} & A_{\text{ext5}} & A_{\text{ext6}} \\ A_{\text{ext7}} & A_{\text{ext8}} & A_{\text{ext9}} \end{bmatrix}}_{(3+2n)\times(3+2n)} \begin{bmatrix} x_1 \\ x_2 \\ x_3 \\ a_1^{(x_1)} \\ \vdots \\ a_n^{(x_1)} \\ a_1^{(x_3)} \\ \vdots \\ a_n^{(x_3)} \end{bmatrix} + \begin{bmatrix} 0 \\ 0 \\ \frac{1}{m}\tilde{f} \\ 0 \\ \vdots \\ 0 \\ \frac{A^{-1}c}{m}\tilde{f} \end{bmatrix} \tag{13.41}$$

whereas

$$A_{\text{ext1}} = \begin{bmatrix} 0 & 1 & 0 \\ 0 & 0 & 1 \\ -\frac{E_0E_1}{m\tilde{R}} & -\frac{E_{01}}{m} & -\frac{E_1}{\tilde{R}} \end{bmatrix} \qquad A_{\text{ext2}} = \begin{bmatrix} 0 & \cdots & 0 \\ 0 & \cdots & 0 \\ & -\gamma\frac{E_{01}}{m}\tilde{\Gamma}^{-1}c^{\mathrm{T}} & \end{bmatrix}$$

$$A_{\text{ext3}} = \begin{bmatrix} 0 & \cdots & 0 \\ 0 & \cdots & 0 \\ & -\gamma\tilde{\Gamma}^{-1}c^{\mathrm{T}} & \end{bmatrix} \qquad A_{\text{ext4}} = \begin{bmatrix} 0 & & 0 \\ \vdots & A^{-1}c & \vdots \\ 0 & & 0 \end{bmatrix} \qquad A_{\text{ext5}} = \begin{bmatrix} -A^{-1}B \end{bmatrix}$$

$$A_{\text{ext6}} = \begin{bmatrix} 0 & \cdots & 0 \\ \vdots & \ddots & \vdots \\ 0 & \cdots & 0 \end{bmatrix} \qquad A_{\text{ext7}} = \begin{bmatrix} -\frac{E_0E_1}{m\tilde{R}}A^{-1}c & -\frac{E_{01}}{m}A^{-1}c & -\frac{E_1}{\tilde{R}}A^{-1}c \end{bmatrix}$$

$$A_{\text{ext8}} = \begin{bmatrix} -\gamma\frac{E_{01}}{m}\tilde{\Gamma}^{-1}A^{-1}cc^{\mathrm{T}} \end{bmatrix} \quad \text{and} \quad A_{\text{ext9}} = \begin{bmatrix} -A^{-1}B - \gamma\tilde{\Gamma}^{-1}A^{-1}cc^{\mathrm{T}} \end{bmatrix}.$$

13.4.2 Finite Element Spatial Discretization

In order to use the presented algorithm in conjunction with a finite element spatial discretization, it is possible to write the matrices $\tilde{A}$ to $\tilde{I}$ in (13.34) for a 2-node finite element of a rod. In this case one gets

$$\tilde{C} = \frac{\rho A \ell}{6}\begin{bmatrix} 2 & 1 \\ 1 & 2 \end{bmatrix} \tag{13.42}$$

for the 'mass matrix' $\tilde{C}$, whereas ρ denotes the material density, A describes the rod's cross section area and ℓ is the element's length. All other matrices can be determined respectively.

13.5 Example

As an example, a viscoelastic rod (length $\ell = 2$ m, cross section area $A = \pi(7.5\text{ mm})^2$), that is initially at rest is considered. Its left end is clamped whereas on the free end an external force $\widehat{F} = 1{,}000$ N is acting in form of a step function in time. Thus, a free vibration is excited, overlayed by a creep process.

The rod is discretized by 30 finite elements. In order to compare the concepts A and B, given in the previous sections, the calculation is carried out using both approaches and compared with a reference calculation which is obtained using

Table 13.1 Material parameters

$E_0 = 2989.53\ \text{N/mm}^2$	$E_1 = 192.92\ \text{N/mm}^2$
$R = 5.276\ \text{Ns}^q/\text{mm}^2$	$q = 0.5$

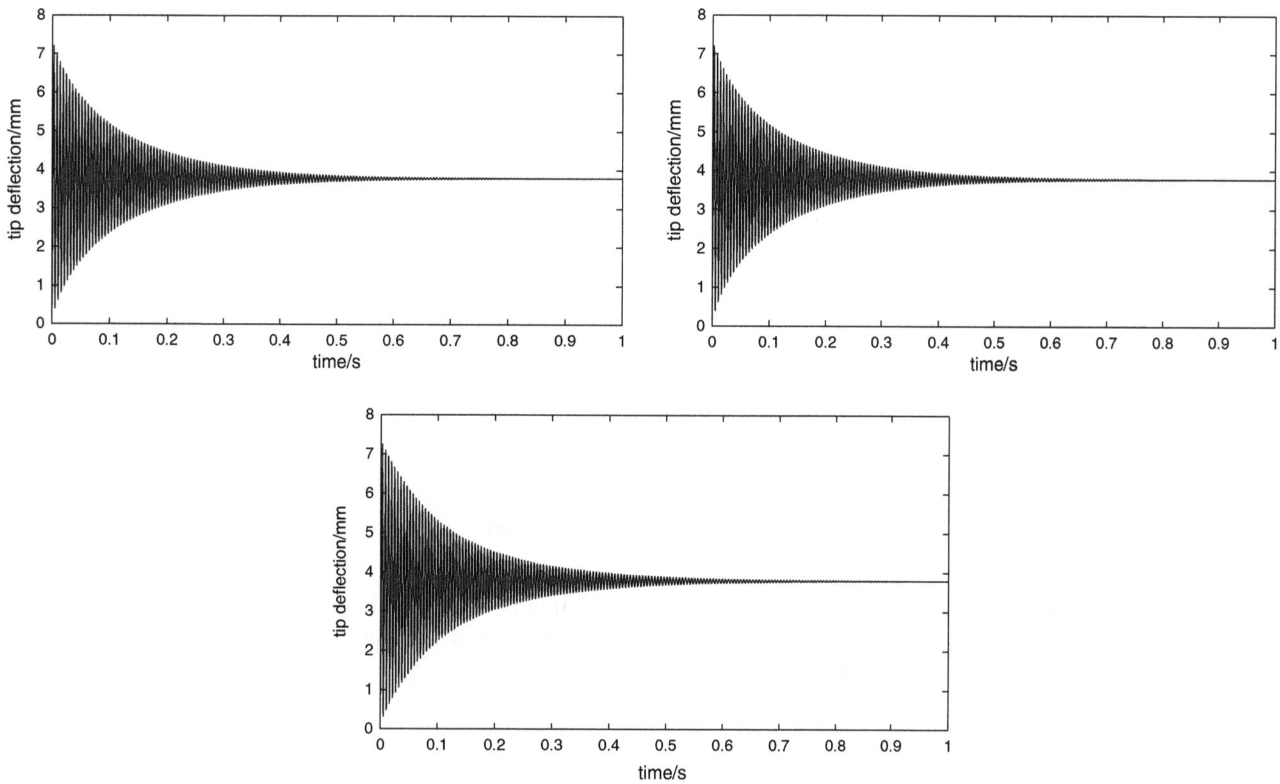

Fig. 13.5 Decaying oscillation of the tip's free end; reference calculation (*upper left*), concept A (*upper right*) and concept B (*lower*)

Table 13.2 Comparison of accuracy, cpu-time and storage requirements of the different concepts

Concept	Reference calculation	Concept A	Concept B
Frequency	184.19 Hz	184.17 Hz	184.30 Hz
Logarithmic decrement	4.05 %	4.02 %	4.22 %
Cpu-time	100 %	3.51 %	1.25 %
Storage	100 %	52 %	779 %

the discretized Grünwald approximation in conjunction with the complete temporal resolution as described in Sect. 13.2. The parameters used in concept A are $i = 100$, $k = 400$ whereas in concept B $n = 11$ finite elements were chosen.

The material under consideration is a thermoplastic polymer (Delrin, DuPont) and its material constants given in Table 13.1 are found by frequency-domain measurements [11]. The displacement of the rod's trip and its neutral position are calculated for a total simulation time of 1 s, using a time step size $\Delta t = 25 \cdot 10^{-6}$ s which leads to 40,000 time steps. The results are shown in Fig. 13.5. A comparison in terms of the computational costs and accuracy is shown in Table 13.2 where the acquired time and memory resources of the reference calculation are taken to be 100 %. The neutral position u_{np} is calculated from 3 subsequent extrema u_{ex} of the free vibration by

$$u_{\text{np}} = \frac{u_{\text{ex},1} u_{\text{ex},3} - u_{\text{ex},2}^2}{u_{\text{ex},1} - 2u_{\text{ex},2} + u_{\text{ex},3}} \tag{13.43}$$

which gives a second-order approximation. The results are shown in Fig. 13.6. The asymptote of the neutral position can be determined analytically by

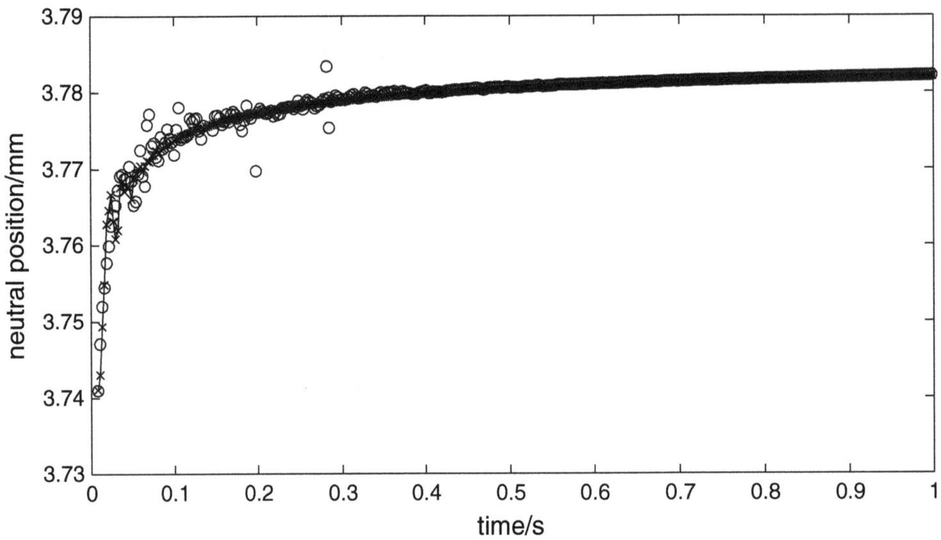

Fig. 13.6 Calculated creep of the neutral position; reference calculation (*solid*), concept A (*crosses*), concept B (*circles*)

$$\lim_{t\to\infty} u_{\rm np} = \frac{\widehat{F}\ell}{E_0 A} = 3.786 \text{ m}m \tag{13.44}$$

The scattering that can be observed in the beginning of the calculation is due to the higher harmonics which are excited by the force step function. Due to the damping they die out quite quickly. For the concept B there also seem to be some numerical reasons for the pronounced distribution.

13.6 Conclusions

Time-domain calculations of structures whose constitutive equations include fractional derivatives lead to high computational costs, especially for large numbers of time steps. In this paper, two different concepts resulting in a drastical reduction of the numerical effort were compared with a costly reference calculation. Therefore, a new algorithm suggested by Singh and Chatterjee was extended to a fractional three-parameter model and included into a structural finite element discretization. Both concepts show very good performance concerning the reduction of calculation time. Also, the results in terms of frequency, creep process and decaying behavior are in good agreement with the reference calculation, where the algorithm by Singh and Chatterjee shows an increased scattering of the creep data. An advantage of this new algorithm is given by the fact that the time integration is unconditionally stable. Therefore, larger time steps are possible. On the other hand, the storage requirements increase drastically. In addition, more general fractional constitutive equations consisting of more than one fractional element are not straightforward to implement.

References

1. Bagley RL, Torvik PJ (1979) A generalized derivative model for an elastomer damper. Shock Vib Bull 49:135–143
2. Bagley RL, Torvik PJ (1983) Fractional calculus — a different approach to the analysis of viscoelastically damped structures. AIAA J 21(5):741–748
3. Bagley RL, Torvik PJ (1983) A theoretical basis for the application of fractional calculus to viscoelasticity. J Rheol 27(3):201–210
4. Bagley RL, Torvik PJ (1986) On the fractional calculus model of viscoelastic behavior. J Rheol 30(1):133–155
5. Caputo M, Mainardi F (1971) Linear models of dissipation in anelastic solids. Rivista del Nuovo Cimento 1(2):161–198
6. Enelund M, Josefson L (1997) Time-domain finite element analysis of viscoelastic structures with fractional derivatives constitutive relations. AIAA J 35(10):1630–1637
7. Koeller RC (1984) Application of fractional calculus to the theory of viscoelasticity. J Appl Mech 51:299–307

8. Lion A (1997) On the thermodynamics of fractional damping elements. Continuum Mech Thermodyn 9(2):83–96
9. Oldham KB, Spanier J (1974) The fractional calculus. Academic, New York
10. Padovan J (1987) Computational algorithms for FE formulations involving fractional operators. Comput Mech 2:271–287
11. Schmidt A, Gaul L (2001) FE implementation of viscoelastic constitutive stress-strain relations involving fractional time derivatives. In: Constitutive models for rubber II. A.A. Balkema Publishers, Tokyo, pp 79–89
12. Schmidt A, Gaul L (2006) On the numerical evaluation of fractional derivatives in multi-degree-of-freedom systems. Signal Process 86(10):2592–2601
13. Singh SJ, Chatterjee K (2006) Galerkin projections and finite elements for fractional order derivatives. Nonlinear Dyn 45(1–2):183–206
14. Yuan L, Agrawal O (2002) A numerical scheme for dynamic systems containing fractional derivatives. J Vib Acoust 124:321–324

Chapter 14
Effect of Pressure on Damping Properties of Granular Polymeric Materials

M. Bek, A. Oseli, I. Saprunov, N. Holeček, B.S. von Bernstorff, and I. Emri

Abstract Reduction of noise and vibration coming from the rail transport activities is an important objective of the environmental policy of the European Union, due to its impact on human and animal health. It has been identified that one of the major sources of noise and vibration in rail transport is from the interaction between the wheel and the rail, the so called rolling noise. One way to mitigate this noise is to attach polymeric damping elements to the rail. By modifying bulk properties of polymeric material we can modify its damping characteristics. In this paper we demonstrated on the example of thermoplastic polyurethane (TPU) the effect of inherent hydrostatic pressure on the time- and frequency-dependent behavior of polymers. For the selected TPU material we found that increasing hydrostatic pressure from 1 to 2,000 bar shifts frequency at which material exhibits its maximal damping properties (G''max) from 37 to 235 Hz. It was also found that change of pressure changes values of storage modulus G' up to 3.5 times (depending on the frequency), while the values of loss modulus G'' are changed up to 5.5 times. Using this property of polymeric materials we designed new generation damping elements composed of glass fiber textile tubes filled with pressurized granulated polymeric materials. Granular material with properly selected multimodal particle size distribution acts as pressurizing agent. At the same time the generated hydrostatic pressure changes frequency dependence of the granular material bulk properties. By modifying material bulk properties we can modify damping characteristics of the new generation damping elements. Applying these damping elements to the rail can substantially reduce vibration amplitudes as well as sound pressure levels, thus reducing exposure of human and animal to noise and vibration.

Keywords Thermoplastic polyurethane • Dissipative granular materials • Effect of pressure • Viscoelasticity • Damping elements

---

M. Bek (✉)
Gorenje gospodinjski aparati, d.d., Velenje, Slovenia

Center for Experimental Mechanics, Faculty of Mechanical Engineering, University of Ljubljana, Pot za Brdom 104, Ljubljana, Slovenia

Institute for Sustainable Innovative Technologies, Pot za Brdom 104, Ljubljana, Slovenia
e-mail: marko.bek@fs.uni-lj.si

A. Oseli
Institute for Sustainable Innovative Technologies, Pot za Brdom 104, Ljubljana, Slovenia

I. Saprunov
Center for Experimental Mechanics, Faculty of Mechanical Engineering, University of Ljubljana, Pot za Brdom 104, Ljubljana, Slovenia

N. Holeček
Gorenje gospodinjski aparati, d.d., Velenje, Slovenia

B.S. von Bernstorff
BASF Aktiengesellschaft, Ludwigshafen, Germany

I. Emri
Center for Experimental Mechanics, Faculty of Mechanical Engineering, University of Ljubljana, Pot za Brdom 104, Ljubljana, Slovenia

Institute for Sustainable Innovative Technologies, Pot za Brdom 104, Ljubljana, Slovenia

H.J. Qi et al. (eds.), *Challenges in Mechanics of Time-Dependent Materials, Volume 2: Proceedings of the 2014 Annual Conference on Experimental and Applied Mechanics*, Conference Proceedings of the Society for Experimental Mechanics Series,
DOI 10.1007/978-3-319-06980-7_14,

14.1 Introduction

Sustainable economic development is one of the main objectives of most industrialized countries, for example the European Union (EU) requires economic growth without additional environmental influences. Reductions in the emission of greenhouse gases that are responsible for global warming and associated to climate change are nowadays a major concern. Around one quarter of all greenhouse gas emissions comes from transport activity [1–3] of which road transport takes the biggest share [4]. For this reason, the goal of environmental policies is to minimize the current difference between road and railroad transport. However, the problem of noise nuisance coming from railroads on population living near railway lines as well as impact on wildlife [5] remains.

Minimizing noise levels can be achieved by implementing different noise control measures: increased damping, reduction of excitations, acoustic shielding or absorption and vibration insulation. In order for noise-control measures to be successful, the dominant source of noise should be identified. It has been found that in many situations rolling noise is the dominant source of railway noise and is caused by wheel and rail vibrations induced at wheel-rail contact area. Wheel and rail roughness generates movement relative to each other. This relative motion generates vibrations, which are transmitted into both structures and from vibrating structures sound is radiated into the air [6].

Several reduction techniques for controlling rolling noise exist, among them are: grinding of rail surfaces, optimizing shape of rail wheels, replacing iron breaks with composite brakes, optimizing rail pads, sound barriers and tuned absorber systems. Currently soft rail pads are being used despite the fact that stiffer ones contribute to the reduction of rail noise. Soft rail pads increase decay rate of rails and it is necessary to apply damping treatments since reducing decay rate is important for reducing noise pollution. For this reason tuned absorber systems were introduced, where cost benefit analysis performed in STAIRRS project showed that for moderate costs, tuned absorber systems gives good results [7], leading to the conclusion that tuned absorber systems can be a very effective way in reducing the noise radiated by the rail.

Many tuned absorber systems have polymeric components to increase their vibration damping ability; however current systems do not utilize the full damping potential of such polymeric materials. As it will be shown on the example of thermoplastic polyurethane (TPU), the inherent (hydrostatic) pressure changes the frequency dependence of polymeric materials. Hence, by proper selection of damping material and pressure to which material is exposed, one can match the frequency range of its maximum damping properties with the resonance frequency of the vibrating structure (rail). In this way we fully utilize damping characteristics of the selected material and maximize the energy absorption properties of the damper.

Using this unique property of polymeric materials enabled us to design and build ultimate adaptive damping elements. For doing this we have additionally utilized patented finding [8] that viscoelastic granular materials with properly selected multimodal size-distribution exhibit fluid-like behavior, while maintaining behavior of the bulk material from which they were made. Hence, they may be used as "pressurizing media" to impose inherent hydrostatic pressure on itself and consequently change its own damping properties.

14.2 Materials and Methods

Within this paper we show on the example of thermoplastic polyurethane (TPU) that by proper selection of the type of the material and proper selection of the hydrostatic pressure, one can shift the maximum of the material damping properties to the frequency range of interest. By doing this, energy absorption properties of a damper could be increased several orders of magnitude. In the case of soft materials this would mean that stiffness of material is increased and at the same time damping capabilities of polymeric material is fully utilized.

We demonstrate this concept via storage modulus, $G'(\omega)$, and loss modulus $G''(\omega)$ measured at different pressures. Material functions in frequency domain were obtained through the process of interconversion from the shear relaxation modulus, $G(t, T, P)$ measured on a specially designed apparatus known as the CEM Measuring System—CMS [9, 10].

14.2.1 Shear Relaxation Measurements

The CMS apparatus [9–11] was developed to study the combined effects of temperature and hydrostatic pressure on mechanical behavior of polymers. The system can measure the volume and the shear relaxation moduli of solid polymer specimens simultaneously subjected to temperatures from −40 to +120 °C and pressures from atmospheric to 5,000 bar [9, 10].

CMS apparatus consists of four main parts: hydraulic system, thermal system, data acquisition system and measuring inserts. Hydraulic system includes a hand pump for pressurizing silicone oil in a pressure vessel. In thermal system, a circulator and thermal bath are used to regulate temperature of specimen. In order to process and record the signals from measuring inserts, data acquisition system is used which combines a carrier amplifier and a computer.

14.2.2 Measuring Principle

The CMS measures four physical quantities: temperature, T(t), pressure, P(t), specimen length, L(t, T, P), and the decaying torque, M(t, T, P), resulting from the initially applied torsional deformation, θ_0, on the sample. Using these quantities measured at constant or varying temperature and pressure, other material functions can be calculated [9, 10]. In this particular case shear relaxation modulus is of interest and is determined by measuring the decaying moment of a specimen exposed to selected constant temperature and pressure boundary conditions. The relaxation modulus curve usually extends over many decades of time, which are practically not possible to characterize at constant temperature or pressure. Keeping this in mind, experiments are performed within a certain time-interval—experimental window. A set of different shear relaxation modulus segments is obtained when experiments are carried out at different temperatures, and/or pressures. The segments are then shifted along the logarithmic time scale to produce master curves using the time–temperature (t–T) and time–pressure (t–P) superposition principles utilizing the closed-form shifting algorithm [14]. The final master curves represent the long-term behavior of the material, at the chosen reference conditions. The mastercurves of storage $G'(\omega)$ and loss modulus $G''(\omega)$ in frequency domain were then obtained from G(t, T, P) through interconversion [11–13, 15].

14.3 Results and Discussion

For clarity reasons the results on storage $G'(\omega)$ and loss modulus $G''(\omega)$ are shown for only two pressures, i.e., 1 and 2,000 bar. The full symbols represent measurements done at lower pressure (P = 1 bar), whereas the empty symbols represent measurements done at higher pressure (P = 2,000 bar). The results for storage modulus G' are shown in Fig. 14.1 and for loss modulus G'' in Fig. 14.2. As expected, the value of both viscoelastic moduli greatly increases as the pressure is increased from 1 to 2,000 bar.

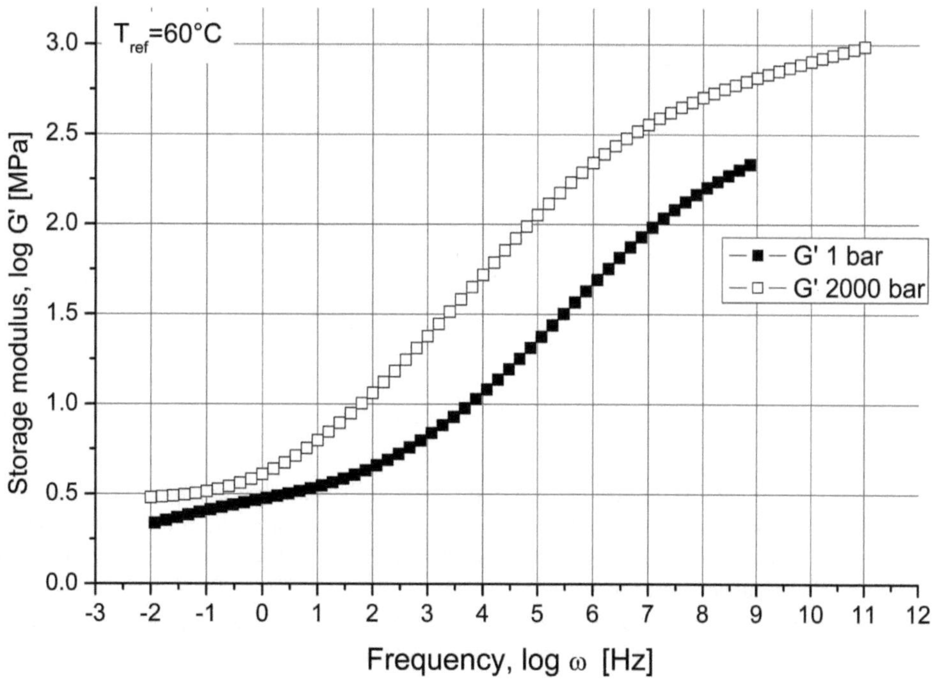

Fig. 14.1 Storage G' at P = 1 bar and P = 2,000 bar

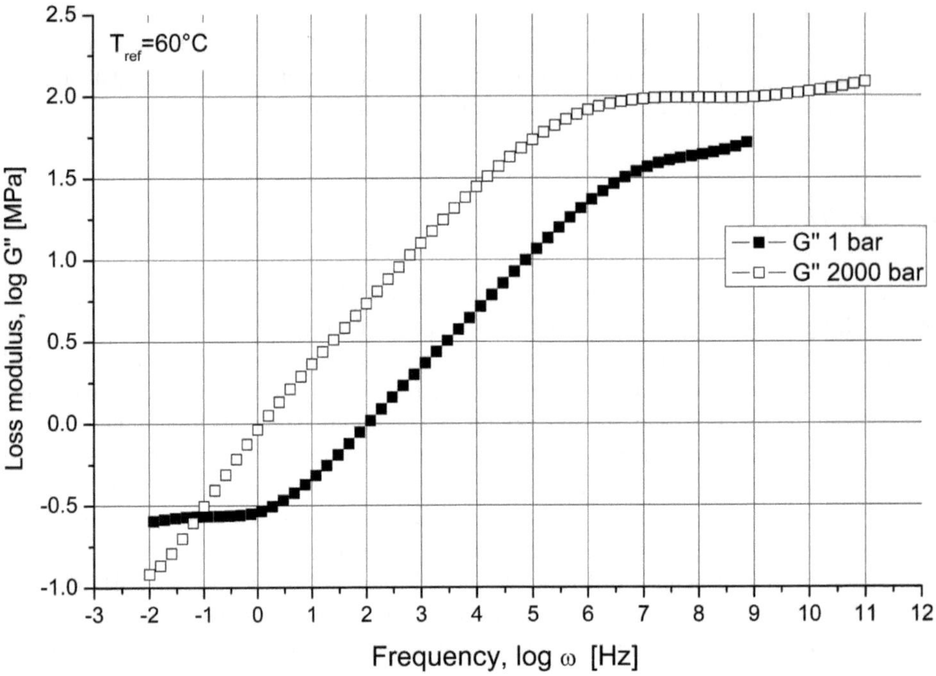

Fig. 14.2 Loss modulus G″ at P = 1 bar and P = 2,000 bar

Fig. 14.3 Storage G′ and loss modulus G″ at P = 1 bar and P = 2,000 bar

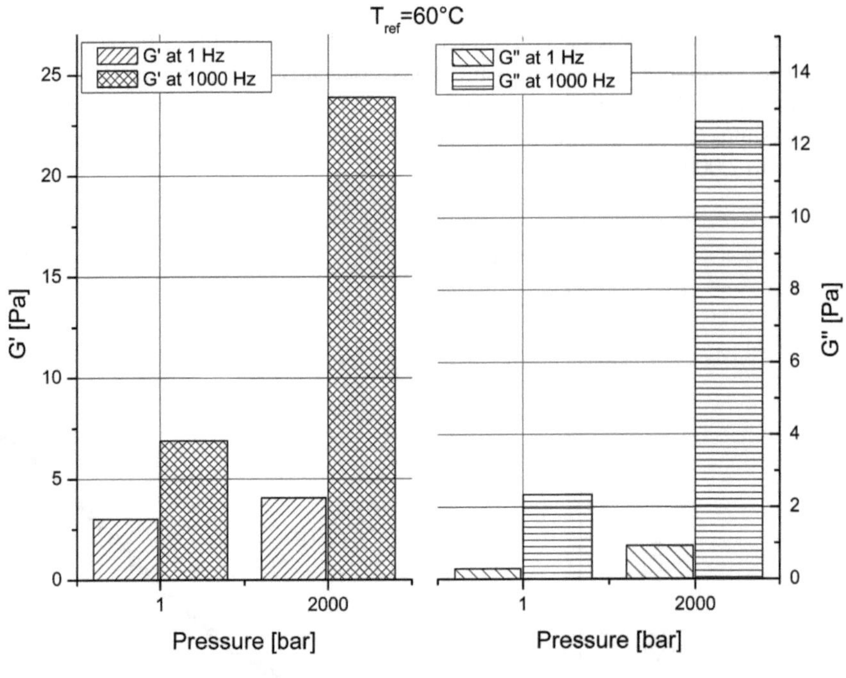

As previously mentioned, we would like to utilize the pressure sensitivity of polymers in the construction of vibroacoustic damping elements for railroad applications. Since the noise and vibration causing the most discomfort are located in the frequency range between 1 and 1,000 Hz, we will examine the effect of pressure on stiffness (G′) and damping (G″) properties of TPU within this frequency window.

From Fig. 14.3 we may clearly see that within the region of interest (1–1,000 Hz), an increase of inherent (hydrostatic) pressure of TPU from 1 to 2,000 bar cause material properties to change for several orders of magnitude. Specifically, at frequency 1 Hz and pressure 1 bar, the storage modulus (representing stiffness) is G′(ω = 1 Hz, P = 1 bar) = 2.99 MPa, whereas at pressure 2,000 bar the storage modulus increases to G′(ω = 1 Hz, P = 2,000 bar) = 4.07 MPa. Hence, material

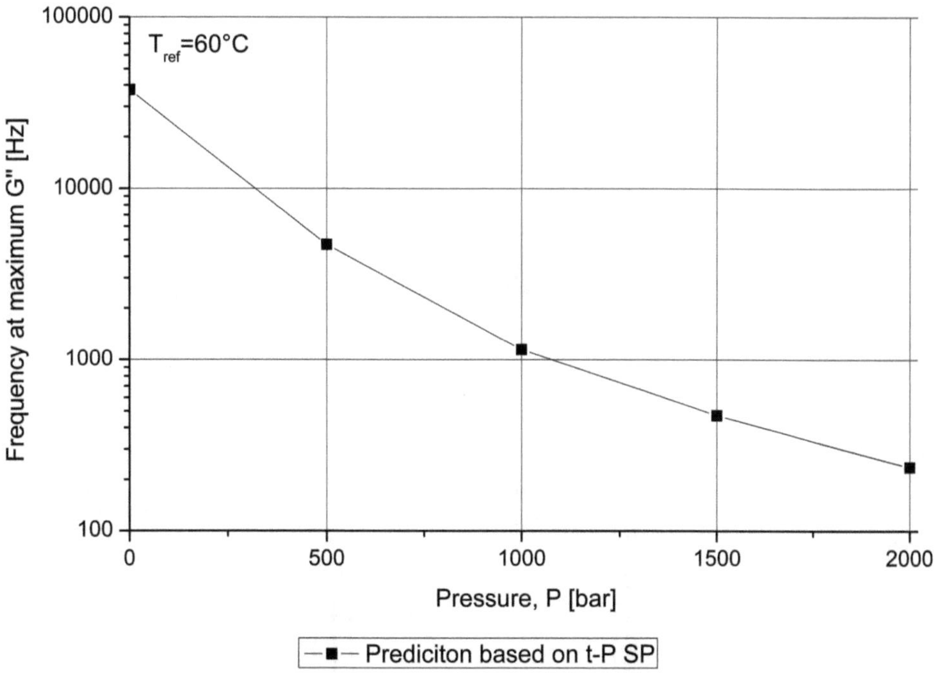

Fig. 14.4 Frequency at maximum damping

becomes 1.4 times stiffer. At the same frequency of 1 Hz the loss modulus (representing material damping characteristics) at pressure 1 bar is $G''(\omega = 1\ \text{Hz}, P = 1\ \text{bar}) = 0.29$ MPa, whereas at pressure 2,000 bar it rises to $G''(\omega = 1\ \text{Hz}, P = 2{,}000\ \text{bar}) = 0.92$ MPa. This means that at elevated pressure the materials ability to dissipate energy increases 3.15 times.

For the higher end of the frequency window, i.e., $\omega = 1{,}000$ Hz, we observe analogous trends. At 1 bar the storage modulus is $G'(\omega = 100\ \text{Hz}, P = 1\ \text{bar}) = 6.89$ MPa, whereas at the pressure $P = 2{,}000$ bar the storage modulus becomes $G'(\omega = 100\ \text{Hz}, P = 2{,}000\ \text{bar}) = 23.89$ MPa, meaning that material stiffness is increased 3.46 times. At the same time the loss modulus at pressure 1 bar is $G''(\omega = 100\ \text{Hz}, P = 1\ \text{bar}) = 2.33$ MPa and at pressure 2,000 bar it becomes $G''(\omega = 100\ \text{Hz}, P = 2{,}000\ \text{bar}) = 12.65$ MPa. Thus, the material ability to dissipate energy has increased 5.41 times.

From Figs. 14.1 and 14.2 one may easily see that by further increasing material inherent pressure we may increase the stiffness and damping properties of TPU up to 100 times!

Using the time–pressure (t–P) superposition principle we can predict how frequency, at which TPU exhibit maximum damping properties (maximal value of loss modulus $G''(\omega)$) changes with pressure.

Figure 14.4 shows frequencies at which loss modulus has maximal values as a function of pressure (from 1 to 2,000 bar). We can see that higher pressure shifts maximal values of damping (maximal values of loss modulus G'') to lower frequencies. For example, frequency at which maximal value of damping at pressure 1 bar occurs is around 37 kHz, whereas at pressure 2,000 bar it moves to as low as 235 Hz.

14.3.1 Damping Elements Based on Dissipative Granular Materials

Based on the presented results, where we showed that existing solutions for structural and vibration control do not and cannot fully utilize damping characteristics of time- and frequency-dependent materials, we designed the ultimate adaptive damping elements system, which we have called DGM System (DGM—Dissipative Granular Materials).

As we have showed, the inherent (hydrostatic) pressure changes frequency dependence of polymeric materials. Hence, by proper selection of damping material and a pressure to which material is exposed one can match the frequency range of its maximum damping properties with the resonance frequency of the selected vibrating structure. In this way we fully utilize damping characteristics of the selected material and maximize the energy absorption properties of the damper.

Granular materials when excited beyond a certain level of stress flow similarly as liquids while maintaining all properties of a bulk material. The macroscopic flow of particles further expands the energy dissipation capability of granular materials. Hence, micro to macro-size multimodal elastomeric granular material may be used as a pressurizing media (similar as air in tires)

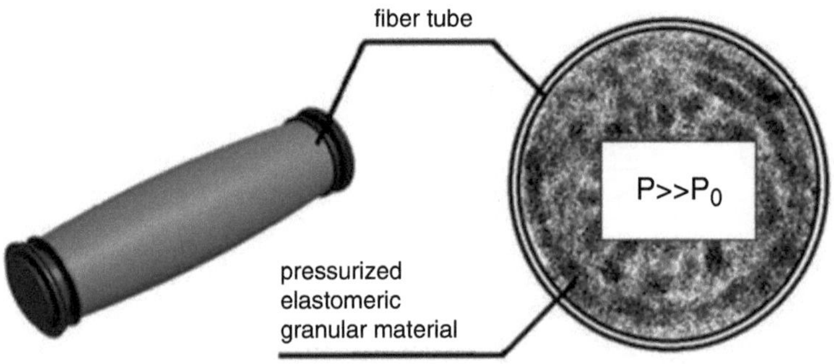

Fig. 14.5 Damping elements consisting of pressurized elastomeric granular material encapsulated by fiber tube

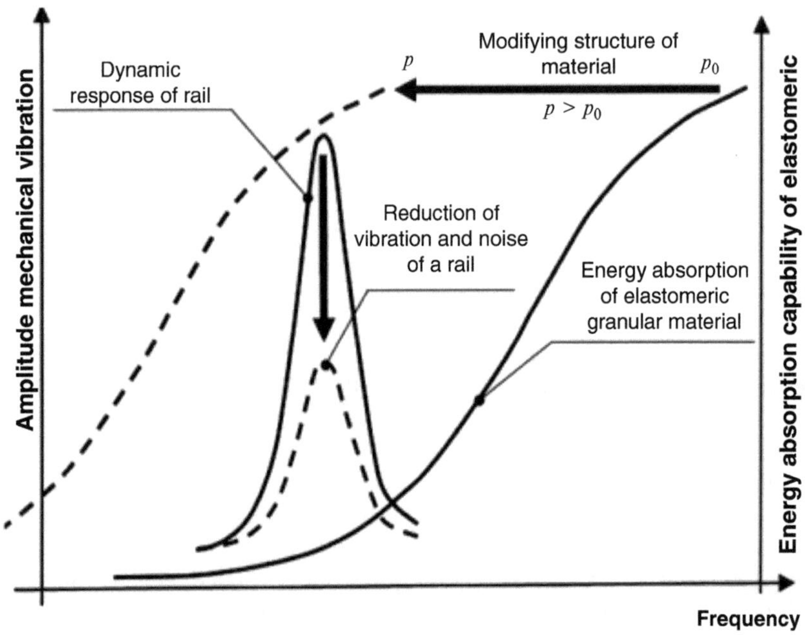

Fig. 14.6 Schematic presentation of working principle

to impose hydrostatic pressure on itself, and change frequency dependence of its own energy absorption properties. With proper adjustment of pressure we also adjust the stiffness of the damping element (again, similar as with air in tires). The proposed particle-filled damping element will provide greater energy dissipation, since when granular materials are deformed there is relative motion of particles; such motion causes energy dissipation through friction. Our proposed solution consists of micro- and macro-sized particles. Smaller particles lead to more surface area per unit volume, which increases the magnitude of frictional dissipation energy caused by particle–particle interaction; while larger particles will allow macroscopic flow, as described above. Hence, our proposed solution utilizes all possible energy dissipation mechanisms and represents an optimal (ultimate) solution for the proposed novel damping system. Such patented damping elements [8] consist of elastomeric granular material, which is encapsulated in a flexible glass fiber tube, as shown in Fig. 14.5.

This design enables us to pressurize the granular material inside the damping element. At higher pressures properties of material shift to lower frequencies, compared to the reference values (Fig. 14.4).

Working principle of such damping element is explained in the diagram in Fig. 14.6, which shows the original dynamic response of the excited structure caused by earthquake and reduced dynamic response of the excited structure when damping elements are applied. The solid line on the right schematically shows damping characteristics of the selected granular viscoelastic material at atmospheric pressure. By self-pressurizing elastomeric granular material to properly selected hydrostatic pressure its frequency characteristic may be adjusted as required, to match the resonance frequency of the structure. This is schematically shown with an arrow showing that line in the right-hand side of Fig. 14.6 changes its frequency position to the dashed line. By modifying hydrostatic pressure to which elastomeric nano-sized granular material is exposed we can adjust the frequency, at which material exhibits maximum energy absorption, such so to match the frequency of the vibrating structure. By doing so, we observe substantial reduction of vibration amplitudes.

14.4 Conclusions

Minimizing noise levels from wheel—rail interaction can be achieved by implementing different noise control measures. Among them, tuned absorber systems can be a very effective way in reducing the noise radiated by the rail. Tuned absorber systems consists of polymeric materials and as it was shown on example of TPU, existing solutions do not and cannot fully utilize damping characteristics of time- and frequency-dependent materials.

In this paper we have demonstrated that by utilizing the basic knowledge on the effect of inherent hydrostatic pressure on the time- and frequency-dependent behavior of polymers it is possible to design and build the ultimate adaptive damping elements system for vibration control applications.

In this process we need an apparatus that allows determination of material functions, as functions of pressure and temperature in time or frequency domain, e.g., G(t). Presented CEM Measuring System is an example of such apparatus.

For the selected TPU material it was found:

- By increasing pressure from 1 to 2,000 bar the frequency at which material exhibits its maximal damping properties was shifted from 37 kHz, at $P = 1$ bar to 235 Hz at $P = 2{,}000$ bar.
- The increase of inherent hydrostatic pressure from 1 to 2,000 bar changes values of storage modulus G' up to 3.5 times (depending on the frequency), while the values of loss modulus G'' are changed up to 5.5 times.

Acknowledgments Operation part financed by the European Union, European Social Fund.

References

1. EC-JRC/PBL (2011) EDGAR version 4.2. http://edgar.jrc.ec.europa.eu/. Accessed 12 May 2012
2. International Transport Forum (2010) OECD. Transport outlook. http://www.internationaltransportforum.org/Pub/pdf/10GHGTrends.pdf. Accessed 10 May 2012
3. International Union of Rail Transport and Environment (2008) Facts & Figures. http://www.uic.org/homepage/railways&environment_facts&figures.pdf. Accessed 10 Jan 2012
4. European Commission (2011) Mobility and transport. http://ec.europa.eu/transport/publications/statistics/doc/2011/pocketbook2011.pdf. Accessed 22 Feb 2012
5. European Commission Directorate General for Research (2002) Habitat fragmentation due to transportation infrastructure. The European review. https://www.milieuinfo.be/productie/beheerplone/nietacm/iene/cost-341/SotA-COST341ER0318.pdf. Accessed 12 Jan 2011
6. Thompson D (2009) Railway noise and vibration: mechanisms, modelling and means of control. Elsevier, Oxford
7. Transport Research Knowledge Centre (2003) Strategies and tools to assess and implement noise—final technical report. http://www.transport-research.info/Upload/Documents/201003/20100310_134523_78957_Final%20Report%20V10_2_.pdf. Accessed 15 Jan 2012
8. Emri I, Bernstorff B (2012) Dissipative bulk and granular systems technology: Anmeldenummer 12006059.5/EP 12006059-2012-08-24. München, Europäisches Patentamt
9. Emri I, Prodan T (2006) A measuring system for bulk and shear characterization of polymers. Exp Mech 46:429–439
10. Kralj A, Prodan T, Emri I (2001) An apparatus for measuring the effect of pressure on the time-dependent properties of polymers. J Rheol 45(4):929–943
11. Knauss WG, Emri I, Lu H (2008) Mechanics of polymers: viscoelasticity: handbook of experimental solid mechanics. Springer, New York, pp 49–95
12. Tschoegl NW, Knauss WG, Emri I (2002) The effect of temperature and pressure on the mechanical properties of thermo- and/or piezorheologically simple polymeric materials in thermodynamic equilibrium: a critical review. Mech Time Depend Mater 6:53–99
13. Tschoegl NW (1989) The phenomenological theory of linear viscoelastic behavior: an introduction. Springer, New York
14. Gergesova MM, Zupančič B, Saprunov I, Emri I (2011) The closed form t–T–P shifting (CFS) algorithm. J Rheol 55:1–16
15. Emri I, von Bernstorff BS, Cvelbar R, Nikonov AV (2005) Re-examination of the approximate methods for interconversion between frequency- and time-dependent material functions. J Nonnewton Fluid Mech 129(2):75–84

Chapter 15
Flow of Dry Grains Inside Rotating Drums

G. De Monaco, F. Greco, and P.L. Maffettone

Abstract In proper ranges of operating conditions, granular materials inside rotating drums display a continuum motion near their free surface. The motion of those discrete systems was studied both experimentally and through Discrete Element Method (DEM) numerical simulations. However, it can be regarded as the flow of a continuum medium, thus allowing a continuum mechanics approach. In our work, we solve the continuum dynamic equations by adopting the visco-plastic JFP constitutive model Jop et al. (Nature 441:727–730, 2006) for the stress tensor, and study the continuous flow of dry grains inside axially rotating cylinders through both 2D and 3D finite volume simulations (FVM). Our preliminary results are in qualitative agreement with some experimental data previously published.

Keywords Rotating drum • Granular flow • Dense flow • Continuum mechanics • Numerical simulation

15.1 Introduction

The flow of granular materials occurs widely in nature as well as in industry, and is an intriguing scientific and technological issue. Granular flows are usually classified in three different regimes [1]: a dense quasi-static regime in which the deformations are very slow and the particles interact by frictional contacts [2]; a gaseous regime in which the flow is very rapid and dilute, and the particles interact by collision [3]; and an intermediate liquid regime in which the material is dense but still flows in a liquid-like way, with its particles interacting both by collision and friction.

In contrast with the solid and gaseous behaviors, for which advanced theoretical frameworks exist, the theory for dense liquid regime is still at a rather early stage of development. Many observed phenomena and main features of liquid-like granular flows are still only tentatively described theoretically; we mention, as examples:

- The presence of a yield criterion: the flow is only possible beyond a critical shear stress;
- The complex dependence of flow response on the flow rate;
- The strong dependence on geometric factors, such as size and shape of the particles and geometric details of the flow.

The above mentioned theoretical approaches to dense liquid-like granular flows are based on direct numerical simulations of the grain system or on descriptions at the continuum mechanics level. For the latter case, of interest here, a constitutive equation for the stress tensor of granular media has been recently proposed [4] and tested in some simple geometries and/or flow conditions. Complex flow situations, on the other hand, have not been systematically tackled within this approach, and the present paper is a novel contribution in this direction.

Among complex granular material flows, one of the most interesting is the flow inside a cylinder rotating about its axis, the so called "rotating drum" problem. This kind of flow has been largely investigated in literature, in part also due to its

G. De Monaco (✉) • P.L. Maffettone
Dipartimento di Ingegneria Chimica, dei Materiali e della Produzione Industriale, Università di Napoli "Federico II",
p.le V. Tecchio 80, 80125 Naples, Italy
e-mail: gaetano.demonaco@unina.it

F. Greco
Istituto di Ricerche sulla Combustione, CNR, p.le V. Tecchio 80, 80125 Naples, Italy

H.J. Qi et al. (eds.), *Challenges in Mechanics of Time-Dependent Materials, Volume 2: Proceedings of the 2014 Annual Conference on Experimental and Applied Mechanics*, Conference Proceedings of the Society for Experimental Mechanics Series,
DOI 10.1007/978-3-319-06980-7_15,

importance in various industrial contexts, and its behaviors can be extremely variegated [5]. The rotating drum is in fact considered a benchmark problem of dense granular flow [5].

Great effort to classify the flow behavior of granular materials in rotating cylinders was spent by Henein et al. [6], who proposed the use of "Bed Behavior Diagrams" to conveniently delineate the different flow behaviors. The most thorough schematization of the possible granular flow configurations was proposed by Mellmann [7], who identified several flow regimes and the transitions among them by means of model calculations. Basically the grain motion in a rotating cylinder can be classified into three categories: the *slipping* motion, at low filling and low friction, where the grains do not flow and move as a bulk; the *rolling* motion, at medium rotational velocities, characterized by the onset of a liquid-like flow; a *cataracting* regime, as the centrifugal acceleration starts to be of the same order of the gravity; and, finally, a *centrifugal* regime.

The *rolling* motion in rotating cylinders can be further subdivided into three regimes [7]. When the rotation velocity is low, the *slumping* flow may occur. This flow consists in a series of successive distinct avalanches. As the rotational speed increases, a transition to fully developed *rolling* takes place, and the discrete sequence of avalanches evolves in a single continuous motion in the upper part of the bed. As the rotational speed further increases, the bed surface begins to arch and *cascading* sets in.

This work is devoted to the study of the fully developed *rolling* flow of a granular medium in half-filled rotating cylinders through numerical simulations. A continuum mechanical level of description is adopted, with the above mentioned constitutive equation [4] for the stress tensor of the granular material.

15.2 Experimental Evidences in the Rolling Regime

Several experimental procedures to study the rolling flow have been devised. One approach consists in looking at the rotating cylinders from one of the lateral transparent wall, measuring the flow depth h and the dynamic angle θ, defined as the average angle of the free surface of the grain phase. Such measurements are usually performed in very narrow cylinders (quasi–2D cylinders), in order to avoid flow in the axial direction. Another approach relies on the use of Magnetic Resonance Imaging (MRI) techniques to investigate the same quantities (h and θ) at the center of wide rotating cylinders. Those approaches are clearly different in many ways. The quasi-2D approach cannot avoid wall effects, while the MRI does not suffer of this limitation. We believe this to be a crucial point: even if is commonly accepted that there is a strong influence of the lateral wall on the flow [8–13], most of the experiments are performed in the worst condition, i.e. by looking at grains near the wall. The MRI drawbacks, on the other hand, lay in the difficulties of the experiment itself, and in that the space and time resolutions of such a technique are quite lower than those of direct measurements. A third approach makes use of the Positron Emission Particle Tracking (PEPT) [14], which again requires very complex apparati.

A review of experimental data obtained by investigating the flow close to the lateral wall can be found in GDR MiDi [1] and Pignatel et al. [8]. GDR MiDi [1] found that the depth of the flowing layer scales with the rotational velocity to the power of 0.5, and that is what one would obtain by hypothesizing a constant shear rate in the flowing layer. The dynamic angle increases with increasing rotation velocity and with decreasing cylinder width.

Felix et al. [15] and Pignatel et al. [8] also report a monotonically increasing scaling of the dynamic angle and the flow depth with the rotational velocity; they found, however, a wide exponent range for the power law dependency of the flow depth on the rotational velocity, with values from 0.15 to 0.68, depending on the ratio $^{D}\!/_{d_P}$ of the cylinder diameter over the particle diameters. The angle is also found to increase with decreasing width, even if an appropriate parameter to collapse the data is not proposed [8]. Pignatel et al. [8] also point out that the wall friction plays an important role in narrow cylinders.

Nakagawa et al. [16] and Yamane et al. [17] used MRI to investigate the rolling flow far away from lateral walls. Their experimental works focused mainly on velocity [18] and density profiles along the depth of the granular material, but reported a single dataset of flow depth versus angular velocity, which scales with a power of 0.4. Yamane et al. [17] also report a linear increase of the dynamic angle with the rotational speed, with the angle being ca. 5° higher near to the wall than at the center of the cylinder at all the angular velocity explored.

As already said, the walls exert a strong influence on tumblers dynamics. Dury et al. [12] measured the difference in dynamic angle at the center (via MRI) and at the wall (via direct visualization), finding a constant difference of ca. 4° regardless of the rotational velocity. Furthermore, they performed DEM simulations to access the profile of the dynamic angle along the axis of wide cylinders. In such a way, they were able to quantify the distance from the wall at which the angle becomes that of the center of the cylinder. Such a distance was found be linear in the drum diameter, following the scale $L = 0.14\, D$. They also found finite size effect for $^{D}\!/_{d_P} < 15$.

Pohlman et al. [10] and Chen et al. [11] studied the 2D velocity profiles on the free surface of wide cylinders both experimentally [10] and numerically [11]. They proved the existence of an axial flow near the walls directed toward the center in the upper part of the flowing layer (upstream), and toward the lateral wall in the lower part of such layer. The axial flow increases with the wall friction and with the flowing length (i.e. with the diameter). The axial flow produces higher streamwise velocities, that can be justified with conservation of mass arguments. They also pointed out that such a flow extend for a distance of $L = 0.5\ D$ from lateral walls.

15.3 On the Description of Granular Liquids Mechanics

The most commonly adopted numerical approach to study dense granular flow is the Discrete Element Method (DEM) (see for example [12, 19, 20]): the granular material is modeled as an assembly of rigid particles and the interactions among them are explicitly considered. This technique has the advantage to reflect the discrete nature of the medium and to access the very local details of the flow. Nevertheless its major drawback is the relatively limited number of particles which can be effectively handled: as an example, we mention a very recent DEM calculation on a system made up of more than one million particles [21]. It should be noticed, however, that a small lab cylinder with a diameter of 20 cm and a width of 40 cm, half-filled with grains having a diameter of 0.5 mm and a void fraction of 0.5, would contain 48 million particles. Few attempts to describe this kind of flow as a continuum has been made by Bonamy et al. [22] and very recently by Huang et al. [23]: they used the Eulerian continuum approach (ECA) coupling the kinetic theory, which holds for more rapid and more diluted flow.

Only recently Jop et al. [4] have developed a continuum description of dense granular flows, based on the local rheology approach proposed by GDR MiDi [1]. The proposed constitutive law, which shares some similarities with classical constitutive equation of visco-plastic fluids such as Bingham fluids, allows a Finite Element Method (FEM) or a Finite Volume Method (FVM) numerical simulation of granular flows.

GDR MiDi [1] proposed (citing the authors) "*a rheology for which stresses and shear rate at a given location in the flow are related through a one-to-one relation*". From dimensional analysis in a shear flow configuration for a granular medium, they find:

$$\frac{\tau}{p} = \mu(I) \tag{15.1}$$

τ being the shear stress, p the pressure, μ the friction factor and I a non-dimensional parameter called the *inertial number*, which is defined as:

$$I = \frac{\dot{\gamma} d_P}{\sqrt{\frac{p}{\rho_P}}} \tag{15.2}$$

where $\dot{\gamma}$ is the shear rate, d_P and ρ_P are the diameter and the density of the (single) particle.

Jop et al. [9] also proposed a functional form of μ in Eq. (15.1):

$$\frac{\tau}{p} = \mu(I) = \mu_s + \frac{\mu_s - \mu_2}{\frac{I_0}{I} + 1} \tag{15.3}$$

Equations (15.1) and (15.2) are nicely confirmed by da Cruz et al. [19], who studied the flow of a dense assembly of dissipative disks subject to a given shear flow by using DEM simulation. They observed steady uniform shear flows, and showed that, in the limit of rigid grains, the shear state is solely determined by the inertial number. Uniform dense flow was found to hold when $I \in [10^{-3}, 10^{-1}]$, and in that range the friction $\mu(I)$ was found to increase towards a plateau, like in Eq. (15.3).

Finally, Jop et al. [4] also proposed a 3D generalization of the 1D constitutive law, which explicitly relies on the assumption that the variation of the solid volume fraction can be neglected. The granular material, in whatever flow configuration, can be described as a continuum fluid having a stress tensor given by:

$$\overline{\overline{\sigma}} = -p\overline{\overline{1}} + \overline{\overline{\tau}} \tag{15.4}$$

$$\overline{\overline{\tau}} = 2\,\eta(II_D,p)\,\overline{\overline{D}} \tag{15.5}$$

$$\eta(II_D,p) = \mu(I)\frac{p}{II_D} \tag{15.6}$$

$\overline{\overline{D}}$ is the rate of deformation tensor, i.e. the symmetric part of the velocity gradient $\overline{\overline{\nabla v}}$, and II_D is the second invariant of $\overline{\overline{D}}$. This model is known as JFP model, after the names of the authors who first proposed it (Jop, Forterre and Pouliquen).

It should be noticed that the viscosity goes to infinity as $\dot{\gamma}$ tends to zero, as it happens for visco-plastic fluids. In the limit of $II_D \to 0$ it is possible to show that the above equations reduce to a yield criterion:

$$II_\tau > \mu(I)\,p \tag{15.7}$$

where II_τ is the second invariant of the viscous stress tensor $\overline{\overline{\tau}}$.

Admittedly, the constitutive law expressed by Eqs. (15.4–15.6) still rests on an empirical ground [9]. Nonetheless, it has been proved that the JFP model quantitatively describes the experimentally determined flow shapes and velocity profiles in the case of a flow on an inclined plane with rough sidewalls [4].

To the best of our knowledge, this is the first work to study the rolling flow with this approach.

15.4 Materials and Methods

In our work, we solve the continuum dynamic equations by adopting the visco-plastic JFP constitutive model [4] for the stress tensor, and study the continuous flow of dry grains inside axially rotating cylinders through both 2D and 3D finite volume simulations (FVM).

We use the open source CFD software package OpenFOAM. It is structured as a library of routines so it is highly modular, and the source code is fully accessible and editable, so it is also very versatile. We use a modified version of the *interFoam* solver, which solves the equations of motion for a two-phases fluid with the Volume of Fluid (VoF) model [24]. VoF model treats the two phases as a single whole fluid (so with uniques velocity and pressure fields) with physical properties varying through space and time. *interFoam* implements the PIMPLE algorithm, which is a blending of the PISO [25] and the SIMPLE [26] algorithms. A thorough description of the numeric of *interFoam*, together with few validation cases, can be found in Deshpande et al. [27].

A sketch of the geometry is shown in Fig. 15.1. The upper half of the cylinder (above the gray plane) is filled with air, the lower one with grains. We varied two geometrical parameter over two levels, obtaining four different cylinders. The first parameter is the diameter ratio $^{D}\!/_{d_P}$, which takes the values of 1,000 and 2,500; the second parameter is the aspect ratio AR, defined as the ratio of the half-width over the diameter of the cylinder: $AR = {}^{W}\!/_{D}$, which takes the values of 0.1 and 1. It should be noticed that for $AR < 0.5$ the walls are expected to affect the dynamic at the center of the cylinder [10]: this is the case for our "narrow" cylinders ($AR = 0.1$), but not for the "wide" ones ($AR = 1$). As the geometry has a symmetry plane, we considered only half of the whole cylinders (the gray cylinder portion in Fig. 15.1). The geometry for the 2D simulations is a circle and can be considered as 3D simulations with $W \to \infty$ and with no axial velocity component. We run simulations at

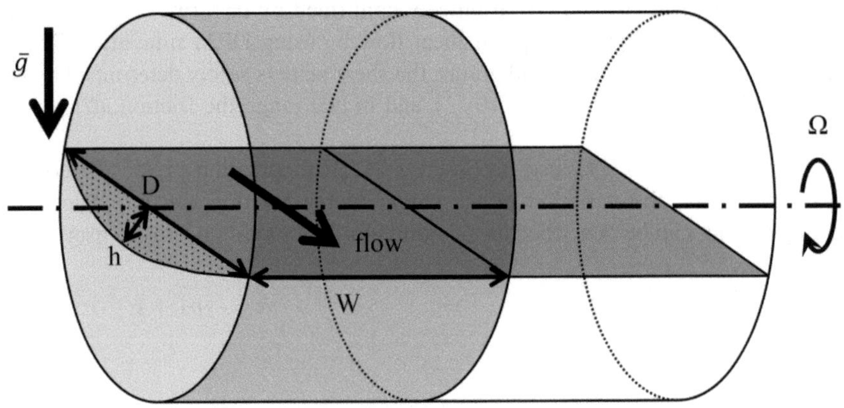

Fig. 15.1 A sketch of the geometry: D is the diameter of the cylinder, W its half-width, h is the depth of the flowing layer, which is the shadowed lens-shaped area, $\overline{g}$ is the gravity, Ω the angular velocity around the axes, which is displayed as a *dash-dotted line*, the plane represent the free surface of the granular phase and the *thick line* is the mainstream flow direction

five different rotational velocities, namely 1, 2, 4, 6 and 10 rounds per minute (*rpm*). For 1 rpm the system is close to the transition to the slumping regime (discrete avalanches), while for 10 rpm it approaches the cascading regime.

We chose to solve the equations of motion in a coordinate system fixed to the rotating cylinder, thus implying that curved walls have zero velocity and a rotating gravity. This choice, along with the implementation of the JFP model, needed a modification of the base solver. The equations which are solved in our version of *interFoam* are:

$$\overline{\nabla} \cdot \overline{v} = 0 \tag{15.8}$$

$$\frac{\partial \rho \overline{v}}{\partial t} + \overline{\nabla} \cdot (\rho \overline{v}\,\overline{v}) = -\overline{\nabla} p^* + \overline{\nabla} \cdot \left[\eta(II_D, p)\left(\overline{\overline{\nabla v}} + \overline{\overline{\nabla v^T}}\right)\right] - (\overline{g}(t) \cdot \overline{r})\overline{\nabla}\rho + \rho\left[2\,\overline{\Omega} \times \overline{v} + \overline{\Omega} \times \left(\overline{\Omega} \times \overline{v}\right)\right] \tag{15.9}$$

$$\frac{\partial \alpha}{\partial t} + \overline{v} \cdot \overline{\nabla}\alpha = 0 \tag{15.10}$$

$$\begin{cases} \rho = \alpha\,\rho_g + (1-\alpha)\,\rho_{air} \\ \eta(II_D, p) = \alpha\,\eta_g(II_D, p) + (1-\alpha)\,\eta_{air} \end{cases} \tag{15.11}$$

$$\eta_g(II_D, p) = \left(\mu_s + \frac{\mu_s - \mu_2}{\frac{I_0}{I} + 1}\right)\frac{p}{II_D} \tag{15.12}$$

Equations (15.8) and (15.9) are the conservation of mass and momentum. The terms $-\overline{\nabla}p^* + (\overline{g}(t) \cdot \overline{r})\overline{\nabla}\rho$ in the right-end side of Eq. (15.9) come from the definition of the modified pressure used by *interFoam* $p^* = p - \rho\overline{g} \cdot \overline{r}$ (the dependence of the gravity on the time comes from the choice of the coordinate system and hereinafter will be omitted):

$$-\overline{\nabla}p + \rho\overline{g} = -\overline{\nabla}p^* - (\overline{g} \cdot \overline{r})\overline{\nabla}\rho \tag{15.13}$$

The other terms, namely the Coriolis force $2\rho\overline{\Omega} \times \overline{v}$ and the centrifugal force $\overline{\Omega} \times \left(\overline{\Omega} \times \overline{v}\right)$, are the fictitious forces which arise because of the non-inertial reference system.

Equations (15.10) and (15.11) express the VoF model: Eq. (15.10) is the simple convection of the volume fraction α of the grains in each cell, which is used to calculate the physical properties in each cell of the domain through Eq. (15.11). Near the interface between the two phases Eq. (15.10) is modified in order to include a compression term for keeping it sharp [28]. ρ_{air} and η_{air} are constants, the density of the granular phase is calculated as $\rho_g = \phi\rho_P$, ϕ being the packing factor and ρ_P the (constant) particle density. $\eta_g = \eta_g(II_D, p)$ is the viscosity of the granular phase, which is calculated in each cell with the local pressure p and the local shear rate $\dot{\gamma}$. Few more words are worth about the packing factor ϕ. It is known that it depends on the inertial number in a linear way [19], but in our calculation we keep it fixed to the value of 0.6. It should be noticed that ϕ and α are both grains volume fraction, but they have completely different meanings: ϕ is the packing factor, that is the fraction occupied by the solid sphere in their 3D arrangement in the bulk; α is the volume fraction of the granular phase considered as a whole fluid (i.e. with its voids), so with a density $\rho_g = \phi\rho_P$, and is used to calculate the physical properties through space and time in each cell. Physical properties vary in time and space because they depend on α.

Equation (15.12) is the implementation of the JPF model, where I is the inertial number defined in Eq. (15.2). μ_s, μ_2 and I_0 are physical properties of the material and can be measured with independent experiments [4]. We chose their values to be equal to those used by Jop et al. [4], namely $\mu_s = \tan(20.9°)$, $\mu_2 = \tan(32.76°)$ and $I_0 = 0.279$. We also regularized the viscosity function:

- Since the constitutive equation diverges as $II_D \to 0$, we bound the viscosity function to a value $\eta_{max} = \rho_g \cdot 100\frac{m}{s^2}$,
- Since the pressure has a relative nature, it might become negative in some cells, so we put to zero every negative viscosity: $\eta_{min} = 0$ (see Eq. (15.12)).

Regarding the boundary conditions for the velocity, we used the OpenFOAM built-in partial slip condition for the walls:

$$\begin{cases} \overline{v} \cdot \overline{n} = 0 \\ \beta\overline{v} + (1-\beta)\dfrac{\partial \overline{v}}{\partial n} = \overline{0} \end{cases} \tag{15.14}$$

and the symmetry condition for the middle plane:

$$\begin{cases} \overline{v} \cdot \overline{n} = 0 \\ \dfrac{\partial \overline{v}}{\partial n} = \overline{0} \end{cases} \tag{15.15}$$

where $\overline{n}$ is the boundary normal and β a slip parameter: for $\beta = 1$ the B.C. becomes a no-slip condition (as the walls have zero velocity in our reference system), for $\beta = 0$ it becomes a perfect slip condition. We fixed $\beta = 0.1$ on all walls. After Jop et al. [4], the pressure level was chosen to be zero at zero time on the initially flat free surface. We imposed a zero gradient boundary condition for the volume fraction α on all domain boundaries.

We measured the dynamic angle at the center of the free surface as the slope of such interface with respect to a plane normal to the gravity. The depth of the flowing layer was measured as the distance from the free surface at which the velocity (in the lab-integral reference frame) became zero, along a line passing through the center of the cylinder and orthogonal to the free surface, just as Felix et al. [15] and Pignatel et al. [8] did.

15.5 Preliminary Results

The preliminary results of our simulations confirm two important features of the rolling flow inside rotating drums:

1. Both the depth of the flowing layer h and the dynamic angle θ increase with the angular velocity, the former following a power law [1, 15], the latter a linear scaling [16];
2. There are strong wall effects on both h and the θ.

Figure 15.2 shows the trends of the flowing layer depth versus the angular velocity at the center (solid line) and near the wall (dashed lines) for cylinders with $AR = 0.1$ (a) and $AR = 1$ (b). The graphs are parametric in $^D/_{d_P}$ (colors) and report the trends of the corresponding 2D cases.

All our cases show power law dependencies of the flow depth with the rotational speed:

$$h \propto \Omega^n \tag{15.16}$$

The narrow cylinders ($AR = 0.1$) show the same exponent n both at the center or near the wall, and they are lower than those of the 2D cases. The wide cylinders ($AR = 1$) show an higher power law index at center than near the wall. Moreover, the trends at the center of the 3D cylinders are close to those of the corresponding 2D cases, at least at higher angular velocity. Some sort of flattening seems to set in at low velocity for the wide cylinders cases. This might be related to the fact that the system is close to the transition to the slumping regime.

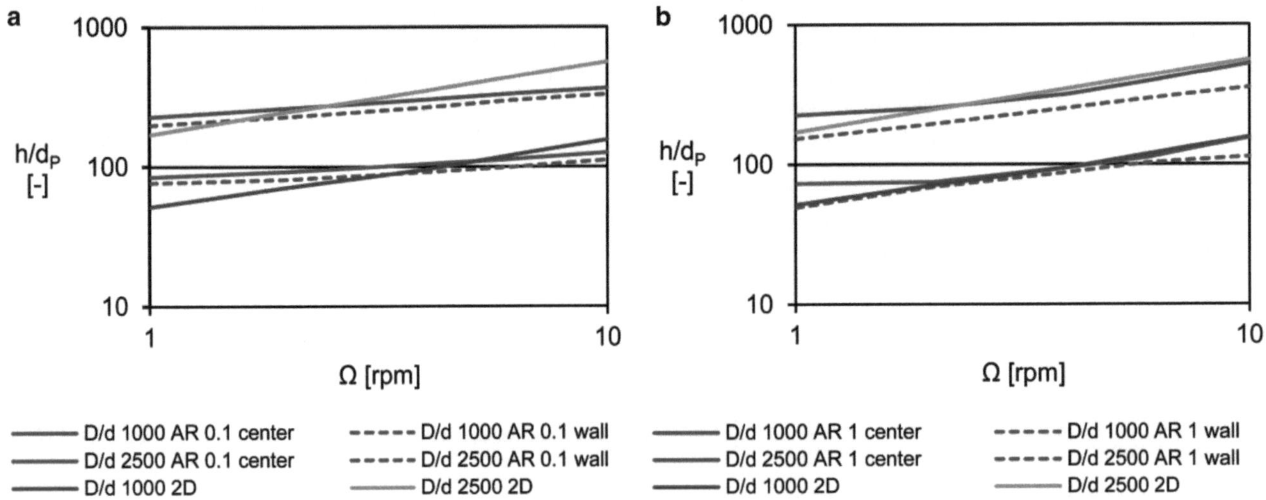

Fig. 15.2 Trends of the flowing layer depth versus the rotational velocity for (**a**) AR $= 0.1$ and (**b**) AR $= 1$ parametric in both $^D/_{d_P}$ and AR; *solid lines* refers to the center of the cylinders, *dashed lines* to the lateral walls; *blue lines* correspond to $^D/_{d_P}$=1,000, *red lines* to $^D/_{d_P}$=2,500 (color figure online)

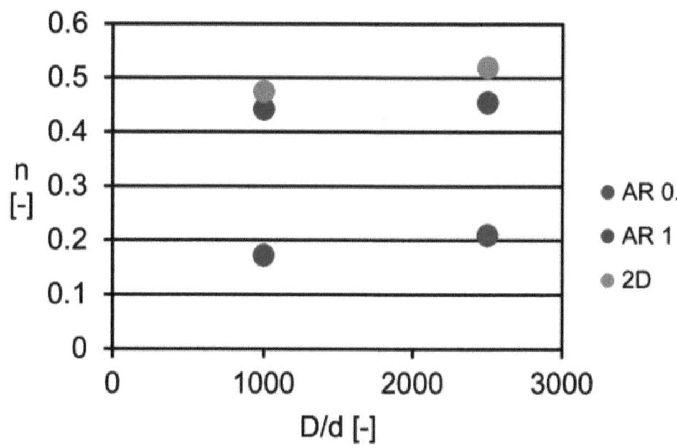

Fig. 15.3 Trends of the power law index of Eq. (15.16) with $^{D}/_{d_P}$, parametric in *AR*

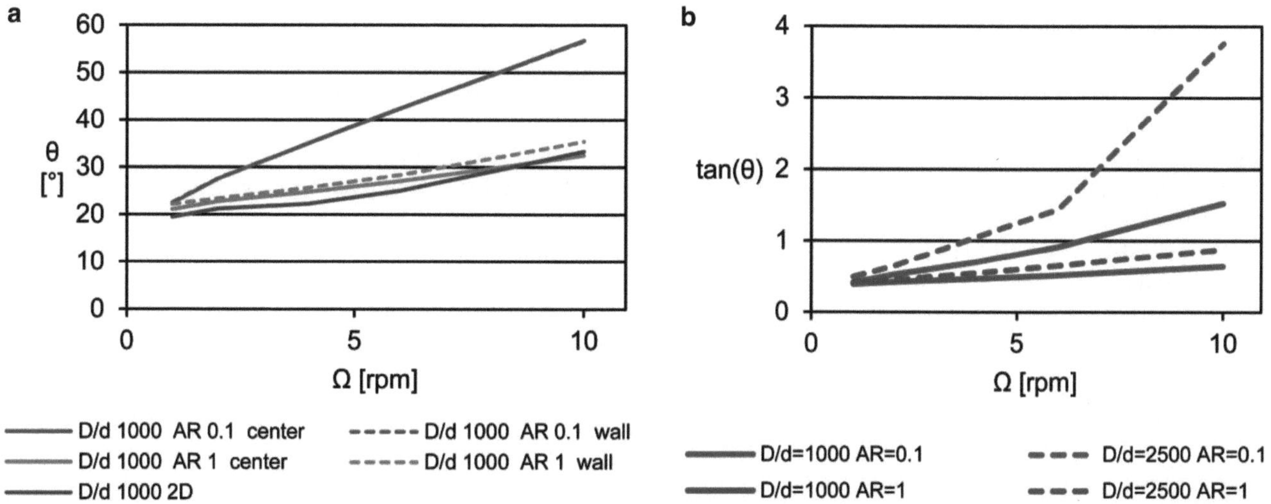

Fig. 15.4 Trends of the dynamic angle versus the rotational velocity (**a**) for $^{D}/_{d_P}$=1, 000, *solid lines* refers to the center of the cylinders, *dashed lines* to the lateral walls; (**b**) measured at the center of the cylinder, *solid lines* refers to $^{D}/_{d_P}$=1000, *dashed lines* to $^{D}/_{d_P}$=2, 500, *red lines* refers to AR = 0.1, *blue lines* to AR = 1 (color figure online)

Our simulations show that the power law indices depend on both $^{D}/_{dP}$ and *AR*: Fig. 15.3 shows their trends (measured at the center of the cylinder) with $^{D}/_{dP}$, parametric in *AR*. For the power law fit at $AR = 1$ we excluded the points at $\Omega = 1$ rpm. In contrast with the result of Felix et al. [15], our results show a weak effect of the $^{D}/_{dP}$ ratio, while the dependence on *AR* is much more relevant: the closer the walls, the lower the indices. The plot in Fig. 15.3 shows that the results of 3D simulations in wide cylinders (red dots) are close to the 2D ones (green dots), although the values of n do not superimpose.

Figure 15.4a shows the trends of the dynamic angle with the angular velocity. For clarity, only results for the cylinders with diameter ratio equal to 1,000 are shown, those obtained for cylinders with $^{D}/_{dP} = 2,500$ are similar. Both the angles measured at the center and near the wall increase linearly with Ω. The trends at the center and near the wall for the narrow cylinders are superimposed: as for the flow depth, the center of the cylinder is not sufficiently distant from the walls to develop a sensibly different behavior. It does happen, instead, for the cases of wide cylinders, where the dynamic angles at the center are slightly lower than those near the wall. Once again, the trends measured at the center of wide 3D cylinders resemble those from the 2D simulations. The slope of the free surface is less sensitive to rotational velocity for wider cylinder, as it is already known from literature [1].

The dynamic angle depends on both the geometric parameters $^{D}/_{dP}$ and *AR*, as shown in Fig. 15.4b (after Midi [1], we report the tangent of the dynamic angle). Increasing the $^{D}/_{dP}$ makes the angle increase at constant *AR*. One might think that this scaling comes from wall effects: the greater lateral surface could be able to better sustain the granular material during the motion. However, this might not hold true anymore when $AR = 1$, as the wall effects are no more relevant at the cylinder center. A possible explanation of the angle increasing $^{D}/_{dP}$ with at $AR = 1$ is the role of inertia: increasing the cylinder diameter determines higher velocities at constant Ω, which might bring up the upstream part of the flowing layer and then the overall free surface. Increasing the *AR*, instead, makes the angle decrease, and this effect is stronger than the one of the diameter ratio: at constant $D/_{dP}$, higher *AR* means farther walls, so they are less effective in sustaining the grains, thus determining lower dynamic angles. This is consistent with the results from MiDi [1].

15.6 Conclusions

In this work the rolling flow of a granular medium in half-filled rotating cylinders is studied through numerical simulations with the JFP constitutive law. The dependences of the flow depth and the dynamic angle on the rotating speed are investigated at different $D/_{dP}$ and *AR* values. Our preliminary results are in fair qualitative agreement with some well-established experimental results for dense granular flow in rotating drum.

The depth of the flowing layer shows a power law dependence on the rotational velocity. The index of the power law depends weakly on the diameter ratio $D/_{dP}$ and strongly on the aspect ratio *AR*, thus suggesting an important effect of the lateral walls on the value of this quantity. This is even more relevant when considering that the vast majority of the available literature data is obtained near the end walls of the cylinder.

The dynamic angle increase linearly with the rotational velocity, and is slightly higher near the wall than at the center of the cylinder. It depends on both the geometric parameters $D/_{dP}$ and *AR*: increasing the $D/_{dP}$ make the angle increase at constant *AR*, while increasing the *AR* make the angle decrease. We conjecture that the former might be an inertial effects, while the latter might be connected with the ability of the lateral wall to better sustain the grains.

More simulations at different value of both parameters are required and have been planned in order to better understand the scaling laws of the flow depth and the dynamic angle in the rolling regime.

We hope also to be able to tackle the complex problem of flow regime transitions (e.g., rolling to slumping transition by decreasing the rotational rate) in the near future.

References

1. GDR MiDi (2004) On dense granular flow. Eur Phys J E 14:341–365
2. Roux J-N, Combes G (2002) Quasistatic rheology and the origins of strain. C R Phys 3:131–140
3. Goldhirsch I (2003) Rapid granular flows. Annu Rev Fluid Mech 35:267–293
4. Jop P, Forterre Y, Pouliquen O (2006) A constitutive law for dense granular flows. Nature 441:727–730
5. Seiden G, Thomas PJ (2011) Complexity, segregation, and pattern formation in rotating-drum flows. Rev Mod Phys 83:1323–1365
6. Henein H, Brimacombe JK, Watkinson AP (1983) Experimental study of transverse bed motion in rotary kilns. Metallurgical Trans B 14B:191–205
7. Mellmann J (2001) The transverse motion of solids in rotating cylinders—forms of motion and transition behavior. Powder Technol 118:251–270
8. Pignatel F, Asselin C, Krieger L, Christov IC, Ottino JM, Lueptow RM (2012) Parameters and scalings for dry and immersed granular flowing layers in rotating tumblers. Phys Rev E 86:011304
9. Jop P, Forterre Y, Pouliquen O (2005) Crucial role of sidewalls in granular surface flows: consequences for the rheology. J Fluid Mech 541:167–192
10. Pohlman NA, Ottino JM, Lueptow RM (2006) End-wall effects in granular tumblers: from quasi-two-dimensional flow to three-dimensional flow. Phys Rev E 74:031305
11. Chen P, Ottino JM, Lueptow RM (2008) Subsurface granular flow in rotating tumblers: a detailed computational study. Phys Rev E 78:021303
12. Dury CM, Ristow GH, Moss JL, Nakagawa M (1998) Boundary effects on the angle of repose in rotating cylinders. Phys Rev E 57:4491–4497
13. Maneval JE, Hill KM, Smith BE, Caprihan A, Fukushima E (2005) Effects of end wall friction in rotating cylinder granular flow experiments. Granular Matter 7:199–202
14. Parker DJ, Dijkstra AE, Martin TW, Seville JPK (1997) Positron emission particle tracking studies of spherical particle motion in rotating drums. Chem Eng Sci 52:2011–2022
15. Felix G, Falk V, D'Ortona U (2007) Granular flows in a rotating drum: the scaling law between velocity and thickness of the flow. Eur Phys J E 22:25–31
16. Nakagawa M, Altobelli SA, Caprihan A, Fukushima E, Jeong E-K (1993) Non-invasive measurements of granular flows by magnetic resonance imaging. Exp Fluids 16:54–60
17. Yamane K, Nakagawa M, Altobelli SA, Tanaka T, Tsuji Y (1998) Steady particulate flows in a horizontal rotating cylinder. Phys Fluids 10:1419–1427
18. Sanfratello L, Caprihan A, Fukushima E (2007) Velocity depth profile of granular matter in a horizontal rotating drum. Granular Matter 9:1–6
19. da Cruz F, Eman S, Prochnow M, Roux J-N, Chevoir F (2005) Rheophysics of dense granular materials: discrete simulation of plane shear flows. Phys Rev E 72:021309

20. Zhu HP, Zhou ZY, Yank RY, Yu AB (2008) Discrete particle simulation of particulate systems: a review of major applications and findings. Chem Eng Sci 63:5728–5770
21. Longmore J-P, Marais P, Kuttel MM (2013) Towards realistic and interactive sand simulation: a GPU-based framework. Powder Technol 235:983–1000
22. Bonamy D, Chavanis P-H, Cortet P-P, Daviaud F, Dubrulle B, Renouf M (2009) Euler-like modelling of dense granular flows: application to a rotating drum. Eur Phys J B 68:619–627
23. Huang A-N, Kao W-C, Kuo H-P (2013) Numerical studies of particle segregation in a rotating drum based on Eulerian continuum approach. Adv Powder Technol 24:364–372
24. Hirt CW, Nichols BD (1981) Volume of fluid (VOF) method for the dynamics of free boundaries. J Comput Phys 39:201–225
25. Issa RI, Gosman AD, Watkins AP (1986) The computation of compressible and incompressible recirculating flows by a non-iterative implicit scheme. J Comput Phys 62:66–82
26. Patankar SV (1980) Numerical heat transfer and fluid flow. Boca Raton, Florida, USA, CRC Press
27. Deshpande SS, Anumolu L, Trujillo MF (2012) Evaluating the performance of the two-phase flow solver interFoam. Comput Sci Discov 5:014016
28. Berberovic E, van Hinsberg NP, Jakirlic S, Roisman IV, Tropea C (2009) Drop impact onto a liquid layer of finite thickness: dynamics of the cavity evolution. Phys Rev E 79:036306

Chapter 16
Statistical Prediction of Tensile Creep Failure Time of Unidirectional CFRP

Yasushi Miyano, Masayuki Nakada, Tsugiyuki Okuya, and Kazuya Kasahara

Abstract The tensile strength along the longitudinal direction of unidirectional CFRP is one of the important data for the reliable design of CFRP structures. This paper is concerned with the statistical prediction of creep failure time under the tension loading along the longitudinal direction of unidirectional CFRP based on the viscoelasticity of matrix resin. It was cleared in this study that the statistical creep failure time under the tension loading along the longitudinal direction of unidirectional CFRP can be predicted by using the statistical static tensile strengths of carbon mono filament and unidirectional CFRP and the viscoelasticity of matrix resin based on Christensen's model of viscoelastic crack kinetics.

Keywords Carbon fiber reinforced plastics • Creep failure time • Statistical prediction • Viscoelasticity

16.1 Introduction

Carbon fiber reinforced plastics (CFRP) has been used for the primary structures of airplanes, ships, automobiles and others, in which the high reliability should be kept during the long-term operation. Therefore, it is strongly expected that the accelerated testing methodology for the long-term life prediction of CFRP structures exposed under the actual environmental temperature, water and others will be established.

The mechanical behavior of matrix resin of CFRP exhibits time and temperature dependence, called viscoelastic behavior, not only above the glass transition temperature T_g but also below T_g. Thus, it can be presumed that the mechanical behavior of CFRP significantly depends on time and temperature [1–6]. We have proposed the formulation for the statistical static strength of CFRP based on the viscoelasticity of matrix resin as shown in our previous papers [7, 8].

The tensile strength along the longitudinal direction of unidirectional CFRP is one of the important data for the reliable design of CFRP structures. The authors developed the test method for the creep and fatigue strengths as well as the static strength at elevated temperatures for the resin impregnated carbon fiber strand (CFRP strand) combined with T300-3000 and epoxy resin [9–11]. Recently the authors have developed the test method for the CFRP strand of T800-12000 and epoxy resin with highly reliable co-cured tab and the temperature dependent tensile strength of this CFRP strand was successfully evaluated [12, 13].

This paper is concerned with the prediction of statistical creep failure time under the tension loading along the longitudinal direction of unidirectional CFRP. The CFRP strands of T800-12000 and epoxy resin for high reliable static and creep tests were employed in this study. First, the static tensile tests for carbon mono filament were carried out at four levels of specimen length by using 50 specimens for each length. Second, the static tensile tests for CFRP strands were carried out at three levels of constant temperature by using 50 specimens for each temperature. Third, the statistical creep failure time at a constant load and temperature was predicted using the statistical results of static tensile strengths at three temperatures and the viscoelastic behavior of matrix resin. Finally, the validity of predicted results was cleared by comparing with the experimental results obtained by the creep tests for CFRP stands of 26 specimens.

Y. Miyano (✉) • M. Nakada • T. Okuya • K. Kasahara
Materials System Research Laboratory, Kanazawa Institute of Technology, 3-1 Yatsukaho, Hakusan, Ishikawa 924-0838, Japan
e-mail: miyano@neptune.kanazawa-it.ac.jp

H.J. Qi et al. (eds.), *Challenges in Mechanics of Time-Dependent Materials, Volume 2: Proceedings of the 2014 Annual Conference on Experimental and Applied Mechanics*, Conference Proceedings of the Society for Experimental Mechanics Series,
DOI 10.1007/978-3-319-06980-7_16,

16.2 Statistical Prediction of Creep Failure Time of Unidirectional CFRP

16.2.1 Formulation for Statistical Static Strength of CFRP Laminates Based on Viscoelasticity of Matrix Resin

We have proposed the formulation for the statistical static strength σ_s of CFRP based on the viscoelasticity of matrix resin as shown in the following equation in our previous paper [8],

$$\log\sigma_s(P_f, t, T) = \log\sigma_0(t_0, T_0) + \frac{1}{\alpha_s}\log[-\ln(1 - P_f)] - n_R \log\left[\frac{D*(t, T)}{D_c(t_0, T_0)}\right] \quad (16.1)$$

where, P_f is the failure probability, t is the failure time, t_0 is the reference time, T is the temperature, T_0 is the reference temperature, σ_0 and σ_s are the scale parameter and the shape parameter on Weibull distribution of static strength, n_R is the viscoelastic parameter, and D_c and D^* are the creep and viscoelastic compliances of matrix resin. The viscoelastic compliance D^* for the static load with a constant strain rate is shown by the following equation.

$$D*(t, T) = D_c(t/2, T) \quad (16.2)$$

The failure probability of unidirectional CFRP under static load with a constant strain rate can be shown by the following equation.

$$P_f = 1 - \exp(-F), \quad \log F = \alpha_s \log\left[\frac{\sigma_s}{\sigma_0}\right] + \alpha_s n_R \log\left[\frac{D_c(t/2, T_0)}{D_c(t_0, T_0)}\right] \quad (16.3)$$

16.2.2 Failure Probability of Unidirectional CFRP Under Static Load

It was cleared in our previous paper that the viscoelastic parameter n_R in Eqs. (16.1) and (16.3) is shown by the following equation for the unidirectional CFRP based on Rosen's model [11].

$$n_R = \frac{1}{2\alpha_C} \quad (16.4)$$

where, α_c is Weibull shape parameter for the tensile strength of carbon fiber mono-filament.

16.2.3 Statistical Creep Failure Time of Unidirectional CFRP

The relationship between the creep failure time and the static failure time can be shown by Fig. 16.1 [14]. This figure shows the creep strength and the static strength versus failure time. The creep strength curve can be obtained by shifting horizontally the static strength curve. The shifting amount log A determined by the slope of the static strength curve is shown by the following equation.

$$\log A = \log(1 + 1/k_R) \quad (16.5)$$

Figure 16.2a shows the slope m_R of the creep compliance of matrix resin against time. Figure 16.2b shows the slope k_R of the static strength of CFRP against failure time. The slope k_R can be obtained from the following equation.

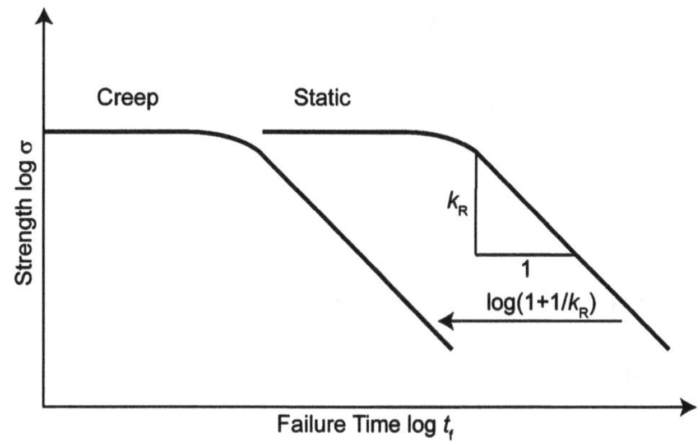

Fig. 16.1 Time shifting between static strength and creep strength

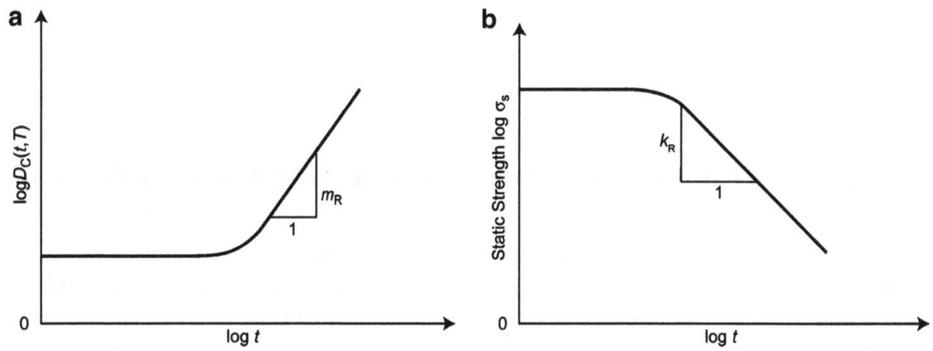

Fig. 16.2 Slope against time. (**a**) Creep compliance of matrix resin. (**b**) Static strength of CFRP

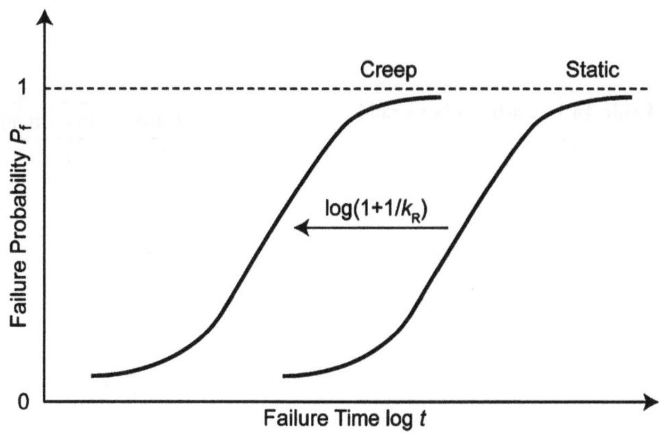

Fig. 16.3 Time shifting between failure probability of static and creep

$$k_R = n_R m_R = \frac{m_R}{2\alpha_C} \tag{16.6}$$

The failure probability against failure time at an arbitrary constant load under static loading (constant strain rate) is predicted by using Eqs. (16.5) and (16.6) as shown in Fig. 16.3 and the creep failure probability against failure time can be predicted by shifting horizontally that of the static loading.

16.3 Experiments

16.3.1 Specimens

CFRP strand which consists of high strength type carbon fiber T800-12000 (Toray Industries Inc.) and a general purpose epoxy resin jER828 (Mitsubishi Chemical Corp.) was molded using filament winding method developed by authors [10]. The composition of epoxy resin and the cure condition of CFRP strand are shown in Table 16.1. The diameter and the gage length of CFRP strands are approximately 1 and 200 mm, respectively. The glass transition temperature T_g of the epoxy resin is approximately 150 °C. The fiber volume fraction of CFRP strand is approximately 50 %. The tensile strength of the CFRP strand σ_s is defined by

$$\sigma_s = \frac{P_{max}}{t_e}\rho \tag{16.7}$$

where P_{max} is maximum load [N]. ρ and t_e are the density of the carbon fiber [kg/m^3] and the tex of the carbon fiber strand [g/1,000 m].

16.3.2 Tensile Strength of Carbon Fiber Mono-Filament and Creep Compliance of Matrix Resin

In our previous paper [13], tensile tests of carbon fiber T800 mono filament for several lengths of filament were conducted at the room temperature to determine the Weibull shape parameter for the tensile strength of the carbon fiber T800 mono filament. The tensile test speed was 1 mm/min. The Weibull distributions of the tensile strength of the carbon fiber T800 mono filament are shown in Fig. 16.4.

The scale parameter β_c* for an arbitrary fiber length L is shown by the following equation based on one dimensional chain model.

Table 16.1 Carbon fiber strand and resin system

Carbon fiber strand	Composition of resin (weight ratio)	Cure schedule
T800-12000	Epoxy: jER828 (100)	100 °C × 5 h
	Hardener: MHAC-P (103.6)	+150 °C × 4 h
	Cure accelerator: 2E4MZ (1)	+190 °C × 2 h

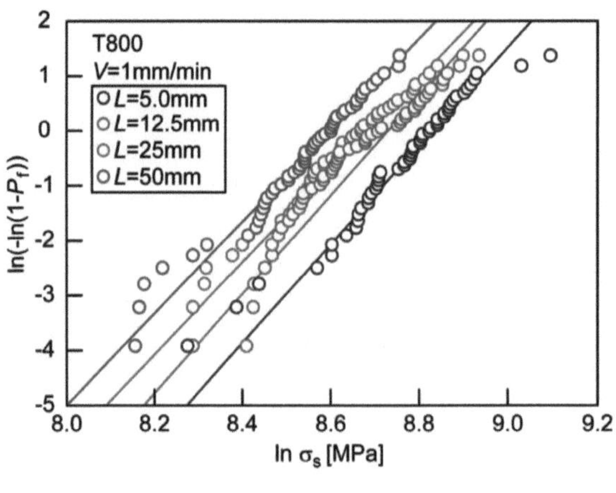

L[mm]	α_c[-]	β_c[MPa]	β'_c[MPa]
5.0	7.8	6,870	6,870
12.5	8.1	6,140	6,123
25.0	8.0	5,890	5,490
50.0	8.0	5,430	5,152

Fig. 16.4 Weibull distributions of the tensile strength of T800 mono filament

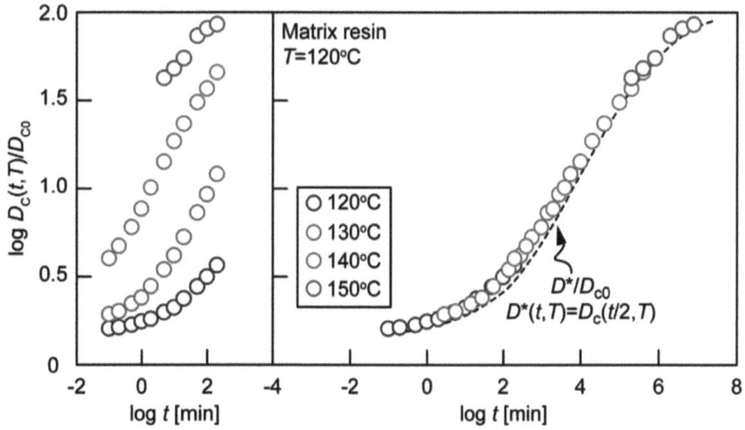

Fig. 16.5 Dimensionless creep compliance of matrix resin at $T = 120$ °C

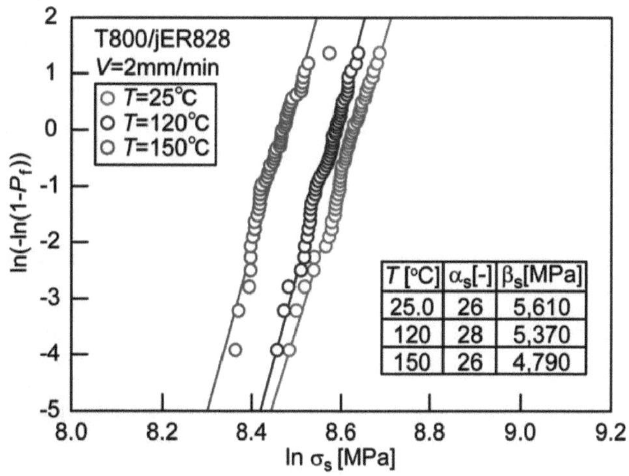

Fig. 16.6 Weibull distributions of static tensile strength of CFRP strand at three temperatures

$$\beta_c^*(L) = \left(\frac{L_0}{L}\right)^{\frac{1}{\alpha_c}} \beta_c(L_0) \tag{16.8}$$

The statistical static strength of carbon fiber T800 shows Weibull distribution based on the one dimensional link model because the shape parameter α_c is constant for various fiber lengths L and the experimental scale parameter β_c agrees well with the predicted one β^*_c for the same fiber length. The shape parameter of carbon fiber T800 can be fixed to $\alpha_c = 8.0$.

The dimensionless creep compliance D_c/D_{c0} measured at various temperatures is shown in the left of Fig. 16.5 and the long-term D_c/D_{c0} at $T = 120$ °C is obtained by shifting horizontally those at various temperatures as shown in the right of Fig. 16.5. The reference temperature and time are selected as $T_0 = 25$ °C and $t_0 = 1$ min in this study. Creep compliance at reference temperature and reference time D_{c0} is 0.33 (GPa)$^{-1}$. The dashed curve is the dimensionless viscoelastic compliance D^* of matrix resin under the constant strain rate at $T = 120$ °C.

16.3.3 Static Strength of CFRP Strand

In our previous paper [13], the static strength tests for CFRP strand were conducted at three levels of temperature, 25, 120, and 150 °C with cross-head speed 2 mm/min. The Weibull distributions of the static strength of CFRP strand at three

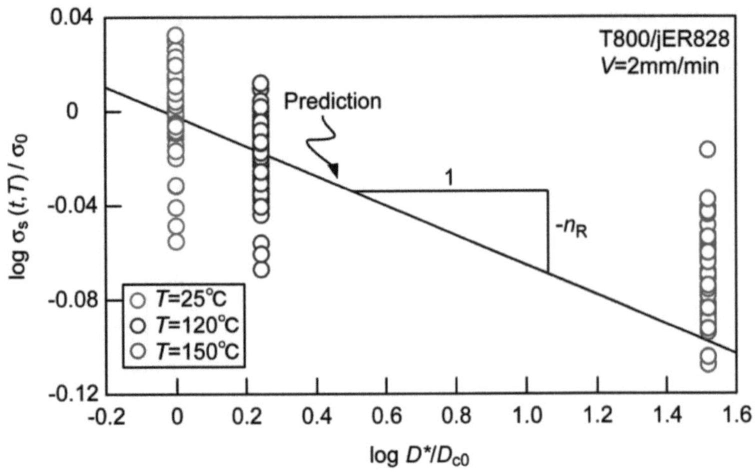

Fig. 16.7 Static strength of CFRP strand versus viscoelastic compliance of matrix resin

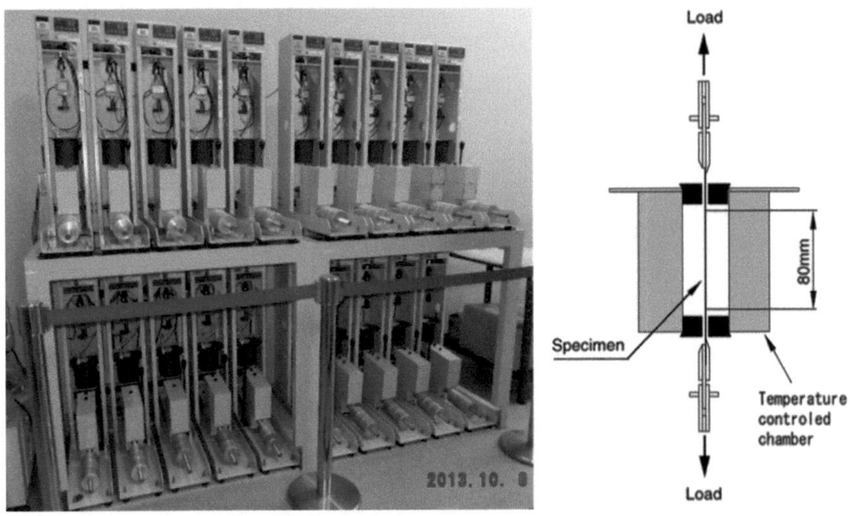

Fig. 16.8 Creep failure testing machine for the CFRP strand specimen

Table 16.2 Condition of creep failure test of T800 CFRP strands

Temperature (°C)	Stress (MPa)	Number of specimens
120	4,564[a]	26

[a]81.4 % of scale parameter of static strength at 25 °C

temperatures are shown in Fig. 16.6. α_s is the shape parameter and β_s is the scale parameter of CFRP strand in this figure. While the scale parameter decreases according to the temperature raise, the shape parameter keeps almost a constant value.

Figure 16.7 shows the dimensionless static strength of CFRP strand σ_s/σ_0 against the dimensionless viscoelastic compliance of matrix resin D^*/D_{c0} at the same time and temperature. The scale parameter of static strength at the reference temperature 25 °C and the reference failure time 1 min σ_0 is 5,608 MPa. The slope agrees well with the solid line predicted by Rosen's model.

16.3.4 Creep Failure Tests of CFRP Strand

Creep failure tests of T800 CFRP stand were conducted by using the specially designed creep failure testing machine shown in Fig. 16.8. The test condition is shown in Table 16.2. The results of the creep failure tests are shown in Fig. 16.9.

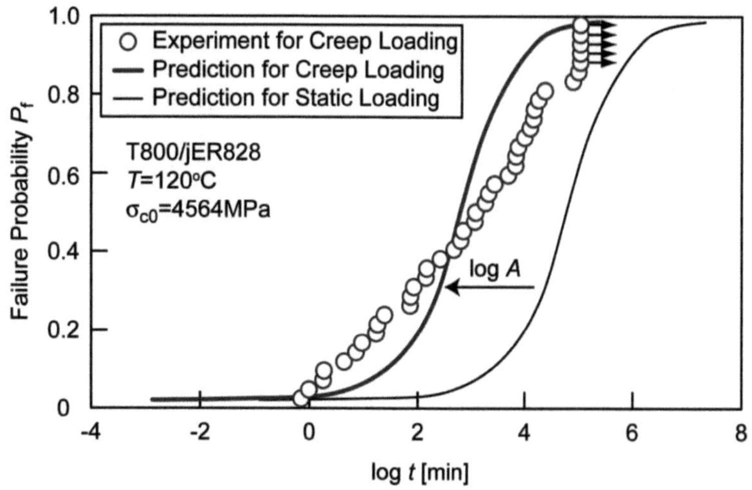

Fig. 16.9 Probability of creep failure time of T800 CFRP strand

Table 16.3 Parameters for statistical creep failure time prediction

Shape parameter of static strength of CFRP strand: α_s	26
Shape parameter of static strength of carbon fiber mono-filament: α_c	8.0
Viscoelastic parameter of matrix resin: n_R	0.063
Slope of viscoelastic compliance of matrix resin: m_R	0.28
Slope of static strength of CFRP strand against failure time: k_R	0.018
Logarithmic time shifting factor: log A	1.8

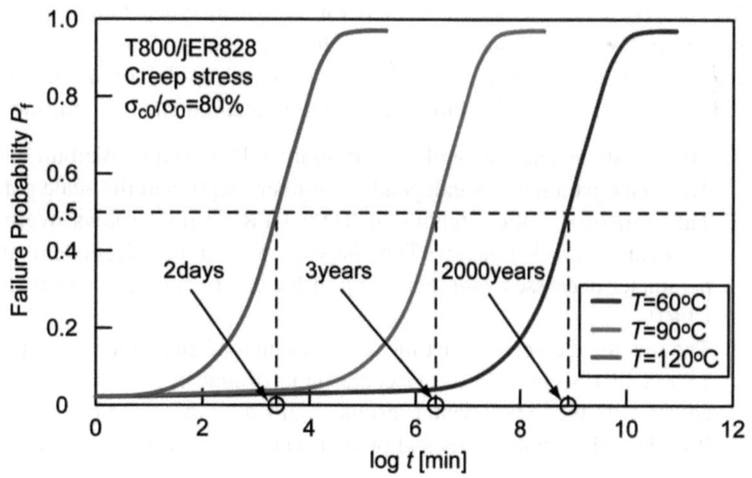

Fig. 16.10 Probability of creep failure time at various temperatures for a constant creep stress

16.3.5 Statistical Prediction of Creep Failure Time of CFRP Strand

As shown in Fig. 16.9, the statistical static failure time were obtained by Eqs. (16.3) and (16.4) and the statistical creep failure time were obtained by shifting horizontally the statistical static failure time by the amount log A calculated by substituting the parameters on Table 16.3 in Eqs. (16.5) and (16.6). The predicted statistical creep failure time agrees well with the experimental data.

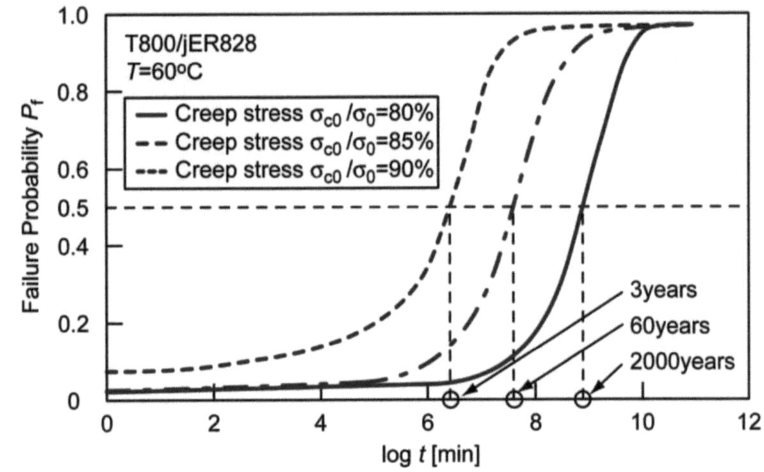

Fig. 16.11 Probability of creep failure time at various creep stresses for a constant temperature

16.4 Probability of Creep Failure Time

The probability of creep failure time at various temperatures for a constant creep stress is shown in Fig. 16.10 and that at various creep stress for a constant temperature is shown in Fig. 16.11 from substituting the parameters of Table 16.3 into Eqs. (16.3)–(16.6). It is cleared from these figures that the creep failure time of CFRP strand drastically change with temperature and creep stress.

16.5 Conclusions

We proposed the prediction method for statistical creep failure time under the tension loading along the longitudinal direction of unidirectional CFRP using the statistical static tensile strengths of carbon mono filament and CFRP strand and the viscoelasticity of matrix resin based on Rosen's model for unidirectional CFRP and Christensen's model for viscoelastic crack kinetics. The applicability of prediction method can be confirmed through the following steps.

1. The statistical static strength of carbon fiber T800 shows Weibull distribution based on the one dimensional link model. The shape parameter is independent of fiber length and the scale parameter decreases with increasing of fiber length.
2. The statistical static strength of T800 CFRP strand shows Weibull distribution based on Christensen's model of viscoelastic crack kinetics. The shape parameter is independent of the creep compliance of matrix resin. The scale parameter decreases with increasing of the creep compliance of matrix resin based on Rosen's model for unidirectional CFRP.
3. The statistical creep failure time at a constant load and temperature predicted using the statistical static tensile strengths of T800 CFRP strand and the viscoelasticity of matrix resin based on Christensen's model for viscoelastic crack kinetics agrees with the experimental results obtained by the creep rupture tests for T800 CFRP stands.
4. It is cleared from the statistical prediction that the creep failure time of CFRP strand drastically change with temperature and creep stress.

Acknowledgments The authors thank the Office of Naval Research for supporting this work through an ONR award with Dr. Yapa Rajapakse and Dr. Kenji Uchino. Our award is numbered to N62909-12-1-7024 and titled "Accelerated Testing Methodology for Long-Term Durability of CFRP Structures for Marine Use". The authors thank Professor Richard Christensen at Stanford University as the consultant of this project.

References

1. Aboudi J, Cederbaum G (1989) Analysis of viscoelastic laminated composite plates. Compos Struct 12:243–256
2. Gates T (1992) Experimental characterization of nonlinear, rate dependent behavior in advanced polymer matrix composites. Exp Mech 32:68–73

3. Miyano Y, Kanemitsu M, Kunio T, Kuhn H (1986) Role of matrix resin on fracture strengths of unidirectional CFRP. J Compos Mater 20:520–538
4. Miyano Y, McMurray MK, Enyama J, Nakada M (1994) Loading rate and temperature dependence on flexural fatigue behavior of a satin woven CFRP laminate. J Compos Mater 28:1250–1260
5. Miyano Y, Nakada M, McMurray MK, Muki R (1997) Prediction of flexural fatigue strength of CFRP composites under arbitrary frequency, stress ratio and temperature. J Compos Mater 31:619–638
6. Sullivan J (1990) Creep and physical aging of composites. Compos Sci Technol 39:207–232
7. Miyano Y, Nakada M, Cai H (2008) Formulation of long-term creep and fatigue strengths of polymer composites based on accelerated testing methodology. J Compos Mater 42:1897–1919
8. Nakada M, Miyano Y (2013) Advanced accelerated testing methodology for long-term life prediction of cfrp laminates. J Compos Mater (2013, on-line published). doi: 10.1177/0021998313515019
9. Miyano Y, Nakada M, Kudoh H, Muki R (1999) Prediction of tensile fatigue life under temperature environment for unidirectional CFRP. Adv Compos Mater 8:235–246
10. Miyano Y, Nakada M, Kudoh H, Muki R (2000) Determination of tensile fatigue life of unidirectional CFRP specimens by strand testing. Mech Time Depend Mater 4:127–137
11. Nakada M, Miyano Y, Kinoshita M, Koga R, Okuya T, Muki R (2002) Time–temperature dependence of tensile strength of unidirectional CFRP. J Compos Mater 36:2567–2581
12. Okuya T, Nakada M, Miyano Y (2013) Reliable test method for tensile strength in longitudinal direction of unidirectional carbon fiber-reinforced plastics. J Reinforced Plast Compos 32:1579–1585
13. Okuya T, Nakada M, Miyano Y (2013) Temperature dependence of longitudinal tensile strength in unidirectional CFRP. J Compos Mater. doi: 10.1177/0021998313512349
14. Christensen R, Miyano Y (2006) Stress intensity controlled kinetic crack growth and stress history dependent life prediction with statistical variability. Int J Fract 137:77–87

Chapter 17
Thermal Crystallinity and Mechanical Behavior of Polyethylene Terephthalate

Sudheer Bandla, Masoud Allahkarami, and Jay C. Hanan

Abstract Polyethylene Terephthalate's (PET) properties are particularly sensitive to its processing history. Processing impacts the extent of crystallinity. Mechanical stretching and melt flow history influence the extent and structure of crystalline domains in the semi-crystalline polymer. Typical processing parameters include the rate of cooling and the amount of stretch. Both influence crystallization differently. Isolating the contribution from stretch or thermal crystallization is valuable for identifying the relationship to mechanical properties. In this work, the influence of thermal crystallinity on the mechanical behavior of PET was observed. Annealed injection molded samples with thermal crystallinity were tested Young's modulus. Using the two-phase composite approach, mechanical behavior of injection molded semi-crystalline PET samples was modeled based on crystallinity from density. Crystallinity measured from different techniques did not always agree.

Keywords Polyethylene terephthalate (PET) • Crystallinity • Young's modulus • Density • Two-phase model

17.1 Introduction

The interdependence of chemistry, processing conditions, microstructure and properties is an important aspect in the engineering of polymers [1, 2]. The ability to control the microstructure through changing the process provides a means for selecting microstructure, and thereby affecting the properties. With wide spread application of semi-crystalline polymers (e.g. PE, PP, and PET), establishing the relationship between microstructure and properties is essential for efficient material use. Modeling the behavior of polymers, in particular for semi-crystalline polymers has received considerable attention. Numerous micromechanical models were suggested considering the semi-crystalline polymer as a composite material, with amorphous and crystalline domains as different phases [3–5]. A number of intrinsic properties like: Young's modulus, elastic and plastic deformations, and texture evolution were studied through this approach [6, 7]. Hitherto there is no single model that is applicable in general for polymers and describes their mechanical properties, due to factors like chemistry, crystallization kinetics, and type of crystallization. For example, Patel and Phillips [8] found that microstructure parameters including the spherulitic radius and lamellar thickness affect the modulus of polyethylene (PE). However, the increase in modulus was observed only up to an aspect ratio of 1,000 (ratio of length to the thickness of the crystal), from controlled crystallization (spherulitic) studies.

Polyethylene terephthalate (PET) is an aromatic semi-crystalline polymer with applications in the field of textiles, packaging, and engineering molding. Crystallinity and molecular orientation affect the modulus of PET [9]. On account of its slow crystallization rate and strain sensitive characteristics, PET can be engineered to have a distinctive microstructure, and thereby distinctive properties. Therefore, correlating the mechanical behavior of PET with its microstructure will help in effectively designing the manufacturing processes. In 1958, Lyons [10] estimated the theoretical modulus for PET at 146 GPa along the molecular axis, based on bond stretching. Nevertheless, the maximum modulus reported was found to be

S. Bandla • M. Allahkarami • J.C. Hanan (✉)
School of Mechanical and Aerospace Engineering, Oklahoma State University, Tulsa, OK 74106, USA
e-mail: Jay.Hanan@okstate.edu

H.J. Qi et al. (eds.), *Challenges in Mechanics of Time-Dependent Materials, Volume 2: Proceedings of the 2014 Annual Conference on Experimental and Applied Mechanics*, Conference Proceedings of the Society for Experimental Mechanics Series,
DOI 10.1007/978-3-319-06980-7_17,

an order of magnitude lower (15.5 GPa for crystalline oriented fibers [11]) than the theoretical value. Rastogi et al. [12] have expanded the two-phase model to a three-phase model, by considering a Rigid Amorphous Phase (RAF) between the amorphous and crystalline phases. They studied the mechanical behavior of PET from glass- and melt-crystallization, under uniaxial compression. Gueguen et al. [13] have proposed a new continuum based three-phase model for estimating the modulus of PET.

Modeling properties requires comprehensive microstructure information (crystal size and percent crystallinity) of the polymer. For the two-phase model, defining the lamellar thickness and length is important and can be obtained through techniques like light scattering, X-ray scattering, thermal analysis [14], and electron microscopy [15]. Crystallinity measurements of PET is subject to the process history and the technique used [16]. Crystallinity measured using XRD, will also include the contribution from molecular orientation. In our earlier work [17], Young's modulus of PET films with different percent crystallinity achieved through stretch crystallization, were reported. The current work focuses on comparing the experimental modulus with micromechanical predictions based for thermally crystallized samples, based on the microstructure information.

17.2 Materials and Methods

Commercially available polyethylene terephthalate (PET) of M_w 37,000 g/mol (0.8 dL/g I.V.), was used for the preparation of two different bulk specimens, as reported earlier [18, 19]. Tensile bars were injection molded using an oil cooled injection molding system, following ASTM D638 Type I specimen (dog bone) dimensions. These were annealed at 190 °C for 10 min and quenched to increase the crystallinity. Amorphous PET cylinders were prepared using a water cooled injection molding system, equipped with a custom mold. All the samples (as received tensile bar, annealed bar, and amorphous cylinder) were tensile tested according to ASTM D638, at 5 mm/min extension rate on a universal testing machine. Multiple samples (at least three) were tested for each condition.

17.3 Density Measurement

Relative densities can be obtained based on Archimedes' principle, using the following equation.

$$\rho = \frac{w}{w - \overline{w}} \times \rho_o \tag{17.1}$$

Where, w—weight of the sample in air, $\overline{w}$—the weight in liquid, and ρ_o—the density of the liquid.

Percent crystallinity was estimated (Table 17.1) using the relative density of PET, by considering the amorphous and crystalline densities at 1.335 and 1.455 g/cc [20].

17.4 X-Ray Analysis

Diffraction patterns were collected on a Bruker D8 Discover micro-X-ray diffraction system, equipped with a 2D area detector. Cu Kα radiation (λ = 1.54184 Å) was used for this analysis. For percent crystallinity measurements, 0.8 mm collimator was used in reflectance, with a scan time of 60 s. Figure 17.1, shows the 2D and 1D diffraction patterns for amorphous and thermally crystallized bulk specimens. With a triclinic crystal structure, PET exhibit four primary reflections from (010), $(\overline{1}10)$, (100) and $(\overline{1}05)$ planes (17.5°, 22.5°, 25.6° and 42.6° 2θ with Cu Kα X-rays) [20]. In the diffraction pattern shown for crystallized specimens, secondary peaks such as $(0\overline{1}1)$ and $(\overline{1}\,\overline{1}3)$ were also observed. Optically

Table 17.1 Measured average density and corresponding crystallinity for PET samples

Sl. no.	Description	Density (g/cc)	Crystallinity (%)
1	Cylinder	1.335	0
2	Tensile bar	1.361	24
3	Annealed tensile bar	1.387	45

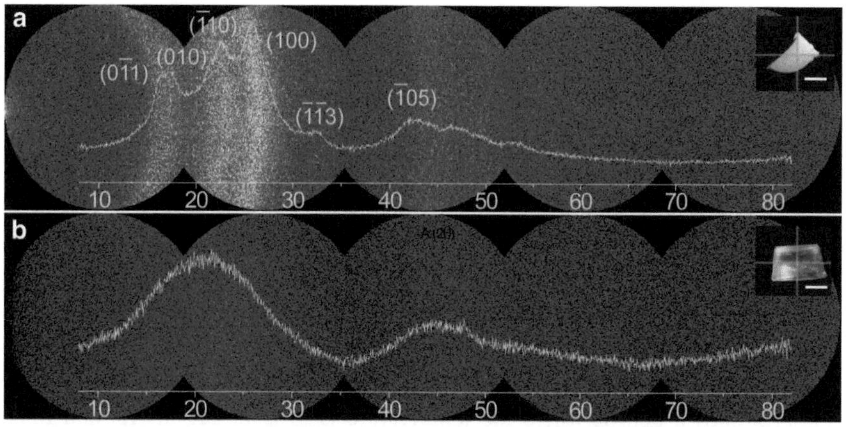

Fig. 17.1 2D and 1D diffraction patterns from (**a**) semi-crystalline and (**b**) amorphous PET, with photographs of the respective specimens shown in the inset (scale—2 mm)

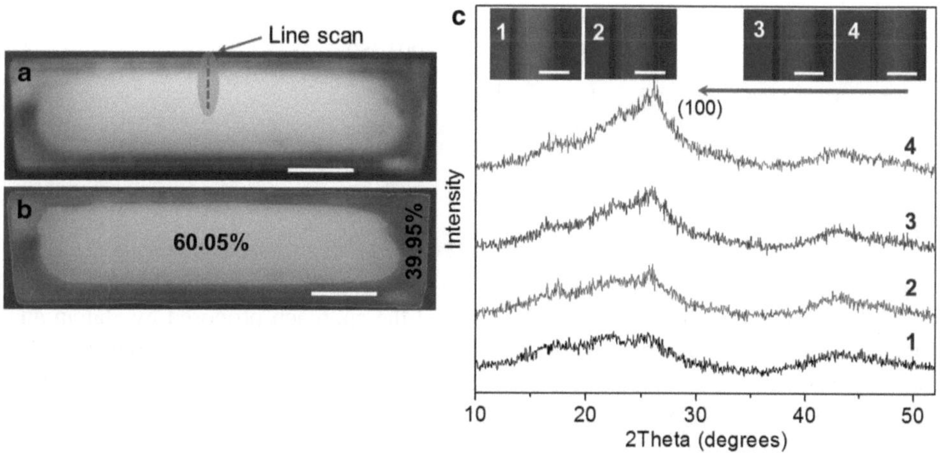

Fig. 17.2 (**a**) Cross-section of the tensile bar with the line scan area pointed, (**b**) distribution of the crystalline and amorphous regions, and (**c**) 1D diffraction patterns along the thickness of the bulk specimen, showing the transition from amorphous to crystalline microstructure (scale—2 mm)

transparent bulk samples, exhibit an amorphous hallo about the 20° 2θ angle. Crystallinity from the diffraction patterns was determined using Eq. 17.2, shown below. Sherrer equation (shown in Eq. 17.3) can be used to estimate the size of the crystallite (β_{hkl}—full width at half maximum for (*hkl*) peak) [21]. By resolving the respective peaks from the crystalline sample, crystallite size with respect to (100) direction was determined to be 2.5 nm (FWHM—2.13° 2θ), which can be lower compared to the actual crystal size, due to peak broadening caused by lattice deformation.

$$X_c\,\% = \frac{A_c}{A_a + A_c} \times 100\,\% \tag{17.2}$$

$$D_{hkl} = \frac{0.89}{\beta_{hkl}\cos\theta} \tag{17.3}$$

Diffraction analysis across the cross-section of the tensile bar before annealing, using a 50 μm beam size (300 s scan time), shows the presence of a crystalline core, surrounded by an amorphous outer layer. Changes in the diffraction peaks, before and after annealing, confirm the change in microstructure. The presence of the diffraction peak corresponding to the (100) plane (shown in Fig. 17.2), suggest the existence of flow induced orientation from injection molding.

Table 17.2 PET crystallinity measured from DSC

Sl. no.	Description	Crystallinity (%)
1	Tensile bar outer layer	14
2	Tensile bar core	33
3	Annealed bar	35

17.5 Differential Scanning Calorimetry (DSC)

Differential scanning calorimetry[1] of as received and annealed tensile bars was also used to analyze for percent crystallinity. For this, percent crystallinity shown in Table 17.2, was determined using Eq. 17.5, where the heat of fusion for a completely crystalline sample (ΔH_c^o) was 140.1 J/g [22]. For the as received tensile bars, the outer layer and inner layer were tested separately. Using the melting point of PET from the annealed sample, the lamellar thickness (I_c) [14] was determined, using Eq. 17.5 shown below, where the temperature (T_m) is in kelvin (K) and thickness is in nanometers. The PET melting temperature was 249.6 °C, based on the DSC thermograms.

$$X_c = \left[\frac{\Delta H_{fusion} - \Delta H_{cold\_crystallization}}{\Delta H_c^o}\right] \tag{17.4}$$

$$\frac{1}{T_m} = \frac{1}{594.5} + \frac{0.001759}{I_C} \tag{17.5}$$

17.6 Micromechanical Modeling

Considering the annealed sample as composite where the amorphous phase is reinforced by the crystalline phase, modulus of the annealed sample was predicted as shown in Table 17.3, using the approach proposed by Halpin and Kardos [3]. The Halpin–Tsai micromechanical model is based on aspect ratio (*l/d*) of the reinforcement. Since the crystallite is spherical with a folded lamellar structure, length of the crystalline phase length (*l*) is equal to the depth (*d*). The crystal is the reinforcement phase and the modulus of the composite was predicted using the following equations:

$$\frac{E_{11}}{E_m} = \frac{1 + 2A_f\eta\phi}{1 - \eta\phi} \tag{17.6}$$

$$\text{Where}, \eta = \frac{E_r - 1}{E_r + 2A_f}, E_r = \frac{E_R}{E_m}, A_f = \frac{l}{d}, \text{and } \phi \text{ is volume fraction.} \tag{17.7}$$

The modulus of the amorphous phase was measured to be 2.5 GPa [19]. For the crystalline phase it was considered as 146 GPa [10]. The percent crystallinity values obtained from density measurements were converted to volume fraction for calculating the modulus.

17.7 Discussion

Figure 17.3, shows the difference in the deformation behavior of PET with increase in crystallinity. Modulus and the tensile strain increased with increase in crystallinity. Decrease in the failure strain of PET after annealed can be attributed to the limited chain mobility from spherulitic crystallite formation.

The numerical model predictions are normally based on the assumption that the reinforcements are oriented along the loading axis. Here there is no orientation and this can explain some of the deviation between the predicted and the experimental values as seen above. In the case of spherulitic crystallization, the crystal will be three-dimensional and

[1] Using a Q-2000 system from TA Instruments.

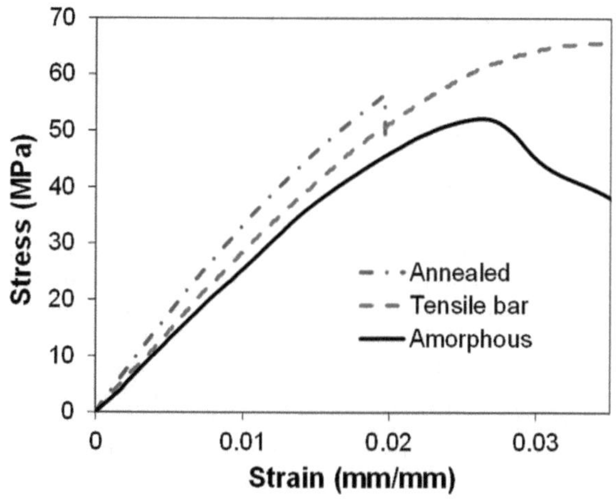

Fig. 17.3 Comparison of stress–strain curves for PET samples

Table 17.3 Comparison of the experimental modulus with numerical predictions using Halpin–Tsai model

Sl. no.	Description	Crystallinity from density (%)	Crystallinity from XRD	Crystallinity from DSC	E_{exp} (GPa)	Standard deviation (GPa)	Halpin–Tsai model, E_{11} (GPa)
1	PET cylinder	0	–	–	2.5	0.1	–
2	Tensile bar	24	–	–	2.7	0.2	4.6
3	Annealed bar	45.5	43.5	34	3.3	0.3	7.8

isotropic, giving rise to a random microstructure with less increase in the modulus. Free volume caused by the crystal amorphous interphase may also reduce the performance.

The measured Young's modulus of the as received tensile bars was 2.7 GPa [18], and for the annealed samples it was 3.3 GPa. Comparing the amorphous and semi-crystalline PET, the increase in crystallinity shows little modulus improvement, even with the presence of a crystalline core. This might be due to the type of crystallization in the injection molded bar. Annealed bars show 43.5 % crystallinity from X-ray diffraction measurements, however DSC measurements show only 35 % crystallinity. This highlights a different type of sensitivity of the measuring techniques.

17.8 Conclusions

Polyethylene terephthalate with different levels of thermal crystallinity were obtained from injection molded samples. Mechanical testing showed that Young's modulus increases with increase in crystallinity, but the increase results in a lower failure strain. Using a two-phase micromechanical model, a poor prediction of the modulus of the tensile bar for different percent crystallinity was obtained. The deviation of the predicted modulus from experimental values, highlight the need for more accurate microstructure information. Further, the current study highlights differences in the amount of crystallinity measured through different techniques.

Acknowledgments This work was possible partially through OSU foundation grants to the Helmerich Research Center and industry sponsorship. Authors would like to acknowledge. Ovation Polymers for the injection molding of the tensile bars, and Dr. Ranji Vaidyanathan for use of the DSC Q-2000 system.

References

1. Krevelen DWV, Nijenhuis KT (2009) Properties of polymers, 4th edn. Elsevier, Amsterdam
2. Bandla S (2010) Evaluation and stability of PET resin mechanical properties. MS, Mechanical and Aerospace Engineering, Oklahoma State University, Stillwater

3. Halpin JC, Kardos JL (1972) Moduli of crystalline polymers employing composite theory. J Appl Phys 43:2235–2241
4. Andrews EH (1974) Morphology and mechanical properties in semicrystalline polymers. Pure Appl Chem 39:179–194
5. Breese DR, Beaucage G (2004) A review of modeling approaches for oriented semi-crystalline polymers. Curr Opin Solid State Mater Sci 8(12):439–448
6. Lee BJ, Parks DM, Ahzi S (1993) Micromechanical modeling of large plastic deformation and texture evolution in semi-crystalline polymers. J Mech Phys Solids 41(10):1651–1687
7. van Dommelen JAW, Parks DM, Boyce MC, Brekelmans WAM, Baaijens FPT (2003) Micromechanical modeling of the elasto-viscoplastic behavior of semi-crystalline polymers. J Mech Phys Solids 51(3):519–541
8. Patel J, Phillips PJ (1973) The Young's modulus of polyethylene. J Polym Sci B 11:771–776
9. Viswanathan A, Wiff DR, Adams WW (1976) An investigation of structure-property correlation in polyethylene terephthalate films. Air Force Materials Laboratory, Wright-Patterson Air Force Base
10. Lyons WJ (1958) Theoretical values of the dynamic stretch moduli of fiber-forming polymers. J Appl Phys 29:1429–1433
11. Ballou JW, Smith JC (1949) Dynamic measurements of polymer physical properties. J Appl Phys 20:493–502
12. Rastogi R, Vellinga WP, Rastogi S, Schick C, Meijer HEH (2004) The three-phase structure and mechanical properties of poly(ethylene terephthalate). J Polym Sci B Polym Phys 42:2092–2106
13. Gueguen O, Ahzi S, Makradi A, Belouettar S (2010) A new three-phase model to estimate the effective elastic properties of semi-crystalline polymers: application to PET. Mech Mater 42(1):1–10
14. Huyskens P, Groeninckx G, Vandevyvere P (1990) What rules the melting point of semi-crystalline polymers? Bulletin des Sociétés Chimiques Belges 99:1011–1017
15. Michler, G. H. (2008). Semicrystalline Polymers. Electron Microscopy of Polymers. H. Pasch. Germany, Springer Berlin Heidelberg: 295–327
16. Farrow G, Ward IM (1960) Crystallinity in poly (ethylene terephthalate): a comparison of X-ray, infra-red and density measurements. Polymer 1:330–339
17. Sudheer B, Masoud A,Jay C. Hanan. Out-of-plane orientation and crystallinity of biaxially stretched polyethylene terephthalate. Powder Diffraction, available on CJO2014. doi:10.1017/S0885715614000190
18. Bandla S, Hanan J (2012) Microstructure and elastic tensile behavior of polyethylene terephthalate–exfoliated graphene nanocomposites. J Mater Sci 47:876–882
19. Bandla S, Hanan JC (2012) Manufacturing tough amorphous thermoplastic-graphene nanocomposites. In: NanoTech, Santa Clara
20. de Daubeny RP, Bunn CW, Brown CJ (1954) The crystal structure of polyethylene terephthalate. Proc R Soc Lond A Math Phys Sci 226:531–542
21. Tong-yin Y, Hai-shan B, Jia-cong H, Wei Z, Qiao-ying G (1981) The double melting peaks of polyethylene terephthalate. Acta Polymerica Sinica 1:461–467
22. Mehta A, Gaur U, Wunderlich B (1978) Equilibrium melting parameters of poly(ethylene terephthalate). J Polym Sci B Polym Phys 16:289–296

Chapter 18
Effect of UV Exposure on Mechanical Properties of POSS Reinforced Epoxy Nanocomposites

Salah U. Hamim, Kunal Mishra, and Raman P. Singh

Abstract Epoxy nanocomposites containing 5 wt% glycidyl Polyhedral Oligomeric Silsesquioxane (POSS) by weight is prepared. Elastic and viscoelastic properties of the neat epoxy and epoxy/POSS nanocomposite has been studied using nanoindentation technique. Elastic modulus of the polymer systems are determined using modified Oliver–Pharr method taking creep rate into account. Creep compliance properties of the studied systems are determined by fitting a standard linear solid model to the creep plot. Both neat epoxy and epoxy/POSS nanocomposite showed an increase in elastic modulus and a decrease in creep compliance as a function of UV exposure time; although, the effect of UV exposure is more pronounced in case of epoxy/POSS nanocomposite. It is speculated that the excess epoxide groups present in the glycidyl POSS is going through further crosslinking during the UV exposure leading to the observed higher rate of property change compared to neat epoxy.

Keywords Glycidyl POSS • Nanocomposite • UV exposure • Elastic modulus • Creep compliance

18.1 Introduction

Epoxy polymers has gained wide variety of applications for its excellent mechanical and thermal properties. To name a few, aerospace, automotive, marine, packaging, micro-electric industries, etc. has been using epoxy polymers in various ways over the last few years. Along with other beneficial characteristics, epoxy polymers also posses excellent chemical resistance, high electrical insulation, and high fatigue resistance making it ideal for different applications.

Environmental degradation can highly affect epoxy polymers. Ultra-violent radiation from sunlight can degrade the mechanical, chemical and functional properties of polymer, and in turn limit its application for outdoor applications. In recent years, considerable amount of research has been conducted to understand the photodegradation mechanism in polymer materials [1–5]. Various inorganic nanoparticles with good barrier properties, such as, clays, graphenes, etc. has been used to mitigate the effect of UV exposure, and at the same time to increase the baseline mechanical properties of the polymers. However, to use inorganic nanoparticles in nanocomposites compatibility between nanoparticles and polymer has to be improved by surface modification using organic substances. Recently, a new hybrid nanoparticle called Polyhedral Oligomeric Silsesquioxane (POSS) came into existence, which has an inorganic cage with functional groups at the corners of the cage [6–10]. The inorganic cage gives the structural rigidity necessary for baseline property improvement of the polymer, while the functional groups provides linkage with the organic crosslinks. Various studies has been conducted to explore the mechanical property improvement of epoxy by reinforcing it with POSS nanoparticles [11–17].

In this paper, the effect of UV exposure on the mechanical properties of POSS reinforced epoxy nanocomposite is explored. Using nanoindentation technique the surface degradation on elastic and viscoelastic properties caused by the UV exposure has been studied.

S.U. Hamim • K. Mishra • R.P. Singh (✉)
School of Mechanical and Aerospace Engineering, Oklahoma State University, Stillwater, OK 74075, USA
e-mail: raman.singh@okstate.edu

H.J. Qi et al. (eds.), *Challenges in Mechanics of Time-Dependent Materials, Volume 2: Proceedings of the 2014 Annual Conference on Experimental and Applied Mechanics*, Conference Proceedings of the Society for Experimental Mechanics Series,
DOI 10.1007/978-3-319-06980-7_18,

18.2 Experimental Details

Neat epoxy and epoxy/POSS nanocomposite specimens were subjected to 24-h cycle of 1 W/m^2 UV radiation at 60 °C for characterizing their environmental durability. The UV radiation was simulated on a QUV/se Accelerated Weathering Tester (Q-Panel Lab Products, Cleveland, Ohio).

18.2.1 Materials

Commercially available diglycidyl ether of bisphenol F (DGEBF) based epoxy polymer EPON 862 (Miller–Stephenson Chemical Company, Inc., Dunbury, Connecticut) with a moderately reactive low viscosity aliphatic amine based curing agent Epikure 3274 was used for this study. Polymer and the curing agent was mixed at a ratio of 100:40 by weight, and was cured for 24 h at room temperature. Post curing of 6 h was conducted at 121 °C. 5 wt% glycidyl POSS (Fig. 18.1) (Hybrid plastics, Hattiesburg, Mississippi) nanocomposite was made by adding 5 wt% POSS, and the same curing cycle for epoxy was followed. For nanoindentation testing, polymer sheets were cut into 20 × 20 mm squares and polished to reduce the surface roughness using a mechanical polisher.

18.2.2 Nanoindentation

An MTS Nanoindenter XP was used to perform the nanoindentation creep experiments on the specimens. A Berkovich pyramidal tip (equivalent half angle, $\alpha = 70.\ 3°$) was used to carry out the nanoindentation experiment in a load controlled environment with 1 mN as the set load. Identical loading and unloading rate of 0.2 mN/s was used for the experiment and a holding segment of 60 s was used to capture the creep behavior of the materials.

The contact stiffness S, extracted from the unloading curve using Oliver–Pharr [18] method was modified using Eq. (18.1), which was proposed by Feng and Ngan [19] to take creep rate into account for calculating elastic modulus.

$$\frac{1}{S_e} = \frac{1}{S} + \frac{\dot{h}_h}{\dot{F}_u} \tag{18.1}$$

$$h_c = h_{max} - \varepsilon \frac{P_{max}}{S_e} \tag{18.2}$$

where, S_e was the modified contact stiffness, and $\dot{h}_h$ and $\dot{F}_u$ were creep rate and unloading rate, respectively. Contact area, h_c was modified using Eq. (18.2) taking modified contact stiffness, S_e into account.

18.2.3 Material Model

Viscoelastic materials are often represented by using a combination of spring and dashpot system. In this paper, a three element model called standard linear solid (SLS) model was used to represent the creep behavior of the materials. SLS

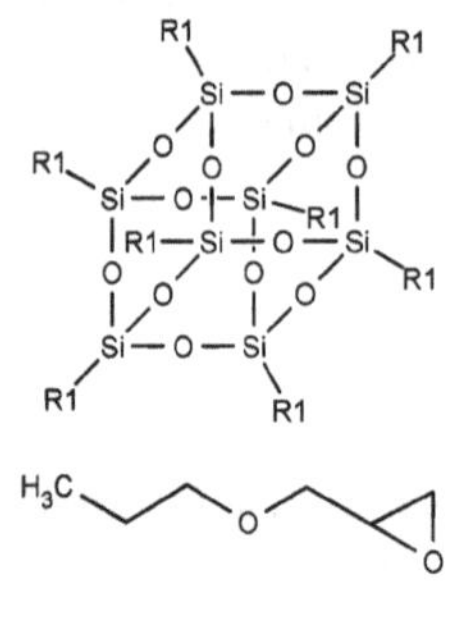

Fig. 18.1 Chemical structure of glycidyl POSS

model was chosen to represent the material behavior because it involved simple mathematical analyses with fewer number of fitting parameters. The basic equation corresponding to the creep compliance for a standard linear solid model is the following

$$J(t) = \frac{1}{E_1} + \frac{1}{E_2}\left[1 - exp\left(\frac{-t}{\tau_c}\right)\right]; \tau_c = \frac{\eta}{E_2} \tag{18.3}$$

where, E_1, E_2 and η are the three parameters representing the springs and the dashpot of SLS model. Although it is a simple model, it is capable of representing both the shot-term (instantaneous) and long-term behavior of the material.

During the nanoindentation experiment a maximum force, P_0 was applied in a finite time, t_0, at a constant rate, r = dP/dt. The relationship between the indenter load and displacement becomes,

$$h^2(t) = \frac{2}{\pi tan\alpha}\left[rt_0\left(\frac{1}{E_1} + \frac{1}{E_2}\right) - \frac{r\tau_c}{E_2}exp\frac{-t}{\tau_c}\left(exp\frac{t_0}{\tau_c} - 1\right)\right] \tag{18.4}$$

Equation (18.4) can be simplified to fit the creep data (h-t) as

$$h^2(t) = C_1 + C_2 exp\left(\frac{-t}{\tau_c}\right) \tag{18.5}$$

18.3 Results

Figure 18.2 shows the stacked load vs. displacement plots offsetted in the Y-axis for both the epoxy and epoxy/POSS systems exposed to different hours of UV exposure. The nature of the nanoindentation experiment was load controlled, and hence, displacements were different in case of the studied material systems. For unexposed epoxy and epoxy/POSS system significant difference in material behavior was observed. It was evident that incorporation of glycidyl POSS made the material system more compliant compared to the epoxy system.

As a result of UV exposure, neat epoxy had undergone subtle change in displacement. On the contrary, UV exposure showed higher impact on the displacement behavior of the epoxy/POSS system. It was observed that, both the displacement values (displacement at maximum load and displacement during the holding segment) were affected due to UV exposure. Figure 18.3 shows the average creep rate for the studied systems during the holding segment of the nanoindentation experiment. The average creep rate in neat epoxy did not change as a function of UV exposure. However, the creep rate was severely affected for epoxy/POSS system. After 48 h of exposure, the creep rate had dropped approx. 67 % for the epoxy/POSS nanocomposite.

Figures 18.4 and 18.5 shows the elastic modulus and creep compliance for the neat epoxy and nanocomposite systems. It was observed that elastic modulus increased almost linearly for epoxy/POSS system as a function of UV exposure; whereas, modulus remained nearly constant for the neat epoxy system. It was also observed that, reinforcing epoxy with glycidyl POSS resulted in a more compliant material response when compared to unexposed neat epoxy. However, after 48 h of UV exposure, the epoxy/POSS nanocomposite specimen showed higher elastic modulus compared to neat epoxy specimen.

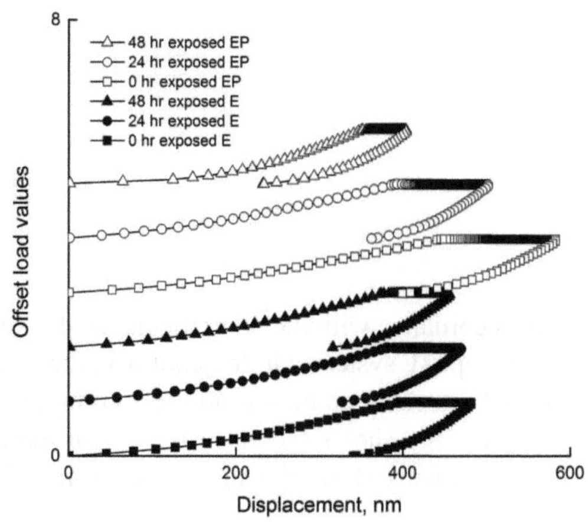

Fig. 18.2 Nanoindentation load vs. displacement plots for the studied systems

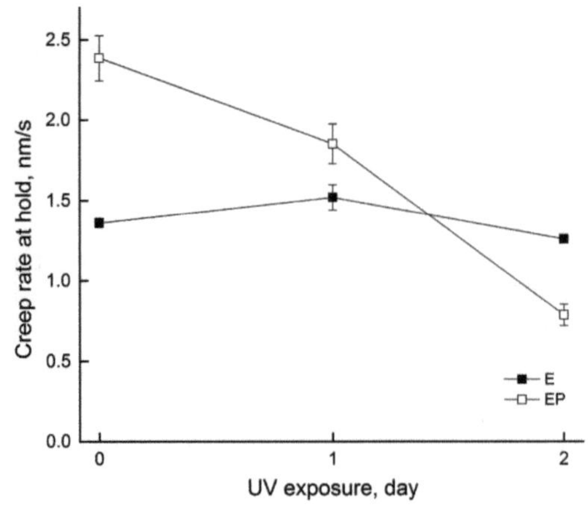

Fig. 18.3 Average creep rate for the studied systems

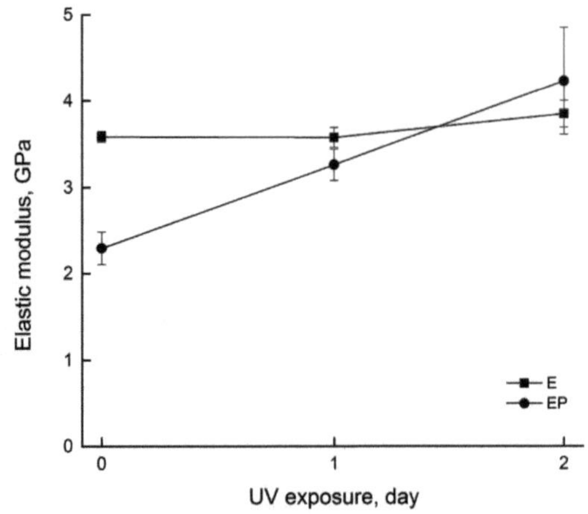

Fig. 18.4 Elastic modulus of the epoxy and nanocomposite systems as a function of UV exposure

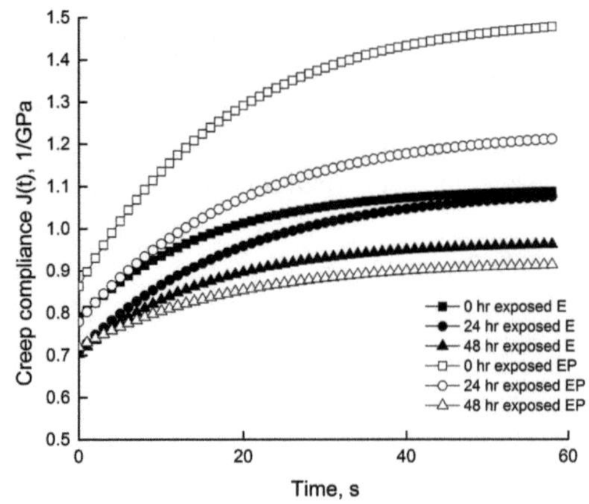

Fig. 18.5 Creep compliance of the epoxy and nanocomposite systems as a function of UV exposure

In accordance with the other mechanical responses, creep compliance showed little change as a function of UV exposure for neat epoxy system. On the contrary, the creep compliance changed dramatically for epoxy/POSS nanocomposite. The material system became less and less compliant as a function of UV exposure time.

Figure 18.6 shows the viscosity (η) for the SLS model as a function of UV exposure for both systems. For the epoxy system viscosity showed insignificant change. However, it increased significantly for epoxy/POSS system.

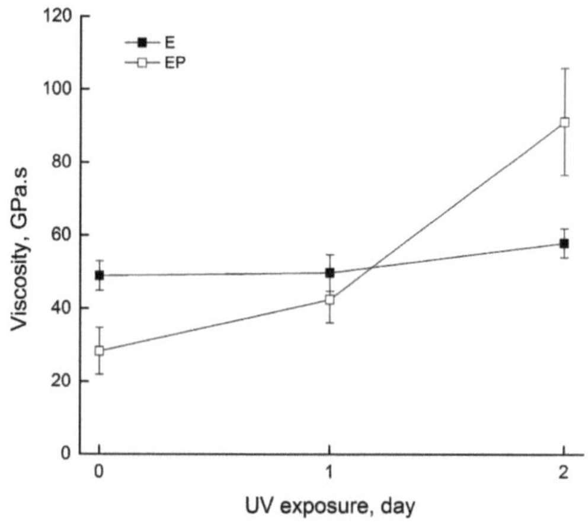

Fig. 18.6 Viscosity of the epoxy and nanocomposite systems as a function of UV exposure

18.4 Conclusion

The effect of UV exposure on the elastic and viscoelastic properties of POSS reinforced epoxy system was studied using nanoindentation technique. Elastic modulus was found by using a modified Oliver–Pharr method proposed by Feng and Ngan, which takes the creep rate into account. Creep measurements taken during the hold segment of the nanoindentation experiment was fitted with three parameter SLS model to find the creep compliance and viscosity of the material systems.

Nanoindentation displacement measurements showed that incorporation of POSS in epoxy made the material more compliant. This observation was well supported by the elastic and viscoelastic properties extracted from the nanoindentation experiments. For the observed properties, it was seen that UV exposure did not significantly affect the neat epoxy. On the other hand, POSS/epoxy system showed huge change after only 48 h of UV exposure. Elastic modulus and viscosity (SLS model) was found to have increased as a result of UV exposure, and creep compliance and creep rate during the hold decreased as function of exposure time.

It is speculated that, the presence of extra epoxide groups in glycidyl POSS may have stayed unreacted making the POSS/epoxy nanocomposite less crosslinked compared to the neat epoxy. Less crosslinked structure could mean higher mobility of the chains while under stress, resulting in a more compliant material. After 48 h UV exposure, however, the elastic modulus and viscosity of the POSS/epoxy system was found higher compared to neat epoxy. Creep compliance and the creep rate values also dropped below the neat epoxy property values. The unreacted epoxide groups that was present in the unexposed nanocomposite system may have undergone further crosslinks due to UV exposure making the structure more rigid for polymer chains to move. As a result, it is possible for the system to show less compliant displacement response reflected in the material properties. However, further investigation is required to ascertain unreacted epoxide groups' presence, and whether or not these were contributing to the increase in crosslink density in the POSS/epoxy nanocomposite leading to the observed properties.

Acknowledgements We gratefully acknowledge that this work is funded in part or fully by a grant through the Oklahoma Nanotechnology Applications Project (ONAP) (Grant no. O9–20) and NASA Experimental Program to Stimulate Competitive Research (EPSCOR) (Grant no. NNXO9AP68A).

References

1. Barkoula NM, Gkikas G, Makri A, Matikas TE, Paipetis A (2010) Effect of the environmental degradation on the viscoelastic response of nano modified epoxies and ctrps. In: 14th European conference on composite materials (ECCM14), Budapest, Hungary
2. Tse KCC, Ng FMF, Yu KN (2006) Photo-degradation of padc by uv radiation at various wavelengths. Polym Degrad Stab 91:2380
3. Kumar BG, Singh RP, Nakamura T (2002) Degradation of carbon fiber-reinforced epoxy composites by ultraviolet radiation and condensation. J Compos Mater 36:2713
4. Woo RSC, Chen YH, Zhu HG, Li J, Kim JK, Leung CKY (2007) Environmental degradation of epoxy-organoclay nanocomposites due to uv exposure and moisture attack: part i. photodegradation. Compos Sci Technol 67:3448
5. Chin JW, Nguyen T, Aouadi K (1997) Effects of environmental exposure on fibre-reinforced plastic (frp) materials used in construction. JCTR 19:205

6. Fina A, Monticelli O, Camino G (2010) Poss-based hybrids by melt/reactive blending. J Mater Chem 20:9297
7. Phillips SH, Haddad TS, Tomczak SJ (2004) Developments in nanoscience: polyhedral oligomeric silsesquioxane (poss)-polymers. Curr Opin Solid State Mater Sci 8:21
8. Achilleos DS, Vamvakaki M (2010) End-grafted polymer chains onto inorganic nano-objects. Materials 20(3):1981
9. Cordes DB, Lickiss PD, Rataboul F (2010) Recent developments in the chemistry of cubic polyhedral oligosilsequioxanes. Chem Rev 110:2081
10. Kuo SW, Chang FC (2011) Poss related polymer nanocomposites. Prog Polym Sci 36:1649
11. Jones IK, Zhou YX, Jeelani S, Mabry JM (2008) Effect of polyhedral-oligomeric-sil-sesquioxanes on thermal and mechanical behavior of sc-15 epoxy. Express Polym Lett 2:494
12. Zhao JQ, Fu Y, Liu SM (2008) Polyhedral oligomeric silsesquioxane (poss)-modified thermoplastic and thermosetting nanocomposites: a review. Polym Polym Compos 16:483
13. Li B, Zhang Y, Wang S, Ji J (2009) Effect of poss on morphology and properties of poly(2,6-dimethyl-1,4-phenylene oxide)/polyamide 6 blends. Eur Polym J 45:2202
14. Mishra K, Singh RP (2010) The mechanics of polymer nanocomposites. In: Nanofocus, Oklahoma City, OK
15. Mishra K, Singh RP (2010) Use of polyhedral oligomeric silsequioxanes(poss) to enhance the fracture toughness of thermosetting resin. In: SEM Technical Conference, Indianapolis, IN
16. Mishra K, Singh RP (2010) Mechanical and chemical characterization of poss incorporated dgeba resin. In: 2010 SAMPE Fall Technical Conference and Exhibition, Salt Lake City, UT
17. Mishra K, Vorrey S, Singh RP (2010) Use of polyhedral oligomeric silsequioxanes to enhance the fracture toughness of thermosetting resin. In: SAMPE 2010 Conference and Exhibition "New Materials and Processes for a New Economy" Seattle, WA
18. Oliver WC, Pharr GM (1992) An improved technique for determining hardness and elastic modulus using load and displacement sensing indentation experiments. J Mater Res 7(6):1564
19. Feng G, Ngan AHW (2002) Effects of creep and thermal drift on modulus measurement using depth-sensing indentation effects of creep and thermal drift on modulus measurement using depth-sensing indentation. J Mater Res 17(3):660

Chapter 19
Overcoming Challenges in Material Characterization of Polymers at Intermediate Strain Rates

William J. Briers III

Abstract Simulations of dynamic loading on polymers often require rate dependent, large-strain material test data to sufficiently calibrate material laws such as nonlinear viscoelastic–viscoplastic. Physical testing at intermediate strain rates creates many challenges because the measurements are susceptible to distortion due to factors including: test specimen variations, fixtures, hydraulic actuator control, instrumentation complexity, data acquisition, electrical noise and post processing. For example, hidden time delays within the total data acquisition system (load cells, displacement sensors, amplifiers/conditioners) create confusion when specimen failure captured by a sensor does not agree in time with images from high speed video. Additionally, noise in measurements can occur due to sample misalignment, fixture and test-stand rigidity, subtleties of a slack adapter (if one is used), and distortions caused by the test environment. After care is taken to minimize structural-born distortions, noticeable levels of noise and oscillations may still exist from these mechanical sources as well as electrical sources such as channel cross-talk of a video camera shutter pulse bleeding onto other sensor channels. Ultimately, prudent use of DSP techniques may be needed to further improve data quality and interpretation. This paper demonstrates success over many such difficulties when performing tensile tests of polymer materials up to 100 s^{-1}.

Keywords Intermediate strain rate • Servo hydraulic • Digital image correlation • DIC • Noise • Digital signal processing • DSP

19.1 Introduction

Computer simulations are used to predict performance of polymer parts during dynamic loads to aid in design. Many polymers have time dependent material properties, therefore material characterization must be completed at multiple strain rates up to and beyond what is expected to be experienced in use. These measurements present many more challenges and opportunities for data corruption than quasi-static (QS) tests. Measurements that may be routine and trivial at low rates require sophisticated equipment, great attention to detail by the experimentalist and competence in digital signal processing (DSP) to produce high quality, meaningful results. Additionally there are currently no well-established standards for these tests so methods vary widely across the intermediate strain rates testing community [1–4]. While a standard may not be available, SAE has formulated some general guidelines [5] which may be helpful for identifying and mitigating sources of noise and measurement error.

To minimize noise and errors, the experimentalist should make the effort to understand every aspect of the process and the potential for problems from specimen preparation, equipment selection and limitations, test setup, data processing, and analysis. The objective of this paper is to provide examples and insight to experimentalists on the challenges of intermediate strain rate testing. All strain rates described in this work are approximate and should be considered as target strain rates for the tests shown. To be clear, the actual strain rate during any given test will show variation throughout the test. While this is an important topic, it will not be further addressed in this paper.

W.J. Briers III (✉)
DuPont Engineering Research and Technology, 974 Centre Rd. CRP702/1019A, Wilmington, DE 19805, USA
e-mail: william.j.briers@dupont.com

H.J. Qi et al. (eds.), *Challenges in Mechanics of Time-Dependent Materials, Volume 2: Proceedings of the 2014 Annual Conference on Experimental and Applied Mechanics*, Conference Proceedings of the Society for Experimental Mechanics Series,
DOI 10.1007/978-3-319-06980-7_19,

19.2 Equipment Synchronization

Characterizing the tensile response of a thermoplastic polymer up to 100 s^{-1} required the use of a servo-hydraulic load frame, force sensors, a sufficiently fast data acquisition system (DAS) with antialiasing protection, and high speed video (HSV) cameras for digital image correlation. The example dataset presented in Fig. 19.1 demonstrates some of the challenges of achieving synchronized data across multiple sensors that include HSV. In the course of this testing, it was noticed that the force response recorded in the data acquisition system (DAS) was indicating failure after the HSV cameras clearly showed a break in the specimen. Figure 19.1a shows a force-time trace of a tensile test at 100 s^{-1} and Fig. 19.1b is a close-up of the peak force with corresponding shutter pulses (rectangular wave form) as recorded from HSV. When the pulses were counted, the peak force occurred at pulse 164 and began to drop at pulse 165, but the break was clearly visible on the HSV at picture 161, Fig. 19.1c. The force was out of sync with the cameras by four pictures (0.022 ms). Initially it was believed to be a problem with the software used to count shutter pulses, but this was verified to be functioning properly by manually counting pulses, which indeed showed more pulses were recorded between the trigger event and the peak force than were on the video. The cameras, trigger, cables, and DAS were examined and found to be performing correctly. Ultimately, the delay was determined to be initiated in the charge mode amplifier between the piezoelectric (PE) charge mode load cell and the DAS. The circuit diagram in the manufacturer's data sheet showed the signal from the sensor was digitized with an alias-protected analog to digital converter, scaled, filtered and converted to analog voltage output. This amplifier had been used for many years but never in a situation when it was critical to be synchronized with the output from other equipment and thus the delay was never previously identified. The time delay was eliminated by using a different load cell that utilized the onboard amplifier in the DAS. Figure 19.2a shows the signal from a piezoresistive (PR) load cell, captured at the same time as the force from the PE load cell conditioned by the external amplifier. The two load cells were mounted in series, allowing simultaneous force measurement. Figure 19.2b is a close-up of the peak forces from both load cells and the shutter pulses from HSV. The PR load cell shows the peak load at the shutter pulse corresponding to the picture of the broken specimen, Fig. 19.2c. The delay between the two load cells was 0.02 ms. This correlated well with the duration of four pictures from HSV cameras (0.022 ms).

The lesson to be learned here is the importance of spending time testing and characterizing each piece of equipment in use, so that limitations can be mitigated or at a minimum, understood.

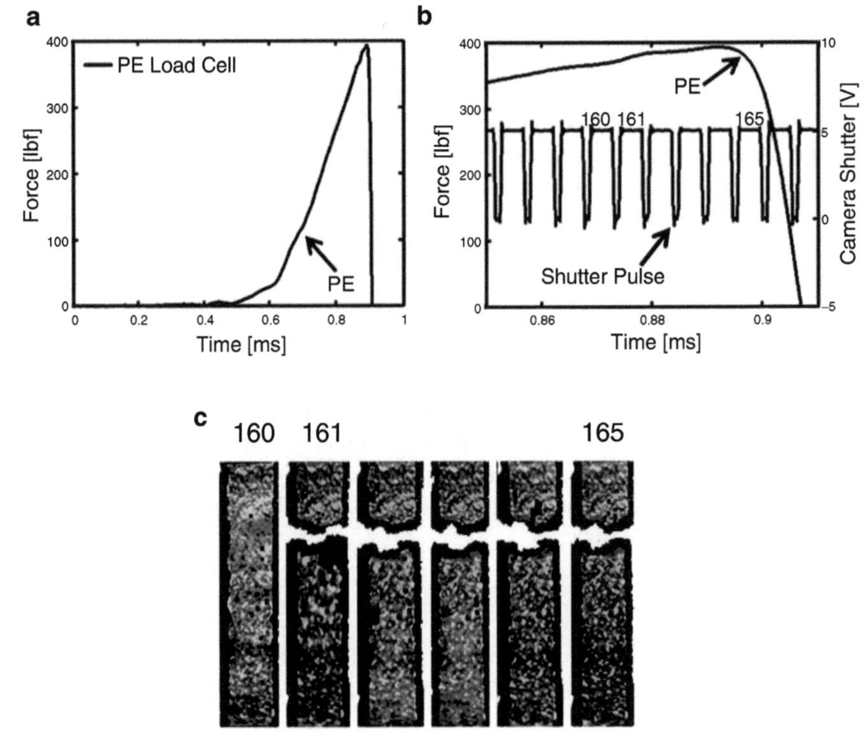

Fig. 19.1 Raw force–time response of PE load cell (a) during tensile test at 100 s^{-1}, close-up of peak force and shutter pluses from HSV (b). Demonstrating incorrect picture count for failure as compared to images from camera (c)

Fig. 19.2 Raw force–time responses of PE load cell (*solid line*) and PR load cell (*dashed line*) showing time delay of PE cell (a) and close-up of peak loads compared to shutter pulse count where response from PR load cell (b). Matches images of specimen failure (c)

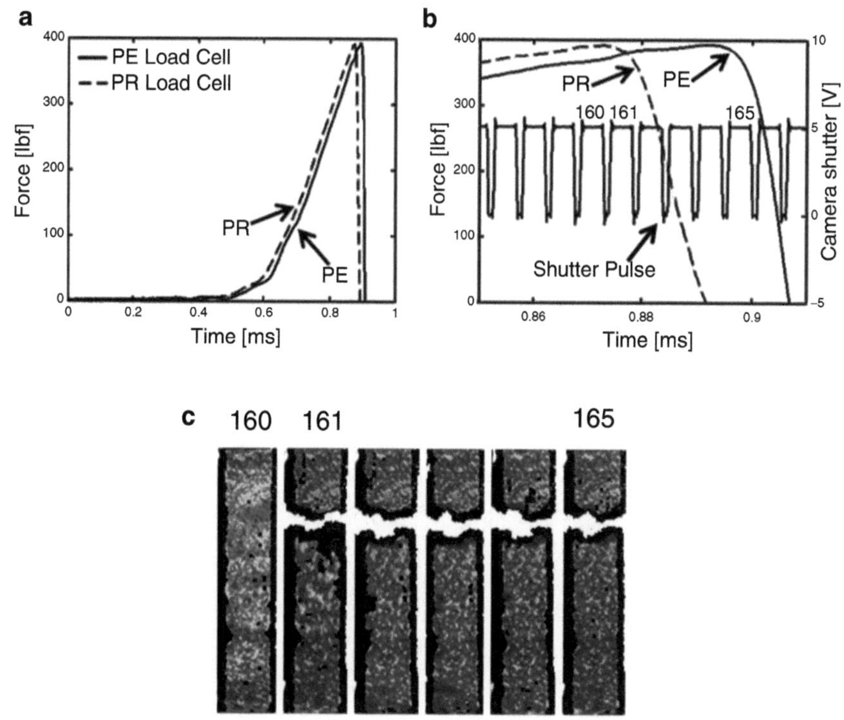

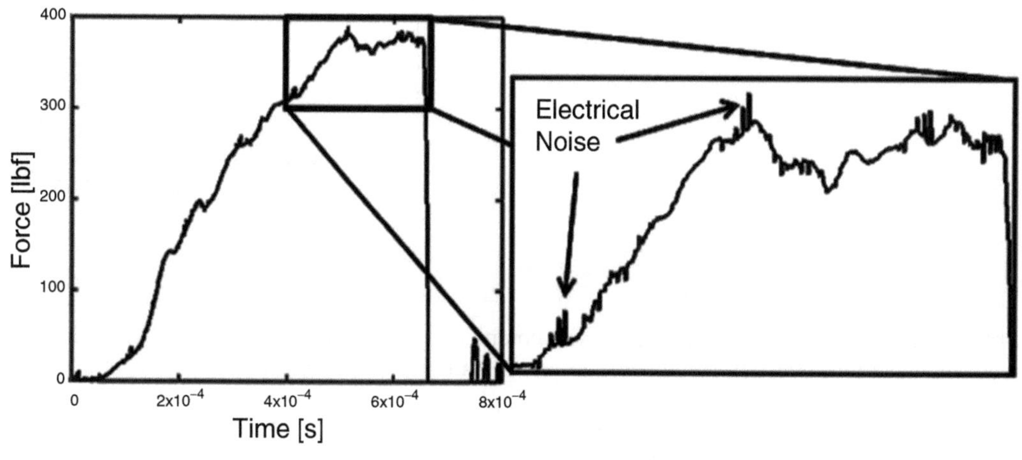

Fig. 19.3 Raw force–time response from a tensile test at 100 s^{-1} showing electrical and vibratory noise

19.3 Noise in Force and Displacement Measurement

Material tests at 10 s^{-1} and above are susceptible to oscillatory distortions due to a host of factors. Figure 19.3 is the force–time response of a tensile test at 100 s^{-1}, which was contaminated by both electrical noise and vibratory noise. The electrical noise correlated with the rising and falling edges of the shutter pulses from the HSV camera. This noise occurred on non-HSV channels by electrical bleed-through (cross-talk) caused by insufficient electrical isolation between channels. The displacement channel was most severely affected by cross-talk, as seen in Fig. 19.4. While the channels of raw data were collected as a function of time, it is common for the test engineer to also look at a force versus displacement plot, especially considering that the data will ultimately be used for material model development. Figure 19.5 presents the results of combining the raw data from Figs. 19.3 and 19.4 into a force–displacement plot; a nearly uninterpretable mess!

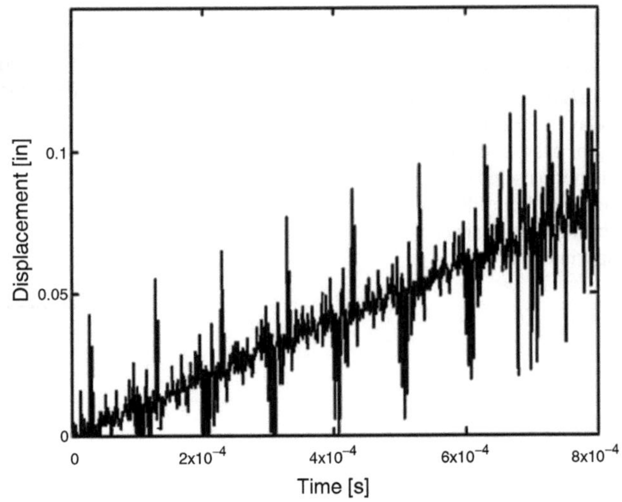

Fig. 19.4 Raw displacement–time response from a tensile test at 100 s^{-1}

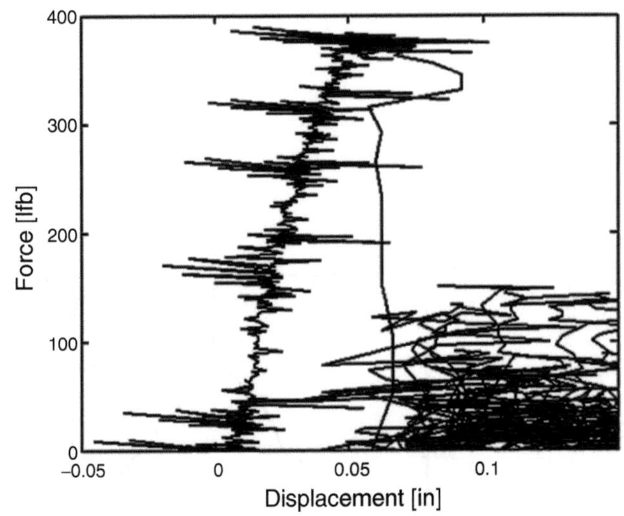

Fig. 19.5 Raw force–displacement response from a tensile test at 100 s^{-1}

Without processing other optical measurements and completing a full analysis protocol, these force–displacement evaluations were intended to guide test settings and to provide a quick sanity check of the test set-up. However, due to data contamination, the raw data plot (Fig. 19.5) appeared completely unusable... or maybe not. As shown next, proper use of advanced DSP techniques can salvage data like this, often resulting in useful and credible results. This data set will be used as an example for how DSP may be used to recover data that has been fouled by noise. To better understand DSP techniques Stearns and David have a useful book [6] which explains various aspects of signal processing algorithms, while Diehl has well explained, practical examples of DSP to solve challenging data problems [7] as well as a valuable training course notebook [8] with many examples, tutorials and explanations of DSP theory written for mechanical engineers.

19.4 Cleaning and Salvaging Data with DSP

When applying DSP to any data, the analyst will typically utilize some type of software to apply digital filters to remove the unwanted frequency content from the raw data. Digital filters come in several forms, but a commonly used workhorse for this type of effort is the lowpass Butterworth filter. This section explores several attempts at cleaning-up the previously presented data using sixth-order Butterworth filters from three different software packages; namely Mathcad® 15 [9], MATLAB® 2013b [10], and Kornucopia® 1.6 [11]. Such DSP tools can be very powerful in cleaning/salvaging data, but these tools can produce undesirable results also, especially if they are not well understood by the user.

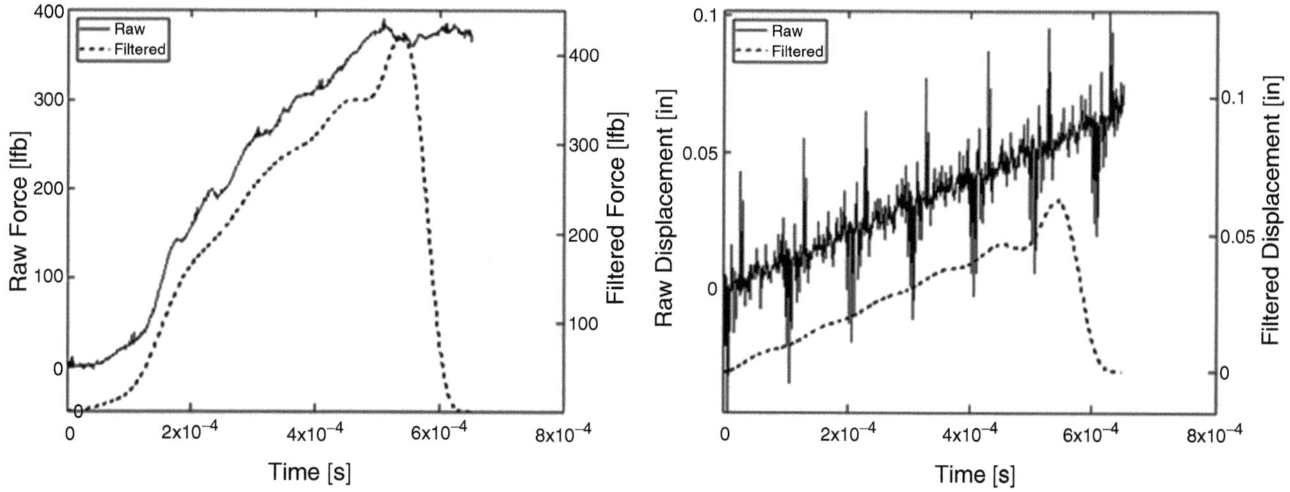

Fig. 19.6 Force–time (*left*) and displacement–time (*right*) responses with a simple, manually created, bi-directional filter using Mathcad® (*dashed lines*)

Applying a lowpass Butterworth filter to data using a single pass filtering approach will result in potentially smoother data, but the data will also be delayed in time by the filter. A simple way to avoid this undesirable distortion is to utilize a bi-directional implementation of the filtering process. With this technique, the data is first passed through the filter (producing a time delay), then this filtered result is reversed and filtered a second time (causing a negative time delay which cancels to first-pass time delay), and then finally this result is reversed again and returned as the filtered result. For the three software products being investigated, MATLAB® and Kornucopia® offer native functions that filter the data bi-directionally. For Mathcad®, the user must manually implement such an algorithm using the single-pass filters offered.

Before the data was filtered, one trimming operation was done to help improve results. The raw force vs time data from Fig. 19.3 has a severe and nearly immediate drop in force when the specimen failed. At this point and beyond, there is no interest in that section of the data. In fact, leaving the post-failed section of data in the dataset that is to be filtered would create undesirable filter distortions around the failure onset location, meaning that it will distort part of the force vs time curve just prior to the onset of failure. To avoid this, the data (both force and displacement) was trimmed at the point where failure begins. The red colored raw data curve in Fig. 19.6 shows the trimmed data to be filtered as described below.

To apply a DSP filter, the user must specify a filter cutoff frequency. For some types of data and problems, the cutoff frequency might be specified by standard procedures or they might be guided by modal analysis or similar. For the data being discussed here, no such information was available. Another common approach to determine cutoff frequency is to compute Discrete Fourier Spectra (so-called FFT) of the force vs time and displacement vs time raw data to see if any obvious spikes or frequency bands appear that might be related to known causes of noise. For this example, these analyses (not shown) were inconclusive and so the cutoff frequency was determined by a simple trial and error process. Different cutoff frequencies were tried, starting with 1/10 of the sampling frequency, and then lowering the cutoff value until a plausibly smooth result was obtained from the filtering process. This is not an exact approach, meaning that no one specific perfect cutoff frequency will be found. For the data analyzed here, a cutoff frequency between 10 and 3 kHz appeared to remove the noise without removing too much of the actual physics believed to be in the data.

The first attempt to clean the data applied a manually created bi-directional filter in Mathcad®. Both the force vs time Fig. 19.6 (*left*) and the displacement vs time Fig. 19.6 (*right*) signals were filtered (dashed lines) using a 10 kHz cutoff frequency and are presented on the secondary y-axis while the raw data (solid lines) are on the primary y-axis. The first two thirds of the filtered response appear reasonable, but the filtered result near the right end of the data (beginning around 5e-4 s) shows a highly undesirable behavior. In both the force and displacement responses, the filtered result near the right end shows a Gibbs phenomenon (the rising bump) and then dives down to a zero amplitude value. This distortion is caused by the way the underlying filtering algorithm in Mathcad® works. In short, it has no end-effect minimization scheme and simply assumes the data begins and ends with zero values. More detailed discussion on the general topic of filter end-effect minimization and filter pre-charging can be found in [7]. Bottom line—this filter-induced distortion is especially bad if the data is to be used to create a material law for finite element analysis.

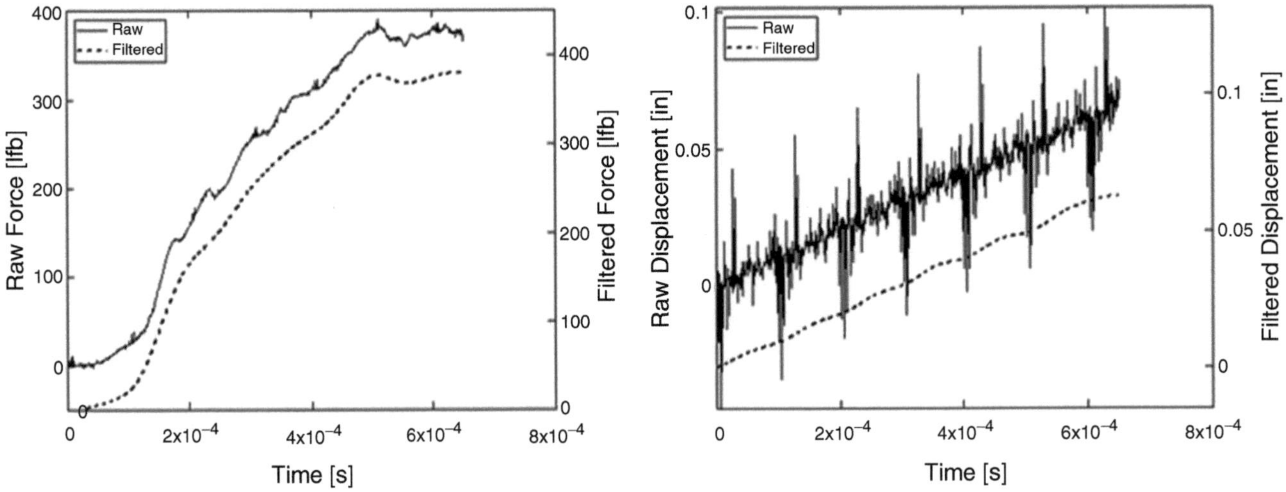

Fig. 19.7 Force–time (*left*) and displacement–time (*right*) responses with MATLAB® 10 kHz filtered responses (*dashed lines*)

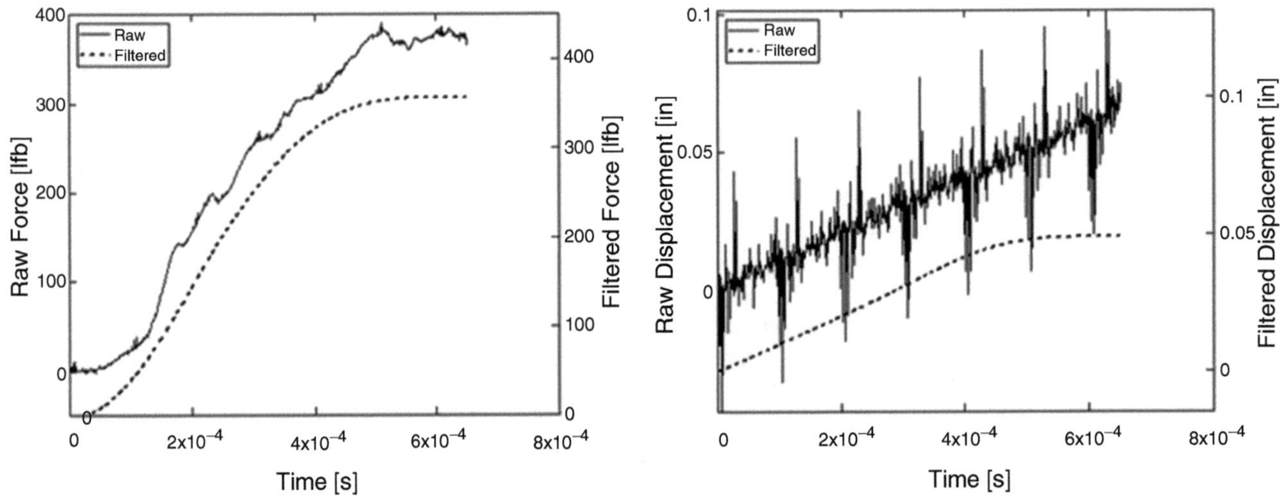

Fig. 19.8 Force–time (*left*) and displacement–time (*right*) responses with MATLAB® 3 kHz filtered responses (*dashed lines*)

The next filtering approach evaluated was MATLAB® via its function "filtfilt". As stated in the MATLAB® documentation, "filtfilt minimizes start-up and end transients by matching initial conditions…" Using a 10 kHz cutoff frequency with filtfilt, Fig. 19.7 (*dashed lines*), appears to have created much improved results that do not suffer from the end-effects that were found with the previous Mathcad® filtering approach (with no end-effect minimization).

The filtered results still show some low-frequency oscillations, so in an attempt to remove these the raw data was filtered using a 3 kHz cutoff frequency. This result shown in Fig. 19.8 (*dashed lines*) is indeed smoother, but has a strange end-distortion on the right side again. The cause of this distortion is related to limitations in MATLAB®'s filtfilt end-effect minimization technique. The way filtfilt attempts to match initial conditions and minimize start-up and end transients is by copying a small section of data from each end of the original data and placing a flipped mirrored version of it onto each end. The length of these artificial copies is related to the number of filter coefficients used in the filter. The idea is that with such extensions and performing filter-precharging after each filter pass of the bi-directional filter, the filter will hopefully yield reasonable results. With a 10 kHz cutoff frequency, the algorithm performed well (Fig. 19.7), but with a 3 kHz cutoff frequency, the algorithm appears to "break down". The problem seen in Fig. 19.8 is caused by the time delay from each filter pass being larger than the filtfilt data extensions. More details on the causes of this type of distortion are found at [8].

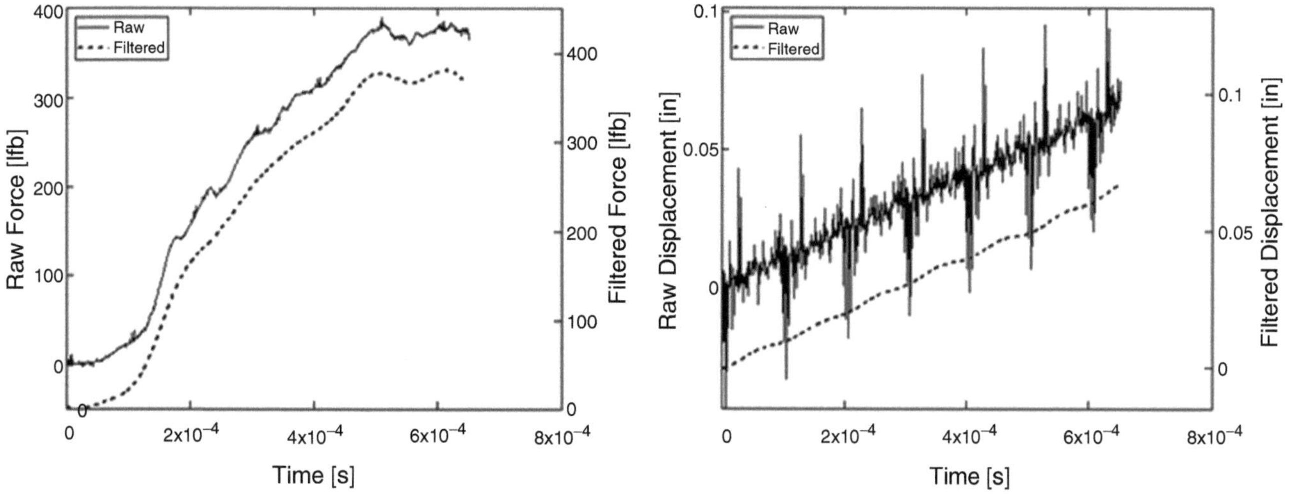

Fig. 19.9 Force–time (left) and displacement–time (*right*) responses with Kornucopia® 10 kHz filter responses (*dashed lines*)

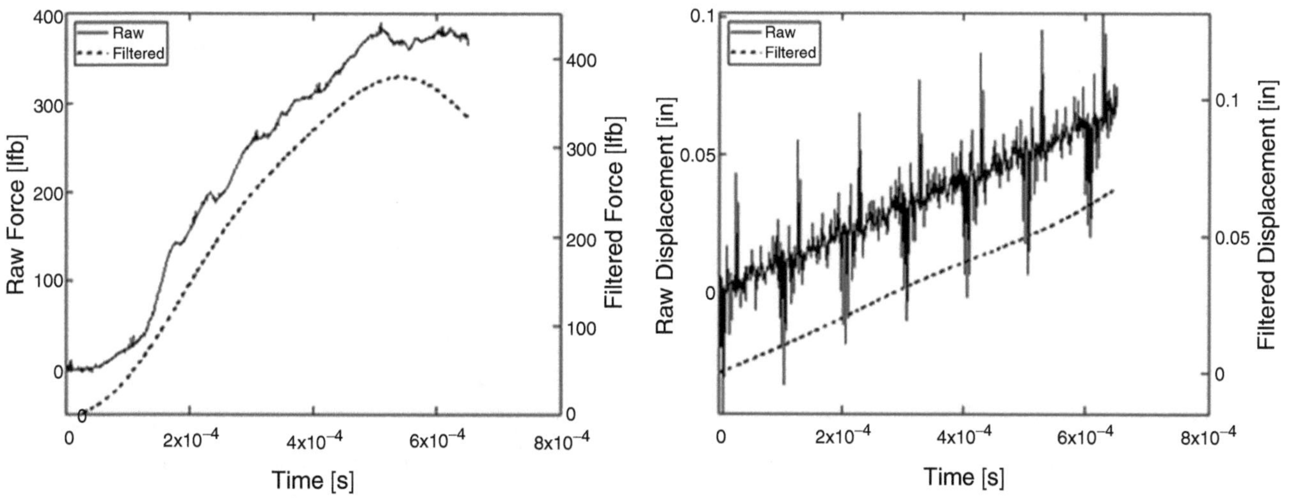

Fig. 19.10 Force–time (*left*) and displacement–time (*right*) responses with Kornucopia® 3 kHz filter responses (*dashed lines*)

The last filtering approach investigated on this data used Kornucopia® via its function "filterXY_k". Like MATLAB®, the filterXY_k function uses a bi-directional filter. However the Kornucopia® function has a variety of user selectable algorithms and options to minimize filter-induced end-effects. Some of the adjustable settings include options for changing start and end assumptions, method of data extension, and delay compensation methods. The default settings employ a unique end-effect minimization algorithm based on an adaptive algorithm. Figure 19.9 presents the Kornucopia® results with a 10 kHz cutoff frequency (*dashed lines*). The filterXY_k function was utilized using its default end-effect minimization algorithm which employs an end condition assumption called "predict" (that is based on an auto-correlation prediction algorithm). The filtered response was very similar to filtfilt at the same cutoff frequency.

When using filterXY_k with 3 kHz cutoff (Fig. 19.10), the results look smooth but the force response near the right end begins to oscillate downward again. This is caused by the auto-correlation-based end assumption. Switching the right end assumption in filterXY_k to "constant" yields the improved results in Fig. 19.11. With this setting, the 3 kHz lowpass Butterworth filter result appears to be credible and will allow for further use of the cleaned data in material model development. The force–displacement plot with adequately filtered data may now prove to be useful compared to the raw data (Fig. 19.12).

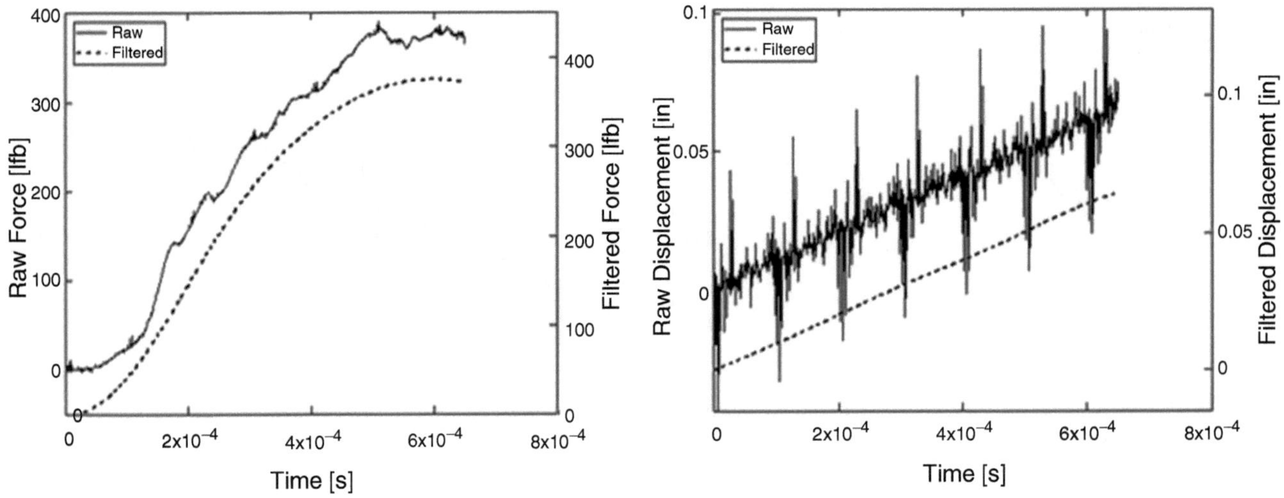

Fig. 19.11 Force–time (*left*) and displacement–time (*right*) responses with Kornucopia® ADV filtered responses (*dashed lines*)

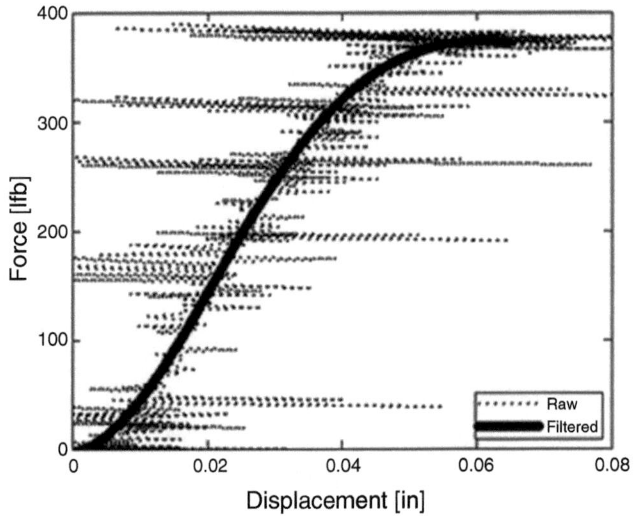

Fig. 19.12 Raw (*dotted line*) and filtered (*solid line*) force–displacement responses from a tensile test at 100 s^{-1}

While the point here is not to say that one software is better than another, the conclusion should be that very different results can be achieved with the tools available in different programs. This is especially important if the user thinks he is essentially performing the same task in each software. The important point here is that the user should be very knowledgeable of the software they are using, and for this type of clean-up work, having a good understanding of DSP is extremely important to the successful use of DSP software.

19.5 Using Improved DAS and Test Set-ups to Mitigate Noise and Distortions

Judicious filtering of the previous data yielded a credible result. This type of effort is often considered as "salvaging" when the initial data has such severe issues. A preferred approach is to address the actual sources of noise and distortion in the data so that cleaner data is collected to begin with.

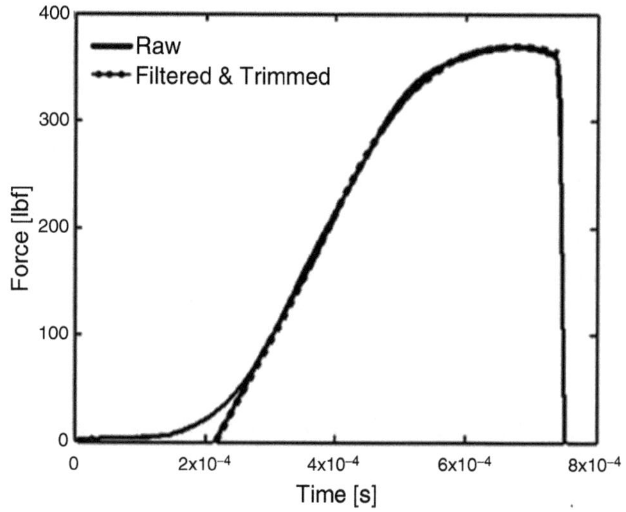

Fig. 19.13 Raw force–time response (*solid line*) and filtered response (*line with dots*) at 100 s^{-1}

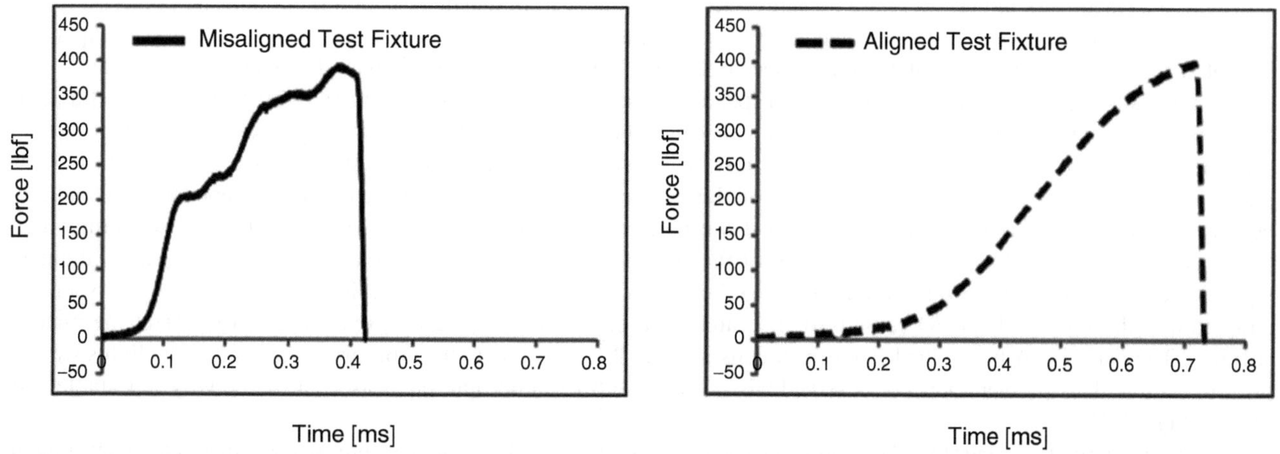

Fig. 19.14 Raw force–time responses during tensile tests at 100 s^{-1} with misaligned test fixture (*left*) and aligned test fixture (*right*)

To mitigate the electrical noise, a different DAS with better isolation between channels was used to prevent cross-talk contamination between channels, especially contamination derived from the HSV framing signal. In addition, this new DAS had improved AAA protection to further ensure data integrity.

The source of mechanical noise was found to be caused by slight misalignment between the grips that hold the specimen. The misalignment wasn't very obvious because the slack adapter, which allows the actuator to accelerate to the desired speed prior to engaging the specimen, had some lateral compliance. Also it was noticed that there was slight run-out in the system which noticeably changed the amount of misalignment when the actuator was manually rotated. The servo hydraulic load frame used for these tests was not equipped with an alignment fixture. The frame had a 1.5 in. diameter hole in the fixed crosshead for attachment of the load train (load cell, grip). A 1.0 in. diameter threaded rod was mounted vertically through the crosshead and held in place with washers and nuts on each end. This rod provided the means of attaching the load train to the crosshead. The difference in diameter between the rod and hole allowed for adjustment, but without an alignment fixture it was difficult to ensure the rod was properly lined up, not only transversely but also angularly though the crosshead. Fixture alignment for this series of tests had to be done by sight and feel because an instrumented alignment tool was not available.

Addressing the causes of noise and distortion at their source results in a much cleaner raw data measurement as seen in Fig. 19.13 (*soild line*). Further cleaning this raw data was accomplished by trimming, filtering (just a small amount) and extending the beginning region of the data with a linear extrapolation Fig. 19.13 (*line with dots*). Clearly there was an improvement in measurement quality due to the new DAS and realignment. The data has much greater integrity and the cleaning effort was obviously a much less daunting task as compared to the efforts required to clean the raw data from Fig. 19.5.

The design of the slack adapter can also be a source of noise during specimen engagement near the beginning of the test. In a series of tensile tests, unrelated to the previous examples, large distortions were observed in the raw force-time response, as seen in Fig. 19.14 (*left*). The test was on glass reinforced nylon with a target rate of 100 s^{-1}, and the oscillations were not significant when the material was tested at lower strain rates. After checking that fixture alignment was correct, and after verifying all mechanical connections such as threaded connectors between components in the load train were properly tightened, high speed video was used to observe the system during the test. The video indicated the slack adapter did not correctly engage and lock upon full extension but bounced, distorting the force signal. This required a redesign of the slack adapter; changing the material of construction and internal geometry of the engagement surfaces. A softer material was chosen to damp the impact and also facilitate momentum transfer between the actuator and the part attached to the specimen grip, minimizing bounce and vibration. Figure 19.14 (*right*) shows the raw response after making design improvements.

19.6 Noise and Distortions Associated with Optical Measurement

Non-contact, optical measurements methods can be extremely useful for intermediate strain rate testing. Like any other sensor or piece of test equipment, optical measurements are susceptible to noise and distortions. The following are a couple examples of challenges in optical measurements.

The five dashed lines in Fig. 19.15 are the stress–strain responses from tensile tests of a thermoplastic polyester resin at 100 s^{-1}. Significant variation in stiffness and elongation at failure were observed in this specific test set, but past experience indicated that this was not typical for these measurements. Upon investigation, it was discovered that a change from the normal test setup proved to be the cause. High speed video cameras require high intensity lights to adequately illuminate the subject. Normally 300 W halogen lights with flexible fiber optic (FO) light pipes were used to illuminate the test specimens, however, at the time of this particular test, these were unavailable and 300 W halogen lights without FO light pipes were selected. The mechanical properties of this material were known to be temperature dependent, so radiative heating from the lights was suspected. A simple test was performed to verify this notion using a fast-response thermocouple (1 mil diameter) mounted in a hole at the center of the specimen gage region. High intensity lights, with and without FO pipes were turned on for various durations up to 1 min. While the FO light pipes had little effect on the specimen temperature, the light without FO caused a temperature rise of 15° in just a few seconds, enough to affect the material stiffness. The variation between the heated specimens in Fig. 19.15 was due to the arbitrary duration that the lights were on prior to testing. This problem was not observed when using the fiber optic light pipes as seen by the very repeatable solid lines in Fig. 19.15.

The lesson learned—be careful when using HSV and high powered lights. The measurement engineer can determine the potential influence of lights by comparing stress vs time response of tests done with and without lights. If the tests with lights have reduced strength or softer response, then temperature of the specimen may be a factor. Also it is worth noting that thermal radiation can affect accuracy of other sensors such as load cells, accelerometers and strain gages. These issues are often mitigated with heat shielding or insulation.

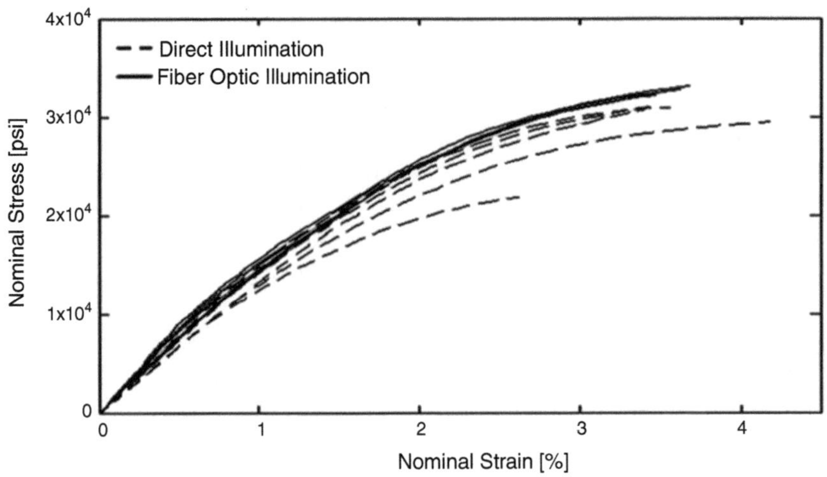

Fig. 19.15 Stress–strain responses at 100 s^{-1} of polymer tensile specimens being illuminated by different approaches. Direct illumination by 300 W Halogen (*dashed lines*) with no control over lighting time and fiber optic illumination with 300 W Halogens (*solid lines*)

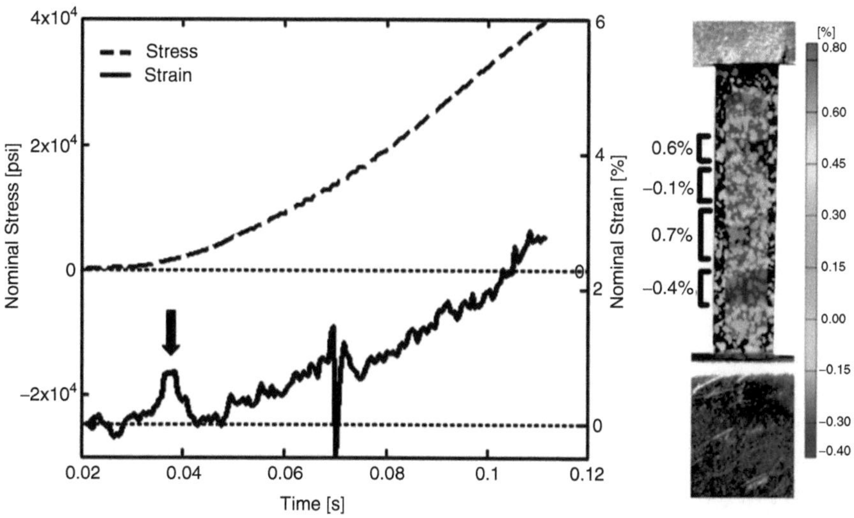

Fig. 19.16 Stress–time (*dashed line*), strain–time (*solid line*) response and DIC overlay (*right*) during tensile test at 1 s^{-1}, showing noise in strain due to non-uniform, dynamic changes in air density between the camera and specimen

While the above example showed a real material response due to improper equipment selection, Digital Image Correlation (DIC) and other optical measurement methods are susceptible to distortions that are not representative of physical reactions in the specimen. For instance, a three dimensional DIC measurement system must be calibrated to account for light refraction between the measurement subject and the cameras' sensors. Sources of refraction include air (or other fluids), transparent materials used as protective shields, camera lens and any optical filters such as polarizers. After the system has been calibrated it is critical that these sources of refraction do not significantly vary.

The following describes an example where dynamic variation in refraction occurred, which in turn caused distortion in the DIC measurement. The test involved measurements on a polymer at −40 °C. The solid line in Fig. 19.16 shows very noisy strain output from DIC and the dashed line is stress, calculated from the load cell output. It is clear the stress response from the specimen did not experience distortions correlated to the strain response, which indicated the problem could be with the optical measurement. The picture in Fig. 19.16 (*right*) shows noticeable variation in the strain field overlaid on the specimen at a single instance in time. This picture occurred in time at the arrow in the strain–time plot. The cause could not be determined by looking at one DIC image at a time, but only became evident when the original images were viewed as a running video (unfortunately not viewable in this paper). Changes in air density caused non-uniform, dynamic optical distortions which were recorded on HSV and perceived as strains in DIC. The effect is similar to "mirage" distortions observed in the air above hot asphalt. As further evidence that air density changes were the cause of the distortions in the optical measurements, it was observed that the metal grip holding the specimen appeared distorted in the video.

The reason behind changes in air density was a heat gun aimed at the window of the thermal chamber to prevent frost. Interestingly, this optical noise was only observed during quasi-static and 1 s^{-1}, but not at 10 and 100 s^{-1}. The speed of lower density air rising up the window was much slower than the two highest strain rates. Thus over the short time duration of the fast tests, the air density appeared constant. This phenomenon could have been recognized prior to testing by performing a DIC noise floor measurement over the expected duration of the test at each strain rate. While noise floor measurements were done routinely, they were performed prior to lowering the chamber temperature and turning on the heat gun.

Lesson learned—the engineer should consider routinely recording the noise floor for each instrument associated with a test, at the test conditions and over the expected duration of the event. Another useful practice is to test a known material, when possible, prior to testing the material of interest. This exercise could elucidate potential measurement problems if the response of the known material does not match historical data.

19.7 Conclusions

Predicting the structural performance of polymer components under intermediate strain rates requires accurate and precise measurements of rate dependent mechanical properties. Obtaining good, credible data from intermediate strain rate testing requires great attention to detail because these tests offer significantly increased potential for noise and distortions as compared to quasi-static measurements. The examples found in this work are meant to highlight some of the complexities of intermediate rate testing and share some experience in overcoming challenges.

The difficulties that can arise in tests at speeds above 10 s^{-1} should not be underestimated. Sources of noise and distortions can occur from many sources including materials, testing hardware and fixtures, sensors, data acquisition systems, data analysis via DSP, or any combination thereof. As demonstrated by the examples presented in this paper, the key to success is to question your data, strive to thoroughly understand causes of noise and distortion, and to mitigate them at the source whenever possible. In the event that such noise and distortions cannot be completely prevented, the prudent use of DSP may be implemented to clean-up measurements. However, unlike classical DSP scenarios on long signals, the user must be aware of the added complexities of applying DPS and filtering to short signals as demonstrated in this paper.

Acknowledgments The author would like to thank; Ted Diehl, for sharing his expertise in DSP and experimental measurements, Alan Wedgewood and Glenn Hake for their on-going support in developing our high speed measurement capabilities. The support of DuPont and many other DuPont colleagues is also acknowledged.

References

1. Xiao X (2008) Dynamic tensile testing of plastic materials. Polym Test 27(2):164–178
2. Boyce BL, Crenshaw TB (2005) Servohydraulic Methods For Mechanical Testing in the Sub-Hopkinson Rate Regime up to Strain Rates of 500 1/s. Sandia National Laboratories Report, SAND2005-5678
3. Zrida M, Laurent H, Grolleau V, Rio G, Khlif M, Guines D, Masmoudi N, Bradai C (2010) High-speed tensile tests on a polypropylene material. Polym Test 29(2010):685–692
4. Perkins L, Lobo H (2005) A novel technique to measure tensile properties of plastics at high strain rates. In: SPE ANTEC proceedings, Boston
5. SAE J2749 (2008) High strain rate tensile testing of polymers. Society of Automotive Engineers International, Warrendale
6. Stearns SD, David RA (1996) Signal processing algorithms in MATLAB. Prentice-Hall, Upper Saddle River
7. Diehl T, Nagaraj B, Carroll D (1999) Using digital signal processing (DSP) to significantly improve the interpretation of abaqus/explicit results. In: Abaqus user's conference, 1999. www.bodietech.com/videos/publications.html
8. Diehl T Analyzing Noisy Data via Filtering and DSP. A DSP short-course. www.bodietech.com
9. Mathcad® 15 Software from *PTC*. www.PTC.com
10. MATLAB® 2013b Software from *Mathworks*. www.mathworks.com
11. Kornucopia® Software from *Bodie Technology Inc*. www.bodietech.com

Chapter 20
Prediction of Statistical Distribution of Solder Joint Fatigue Lifetime Using Hybrid Probabilistic Approach

Hyunseok Oh, Hsiu-Ping Wei, Bongtae Han, Byung C. Jung, Changwoon Han, Byeng D. Youn, and Hojeong Moon

Abstract We propose a novel hybrid approach to examine uncertain sources that cause variations in solder joint lifetimes and to predict statistical distributions of solder joint lifetimes under actual operating conditions. The uncertainty in input variables of a life prediction model is propagated to the output by an approximate integration method. The variations in the output are statistically compared to experimentally-measured lifetimes. A set of input variables that minimize the discrepancy between predicted and experimental results is obtained through the optimization technique. The proposed approach is implemented for chip resistor assemblies, and the fatigue life of solder joints under the actual operating conditions is predicted after calibration.

Keywords Solder joint • Accelerated life testing • Lifetime prediction • Statistical model validation • Uncertainty

20.1 Introduction

Solder joint can fail under operating and environmental loading conditions such as thermal cycling, thermal shock, vibration, and humidity. Reliability assessment of solder joints is important to avoid intermittent and/or catastrophic failure of electronics systems. Accelerated life testing has been used to evaluate their reliability of solder joints in electronic packaging [1]. For example, the most widely accepted accelerated thermal cycling condition of chip resistor assemblies is −55 to 125 °C.

Experimental data from accelerated life testing are analyzed to extract useful information such as mean cycles to failure. Weibull and lognormal distributions are widely used in analyzing experimental data from thermal cycling tests of solder joints. After fitting experimental data to a functional form of statistical distributions, parameters of statistical distributions are obtained. With a set of estimated parameters, one can extract information about the reliability of solder joints. Previous studies [2–4] used the mean cycles to failure (or the characteristic life) to characterize the durability of solder joints. This is critical when comparing the reliability of different designs for enhanced reliability. Nevertheless, understanding variations in solder joint life is equally important in increasing yield rates, determining warranty periods, and qualifying products.

The ultimate goal of the reliability assessment of solder joints is to predict their reliability under actual operating conditions. Several approaches including stress-based, strain-based, and energy-based methods were developed to predict

H. Oh (✉) • H.-P. Wei • B. Han
Department of Mechanical Engineering, University of Maryland, College Park, MD 20742, USA
e-mail: hsoh@umd.edu

B.C. Jung
Korea Institute of Machinery and Materials, Daejeon 305-343, Republic of Korea

C. Han
Components & Materials Physics Research Center, Korea Electronics Technology Institute, Seongnam 463-816, Republic of Korea

B.D. Youn
Department of Mechanical and Aerospace Engineering, Seoul National University, Seoul 151-742, Republic of Korea

H. Moon
Semiconductor Business, Samsung Electronics, San #16, Banwol-Dong, Hwasung-City, Gyeonggi-Do, Republic of Korea

H.J. Qi et al. (eds.), *Challenges in Mechanics of Time-Dependent Materials, Volume 2: Proceedings of the 2014 Annual Conference on Experimental and Applied Mechanics*, Conference Proceedings of the Society for Experimental Mechanics Series,
DOI 10.1007/978-3-319-06980-7_20, © The Society for Experimental Mechanics, Inc. 2015

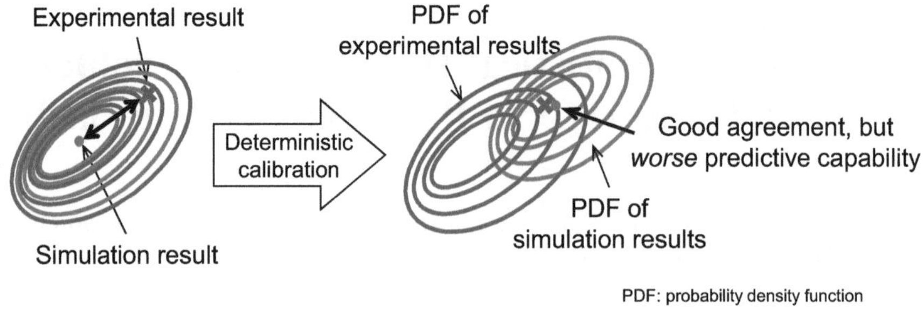

Fig. 20.1 Caveat of the deterministic model calibration approach

fatigue life of solder joints [5]. Yet, the life prediction is still challenging due to inherent manufacturing variability, uncertain material properties, etc. These uncertainties should be considered to improve the predictive capability of fatigue life models.

Recently, model calibration methods were used to enhance the model prediction. Model calibration is the process of adjusting unknown model variables in the predictive model. Although deterministic model calibration is a proper method to minimize mismatch between results from predictive models and experiments, ad-hoc adjustment of model variables can degrade the predictive capability of the model. Figure 20.1 [6] illustrates this caveat of the deterministic model calibration approach using a single experimental result. The red and blue solid curves are statistical distributions calculated by predictive modeling and obtained by experiments, respectively. If model calibration is conducted in such a way that the red dot calculated using a predictive model matches the blue cross obtained from an experiment, the predictive capability is actually compromised after model calibration. This can be prevented if the model calibration is performed on statistical metrics such as mean and variance with multiple experimental results.

This study aims to develop a methodology that statistically calibrates uncertain variables in fatigue life models for solder joints by utilizing experimental data from accelerated thermal cycling testing. The proposed methodology is implemented for chip resistor assemblies, and the fatigue life of solder joints under the actual operating conditions is predicted after calibration.

20.2 Methodology

A lifetime predicted by a fatigue life model can be described as:

$$\hat{y}\left(x_{known}, x_{unknown}\right) = y\left(x_{known}, x_{unknown}\right) + \varepsilon \tag{20.1}$$

where $\hat{y}$ and y are predicted and true lifetimes; x_{known} and $x_{unknown}$ are the known and unknown variable vectors, respectively; ε is the error between the predicted and true lifetimes. The objective is to calibrate the unknown variables in fatigue life prediction models using experimentally measured lifetimes. The error should be minimized after calibrating life prediction models.

Figure 20.2 shows the outline of the methodology proposed in this study. "Unknown" variables to be calibrated should be selected by the user. Examples of unknown variables are dimensions of solder joints, material properties, and thermal cycling profiles. Unknown variables can be expressed using a functional form of statistical distributions. It is to be noted that an initial statistical distribution of unknown variables should be assigned. This can be done by reviewing published papers, industry standards, and experts' opinion. The known and unknown variables are inputs to life prediction models.

Several methods are used to propagate uncertainty of inputs to outputs of the life prediction models. Random sampling methods such as Monte Carlo sampling are the most accurate but computationally expensive. Response surface methods are very popular since they reduce the number of runs by design-of-experiment sampling. However, their results depend on the accuracy of a response surface. Their computational cost increases exponentially as the number of variables increases. Recently, approximate integration methods were developed as an efficient uncertainty propagation approach [7]. In this study, the eigenvector dimensional reduction (EDR) method based on approximate integration was used [8–10].

The statistical distribution of fatigue life calculated by the EDR method is compared to the experimentally measured lifetime data. The objective function used in the comparison is the likelihood function. When the likelihood value converges, the model calibration terminates. Otherwise, statistical distributions of unknown variables are updated and their uncertainty is propagated. The iterations are conducted until the likelihood value converges. Having a life prediction model with calibrated variables, statistical distributions of cycles to failures can be predicted under different loading conditions.

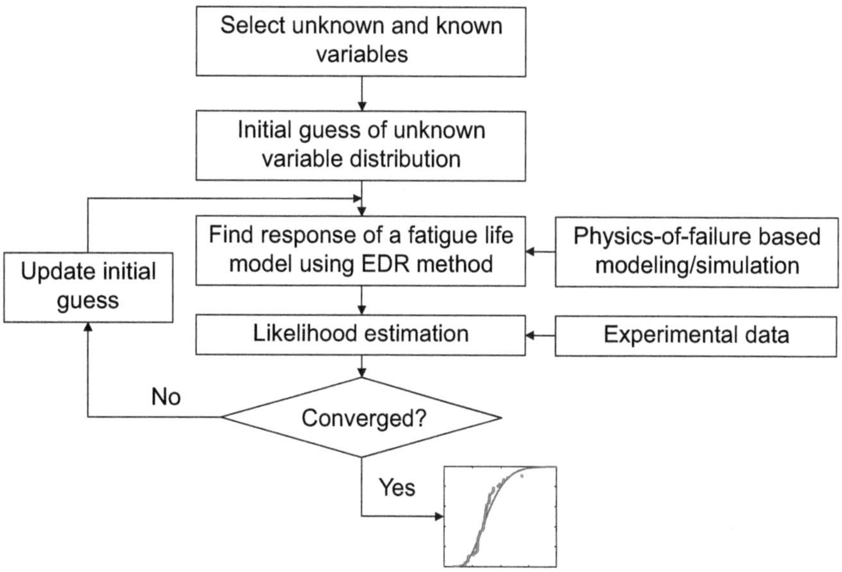

Fig. 20.2 Proposed hybrid approach using physics-of-failure model and statistical model calibration

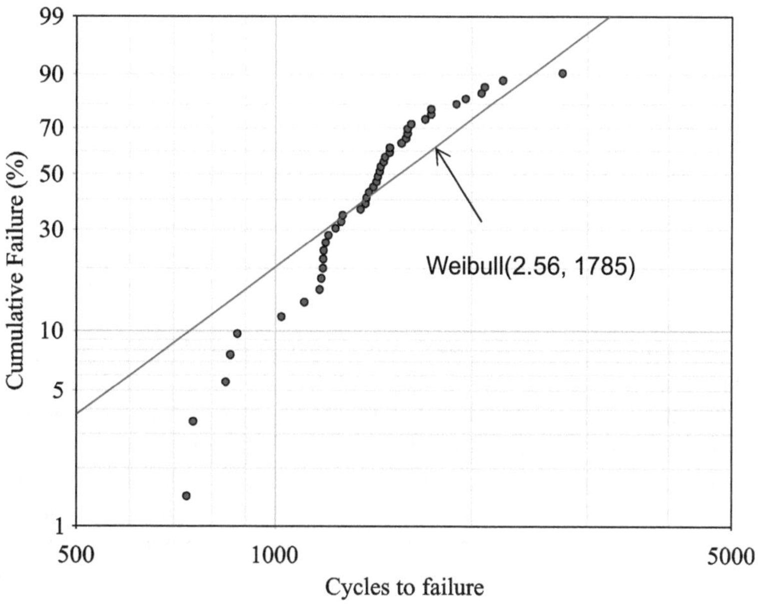

Fig. 20.3 Parameters of Weibull distribution calculated by maximum likelihood estimator

20.3 Implementation

A data set from thermal cycling tests [11] was used to demonstrate the methodology developed in this paper. The chip resistor is mounted on a printed circuit board using the eutectic tin-lead solder. Chip resistor assemblies were subjected to temperature excursions from −50 to +125 °C. A summary of the experimental results is shown in Fig. 20.3. The details can be found in [11].

The relationship between strain energy density (ΔW_{ave}) and cycles to failure (N) has been known as [12]:

$$N_0 = K_1(\Delta W_{ave})^{K_2} \tag{20.2}$$

$$N_p = \frac{a}{K_3(\Delta W_{ave})^{K_4}} \tag{20.3}$$

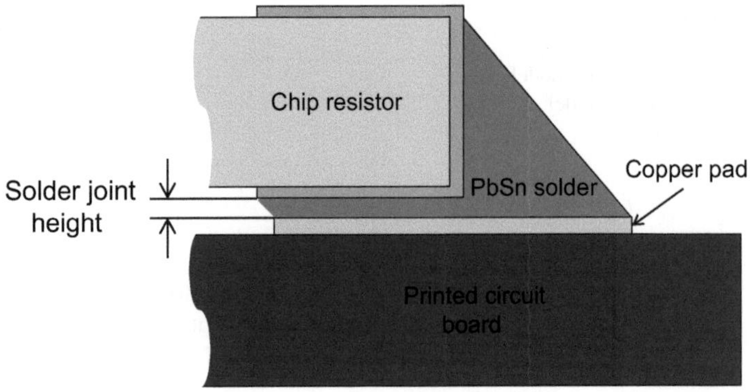

Fig. 20.4 Unknown variable to be calibrated: solder joint height

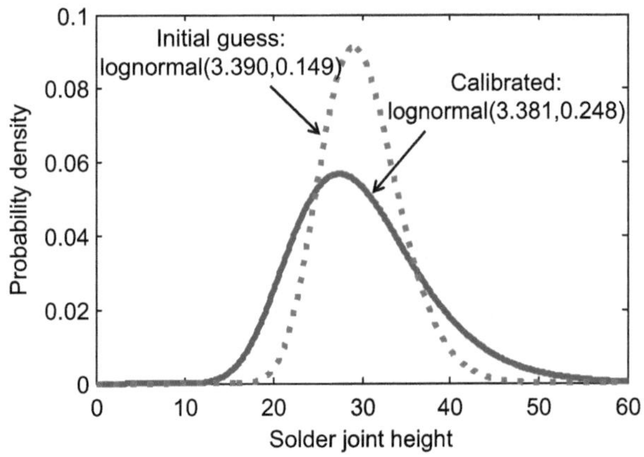

Fig. 20.5 Solder joint height before and after model calibration

where N_0 is the cycles to initiate cracks; N_p is the cycles to propagate cracks until failure; a is the crack size; K_1, K_2, K_3 and K_4 are the empirical constants. The total lifetime for chip resistor assemblies is the sum of N_0 and N_p. It has been known that N_0 for chip resistor assemblies is much smaller than N_p [13]. Therefore, only Eq. (20.3) was used to estimate solder joint lifetimes.

The unknown variable considered in this study is the solder joint height in the chip resistor assemblies (Fig. 20.4). It is assumed that the solder joint height follows a lognormal distribution. The initial guesses of shape and scale parameters of the lognormal distribution are 0.1980 and 3.3816, respectively. The known variables were K_3 and K_4. The details about how to determine statistical distributions for K_3 and K_4 are beyond the scope of this paper, which will be discussed in a separate article. The statistical distribution for K_3 was a lognormal distribution with the shape and scale parameters of 0.00362, 0.02822, respectively. The statistical distribution for K_4 was a normal distribution with the mean and standard deviation of 2.496 and 0.05239, respectively.

The calculation of strain energy densities of the solder relied on Anand's model. The constants for Anand's model in [14] were used in the analysis. Material properties for chip resistor assemblies used in the analysis can be found in [11]. A commercial FEA package (ANSYS^{TM}) was used for the calculation.

20.4 Result

Figure 20.5 shows the solder joint heights before and after model calibration. The scale parameter of the lognormal distribution decreased from 3.390 to 3.381. The shape parameter of the lognormal distribution increased from 0.149 to 0.248. This indicates that the variation of solder joint height is larger than the initial guess. The interval of the calibrated solder joint was 18.1 to 47.8 with the 95 % confidence level.

Figure 20.6 presents the prediction of cycles to failure of solder joints used in portable electronics. The calculation was conducted by the calibrated life prediction model. The failure probability of 1 % corresponded to 7,945 cycles to failure. If one million chip resistors are used in portable electronics, 10,000 units may failure before 7,945 cycles.

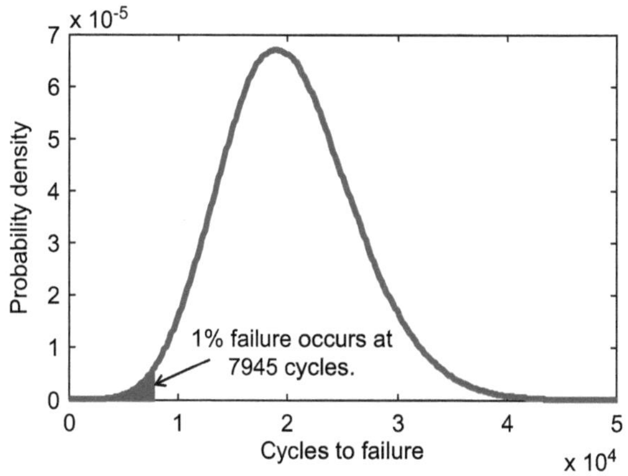

Fig. 20.6 Reliability estimate of chip resistors under a portable electronics usage condition (−40 to 70 °C)

20.5 Conclusions

A novel hybrid approach was proposed and implemented to calibrate uncertain variables in solder joint life prediction models and to predict statistical distributions of solder joint lifetimes under actual loading conditions. The inherent manufacturing variability (i.e., solder joint height) of chip capacitor assemblies was determined statistically from the accelerated test results using the proposed approach. The failure probability of chip resistors under an actual portable electronics usage condition was estimated using the statistical life prediction model. The proposed approach can be potentially used to increase yield rates, to determine warranty periods, and to qualify products.

Acknowledgment This work was supported by the International Collaborative Energy Technology R&D Program of the Korea Institute of Energy Technology Evaluation and Planning (KETEP) granted from the Ministry of Trade, Industry & Energy of Korea (No. 20118520020010).

References

1. Performance test methods and qualification requirements for surface mount solder attachments: IPC-9701, Northbrook, 2002
2. George E, Das D, Osterman M, Pecht M (2011) Thermal cycling reliability of lead-free solders (SAC305 and Sn3.5Ag) for high-temperature applications. IEEE Trans Device Mater Reliab 11:328–338
3. Chai F, Osterman M, Pecht M (2012) Reliability of gull-wing and leadless packages subjected to temperature cycling after rework. IEEE Trans Device Mater Reliab 12:510–519
4. Salmela O, Nierninen T, Saerkka J, Tarnmeranaa M (2008) Reliability analysis of lead-free solder castellations. IEEE Trans Compon Packag Technol 31:13–22
5. Ridout S, Bailey C (2007) Review of methods to predict solder joint reliability under thermo-mechanical cycling. Fatigue Fract Eng Mater Struct 30:400–412
6. Youn BD, Jung BC, Xi Z, Kim SB, Lee WR (2011) A hierarchical framework for statistical model calibration in engineering product development. Comput Methods Appl Mech Eng 200:1421–1431
7. Rahman S, Xu H (2004) A univariate dimension-reduction method for multi-dimensional integration in stochastic mechanics. Probab Eng Mech 19:393–408
8. Hu C, Youn BD (2011) Adaptive-sparse polynomial chaos expansion for reliability analysis and design of complex engineering systems. Struct Multidiscip Optim 43:419–442
9. Wang P, Youn BD, Xi Z, Kloess A (2009) Bayesian reliability analysis with evolving, insufficient, and subjective data sets. J Mech Des 131:111008
10. Youn BD, Xi Z (2009) Reliability-based robust design optimization using the eigenvector dimension reduction (EDR) method. Struct Multidiscip Optim 37:475–492
11. Han C, Han B (2014) Board level reliability analysis of chip resistor assemblies under thermal cycling: a comparison study between SnPb and SnAgCu. J Mech Sci Technol 28:879–886
12. Darveaux R (2000) Effect of simulation methodology on solder joint crack growth correlation. In: Proceedings of the 50th electronic components and technology conference. Las Vegas, pp 1048–1058
13. Lu H, Bailey C, Dusek M, Hunt C, Nottay J (2000) Modeling the fatigue life of solder joints for surface mount resistors. In: Proceedings of international symposium on electronic materials and packaging, Hong Kong, pp 136–142
14. Ghorbani HR, Spelt JK (Nov 2007) An analytical elasto-creep model of solder joints in leadless chip resistors: part 2—applications in fatigue reliability predictions for SnPb and lead-free solders. IEEE Trans Adv Packag 30:695–704

Chapter 21
Effect of Moisture and Anisotropy in Multilayer SU-8 Thin Films

C.J. Robin and K.N. Jonnalagadda

Abstract Biocompatible polymers, like SU-8, have become a vital part of Microelectromechanical System (MEMS) devices, especially as a structural layer in biosensors and micro-fluidics. In the present study, effect of moisture and anisotropy on the mechanical response of 2 micron thick SU-8 polymer films was investigated using optical in-situ microtensile experiments. The tensile specimens were fabricated using soft lithography, which consisted of deposition, patterning and release of the film from the silicon wafer. A novel methodology was followed to capture optical images of the patterned SU-8 surface during the experiment, for strain calculations using Digital Image Correlation (DIC). Tensile properties were derived from the uniaxial stress–strain curves. The tensile strength and fracture strain of the multilayer films increased by more than 10 and 5 %, respectively, as compared to the single layer film. Some experimental results along with Fourier Transform Infrared (FTIR) spectroscopy data is shown to support the observations from our experiments.

Keywords SU-8 • Multilayer • Digital image correlation • Lithography • Fourier transform infrared spectroscopy

21.1 Introduction

Polymers, especially biocompatible polymers, have become the drivers in the field of Microelectromechanical Systems (MEMS). SU-8 resin, a UV sensitive thermosetting epoxy polymer (formulated by IBM) [1], has been employed in MEMS applications as a structural material because of its bio-compatibility and ability to form high aspect ratio features. SU-8 polymer has a high functionality owing to its eight epoxy rings that participate in cross-linking during UV processing and thermal baking. Mechanical response of SU-8 thin films is of interest because of their susceptibility to different loading conditions, particularly in-plane loading, in various MEMS applications. Studying these films using uniaxial and biaxial experiments can help in the design of reliable MEMS devices [2, 3]. In the previous thin film experimental studies [4–7], optical methods like Digital Image Correlation (DIC) has been used in combination with AFM and in-situ optical techniques to provide accurate measurement of full field displacement and strain fields. Here, we have utilized a modified optical in-situ tensile method to extract the mechanical properties of single and multi layer SU-8 thin films [8]. Initial investigations revealed two aspects of interest in the processing of these polymer films, viz. moisture and spin-coating.

Moisture is an unavoidable environmental condition in most applications like micro resonators and micro fluidic devices. Polymers are sensitive to humidity and are therefore used as sensitive material in humidity sensors. Feng and Farris [2] studied the residual stress variation when SU-8 membranes are kept in a humid environment and found it to decrease with increase in cross-linking density. In another study, the processing liquids itself were found to affect the SU-8 structures due to the presence of voids and holes. The processing liquids like IPA and PGMEA were found to displace water resulting in high residual stresses [9]. Although SU-8 is affected by the prolonged exposure to, water or moisture [9–11], there is no literature available on the variation in mechanical properties of SU-8 due to moisture, or deterministic evaluation of the length of exposure time to moisture that can potentially have effect on mechanical behavior of SU-8.

Spin-coating is the first step in thin film fabrication, which is responsible for the thickness and uniformity in the final specimens. However, it is necessary to know if the spinning causes any anisotropy in the mechanical property of the coated films.

C.J. Robin (✉) • K.N. Jonnalagadda
Department of Mechanical Engineering, Indian Institute of Technology Bombay, Mumbai 400076, India
e-mail: cjrobin23@iitb.ac.in

H.J. Qi et al. (eds.), *Challenges in Mechanics of Time-Dependent Materials, Volume 2: Proceedings of the 2014 Annual Conference on Experimental and Applied Mechanics*, Conference Proceedings of the Society for Experimental Mechanics Series,
DOI 10.1007/978-3-319-06980-7_21,

The shear flow during spin-coating has been found to cause anisotropy in SU-8 composites [12–14]. Most of the orientation studies as a result of spin-coating were on aromatic long-chain polymers like polyimide [15] and poly(arylene ether)s [16]. It is necessary to study if there is an in-plane orientation of the polymer chains, as they can play an important role in the mechanical response, and thereby, reliability of the MEMS devices where SU-8 thin films is used as a structural or device layer. In an earlier work [8], we have studied anisotropy in spin coated SU-8 thin films. In this study, the molecular orientation in single and multilayer SU-8 thin films with respect to the position on the wafer has been studied in conjunction with FTIR spectroscopy.

21.2 Experimental Details

21.2.1 Sample Preparation

The specimens were prepared with two formulations of SU-8 (single layer—2002 and multilayer—2000.5), through the process flow shown in Fig. 21.1a. The specimen layer of SU-8 was spin- coated (3,000 rpm) over a sacrificial oxide layer, of thickness 500 nm, on a silicon wafer. The multi layer specimen of SU-8 2000.5 was fabricated in eight steps to give similar thickness of 2 μm. The prebaking consisted of two-steps 2 min at 70 °C and 3 min at 90 °C. This was followed by UV exposure of 250 mJ/cm^2, with pattern for the specimens and post exposure bake of 3 min at 70 °C and 5 min at 90 °C. The polymer films were then developed for 1 min and hardbaked for 15 min at 200 °C. The support layer of SU-8 2100 was then patterned through similar steps with the difference in the baking times (15 min at 70 °C and 75 min at 90 °C) and exposure (300 mJ/cm^2). Finally, the specimen dies were released by removal of the sacrificial layer, achieved through buffered Hydrofluoric acid (BHF) etching. The final dog bone specimen had a gauge length of 800 μm and width of 400 μm. The thickness was measured utilizing a profilometer and found to be 1.93 μm for SU-8 2000.5 and 1.96 μm for SU-8 2002.

21.2.2 Effect of Moisture

To study the effect of moisture, tensile experiments were conducted on the SU-8 2000.5 specimens, which were exposed to water for different periods. Experiments were conducted using a micro-tensile testing setup developed in our laboratory (Fig. 21.1b) based on earlier designs [4, 5, 8]. An optical system (Olympus microscope with 100X magnification and Sony monochrome CCD camera) was used to capture the images of the specimen during in-situ experiments. These images were used for extracting the full field data in the direction of loading, through DIC. Tensile experiments were carried out at different strain rates, varying from 1×10^{-5} to 1×10^{-3}/s. ATR-FTIR spectroscopy was used to determine the amount of moisture and structural changes in the material. The spectra also help in differentiating the solvent and absorbed moisture,

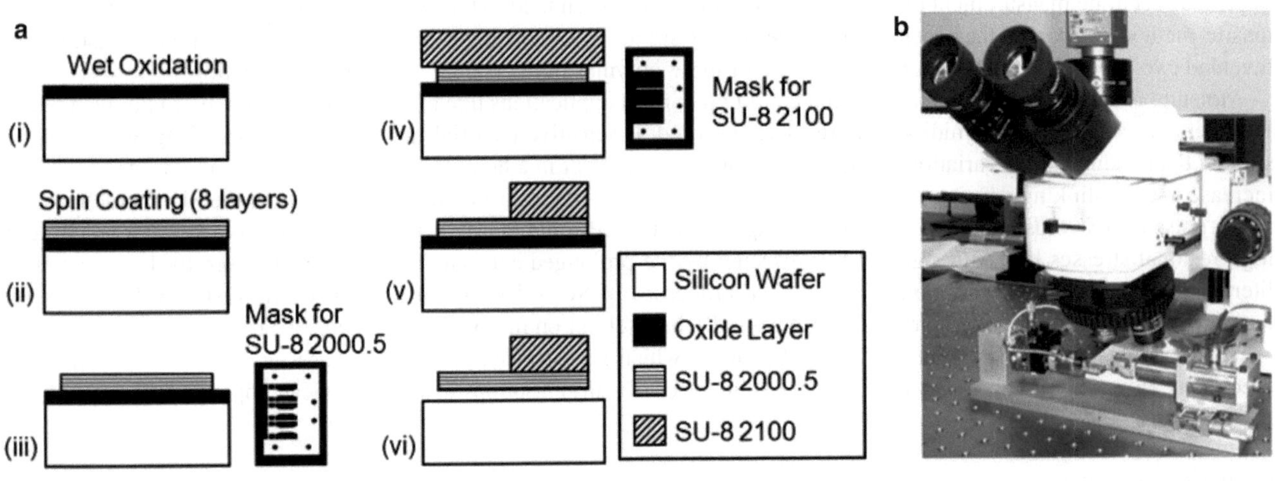

Fig. 21.1 (**a**) Process flow for fabrication of SU-8 thin films (**b**) In-situ optical micro-tensile setup used in the experiments on transparent polymeric films [8]

which has significant effect on the mechanical response of SU-8 polymer [17]. The spectra were recorded in the MIR range from 400 to 4,000 cm^{-1} with air taken as the reference background.

21.2.3 Spin Coating Anisotropy

Orientation studies in spin-coated SU-8 thin films were conducted for single layer SU-8 2002 films [8]. In this work, the effect has been studied in multilayer films by conducting tensile experiments on tangential and radial specimens, named in relation to their positions on the Si wafer. Radial specimens have their gauge length parallel to the radius of the wafer while for tangential specimens they are perpendicular. The molecular structure and its correlation to the mechanical behavior of the spin-coated film were investigated utilizing a Polarizer along with the standard FTIR spectroscopy setup [18].

21.3 Results and Discussion

Being a transparent polymer, (Fig. 21.2a) SU-8 cannot be directly used for microtensile experiments, since DIC requires a patterned/speckled surface. For this purpose, the film surface was patterned using copper oxide nanoparticles, which were randomly dispersed on the film surface by blowing on the particles and allowing it to settle on the specimen (Fig. 21.2b). This patterned specimen was then used to generate the DIC compatible images during the experiments.

The specimens were loaded under uniaxial tension until failure. Fig. 21.2c shows the displacement contours generated by DIC analysis (in the 'u' direction), which were super imposed on the patterned specimen. The specimens for analysis were dipped in water for different lengths of time. These specimens were then loaded under tension at a strain rate of 2×10^{-3}/s. Stress–strain curves were then extracted from the experimental data obtained from the load cell and DIC analysis (Fig. 21.3). It was observed that the material response has changed with the presence of water/moisture. The diffusion of water into the voids in the thin film [18] results in a lower stiffness as observed in other polymers [14, 19].

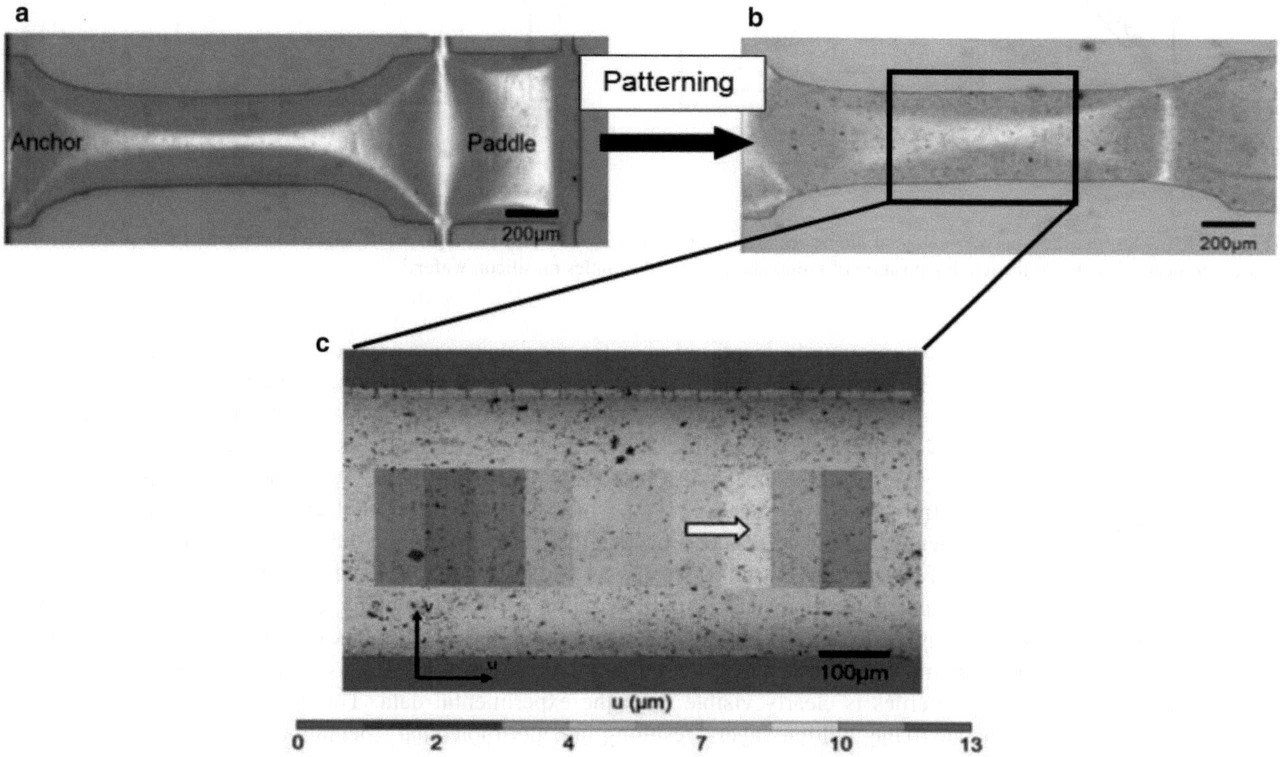

Fig. 21.2 (**a**) Released specimen (**b**) Patterned specimen (**c**) Full field displacement contours along the loading axis [8]

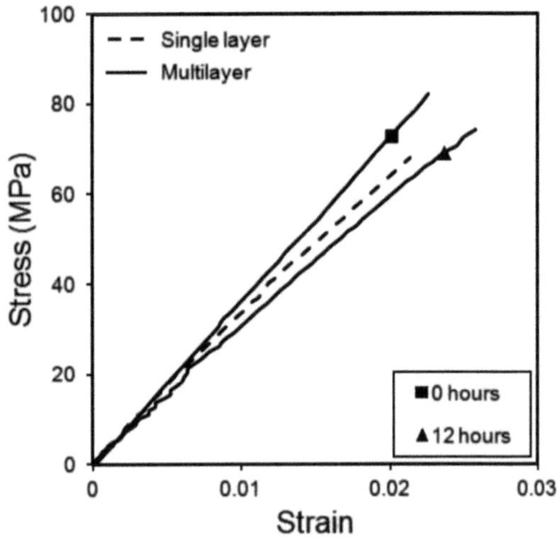

Fig. 21.3 Tensile stress–strain curves of single layer and multilayer SU-8 films dipped in water for 0 and 12 h

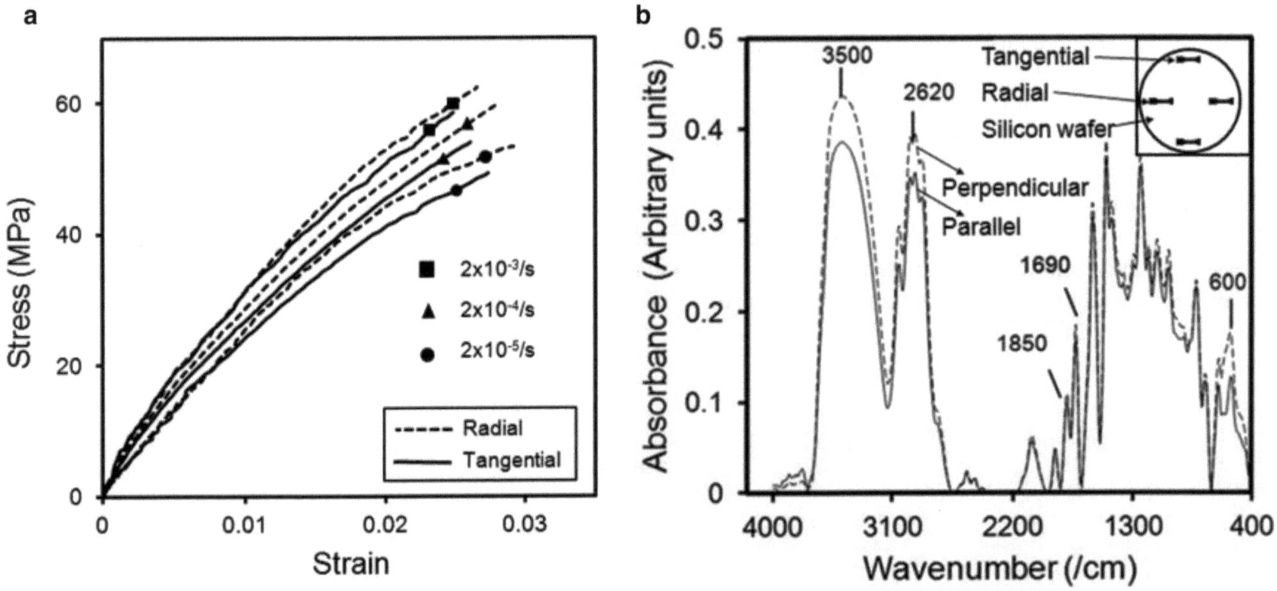

Fig. 21.4 (**a**) Stress–strain curves for radial and tangential samples at various strain rates (**b**) FTIR spectra for parallel and perpendicularly polarized beams [8] (Inset: the relative position of radial and tangential samples on silicon wafer)

The elastic modulus and the fracture strength decreased when the specimens were exposed to water. This suggests that the material undergoes a structural change in the presence of water and not mere absorption, resulting in the swelling of the polymer. Thus, even an optimum processing condition does have an effect on the residual solvent present in the sample causing the internal stresses in SU-8 epoxy and, therefore, do not completely eliminate its effect on mechanical properties.

As observed with most polymers [13, 20] the mechanical response of SU-8 is affected by the molecular/chain orientation. The anisotropy is clearly visible from the curves (see Fig. 21.4), which show higher strength and moduli for radially oriented samples than the tangentially oriented samples. This confirms that there is a variation in the mechanical behavior depending on the position and orientation of the samples on the wafer. It was also observed that the difference in flow stress increases with the strain at all strain rates studied, which means the radial samples are stronger at any strain. In addition, this shows the orientation is not brought about by the loading of the specimens but due to the prior processing steps. The thin film strength was significantly improved in multilayer films as opposed to single layer films [8] as seen from the Fig. 21.3. The enhancement in the tensile properties is clearly visible from the experimental data. This is attributed to the improved cohesion assisted shear flow due to the multiple layers, resulting in better molecular orientation.

Polarized FTIR spectra were generated with respect to the gauge length of the specimens. The parallel beam was incident perpendicular and perpendicular beam parallel to the gauge length of the specimen giving the peaks as shown in Fig 21.4 b. The IR absorption spectra give the spatial orientation and relative intensity of the chemical bonds in the material under inspection. The peaks of 600, 910 and 3,900 cm^{-1} are significant. The dichroic ratio (R) in polarized FTIR spectra is defined as the ratio of absorbance intensities, i.e., parallel beam to the perpendicular beam. It was found to be greater than 1.5 in the vibrational bands namely, 3,700–4,000, 1,920–1,950 and 1,685–1,700 cm^{-1}, which confirmed the presence of orientation [8].

21.4 Conclusion

Multilayer films were found to have better tensile properties than single layer films. The tensile strength and fracture strain of the multilayer films were ~6 % higher than the single layer films. Effect of moisture was found to be significant in the mechanical response of the SU-8 thin film. The polar groups in the material were found to have an affinity for water molecules under long exposure to moisture. It was found through ATR-FTIR spectroscopy that there is a structural variation in SU-8 thin film specimens resulting in reduction of the stiffness of the material. The investigation of anisotropy in multilayer films gave interesting results, suggesting molecular orientation due to spin coating process. Microtensile experiments in conjunction with polarized FTIR study was used to support this observation. The mechanical properties like tensile strength and modulus changed with the position of the film on the wafer. This aspect can be of high value to MEMS applications. It was confirmed that there is anisotropy in the plane of the films through polarized FTIR experiments.

Acknowledgments The work was supported by Department of Science and Technology (DST), India. The authors would like to thank Center for Excellence in Nanoelectronics, IIT Bombay, for providing the facilities for the specimen fabrication. The authors are also grateful to sophisticated analytical instruments facility (SAIF) for the FTIR spectroscopy facility.

References

1. Lorenz H, Laudon M, Renaud P (1998) Mechanical characterization of a new high-aspect-ratio near UV-photoresist". Microelectron Eng 41/42:371–374
2. Feng R, Farris RJ (2002) Influence of processing conditions on the thermal and mechanical properties of SU8 negative photoresist coatings. J Micromech Microeng 13(1):80–88
3. Wouters K, Gijsenbergh P, Puers R (2011) Comparison of methods for the mechanical characterization of polymers for MEMS applications. J Micromech Microeng 21:115027
4. Jonnalagadda KN, Chasiotis I, Yagnamurthy S, Lambros J, Pulskamp J, Polcawich R, Dubey M (2009) Experimental investigation of strain rate dependence of nanocrystalline Pt thin films. Exp Mech 50:25–35
5. Jonnalagadda KN (2008) Failure and fracture of thin film materials for MEMS. Ph.D. thesis
6. Sutton MA, Wolters WJ, Peters WH, Ranson WF, McNeill SR (1983) Determination of displacements using an improved digital image correlation method. Image Vis Comput 1:133–9
7. Sharpe WN (2008) Springer handbook of experimental solid mechanics. Springer (SEM), Boston
8. Robin CJ, Vishnoi A, Jonnalagadda KN (2014) Mechanical behavior and anisotropy of spin-coated SU-8 thin films for MEMS. J Microelectromech Syst 23(1):168–180
9. Wouters K, Puers R (2010) Diffusing and swelling in SU-8: insight in material properties and processing. J Micromech Microeng 20:095013
10. Blanco VM, Carballo, Melai J, Salm C, Schmitz J (2009) Moisture resistance of SU-8 and KMPR as structural material. Microelectron Eng 86:765–768
11. Schmid S, Kuhne S, Hierold C (2009) Influence of air humidity on polymeric microresonators. J Micromech Microeng 19:065018
12. Xu X, Thwe MM, Shearwood C, Liao K (2002) Mechanical properties and interfacial characteristics of carbon-nanotube-reinforced epoxy thin films. Appl Phys Lett 81:2833–2835
13. Ward IM (2004) Introduction to the mechanical properties of solid polymers, 2nd edn. Wiley, Chichester
14. Young RJ (1991) Introduction to polymers, 2nd edn. Chapman and Hall, London
15. Lin L, Bidstrup SA (1993) Processing effects on optical anisotropy in spin-coated polyimide films. J Appl Polymer Sci 49(7):1277–1289
16. Jiang J, Callender CL, Blachetiere C, Jacob S, Noad JP, Ding J, Qi Y, Day M (2006) Optimizing fluorinated poly(arylene ether)s for optical waveguide applications. Opt Mater 28(3):189–194
17. Keller S, Blagoi G, Lillemose M, Haefliger D, Boisen A (2008) Processing of thin SU-8 films. J Micromech Microeng 18(12):125020
18. Peter K, Vollhardt C, Schore NE (2006) Organic chemistry. W. H. Freeman, New York
19. Herrera-Gómez A, Velázquez-Cruz G, Martín-Polo MO (2001) Analysis of the water bound to a polymer matrix by infrared spectroscopy. J Appl Phys 89:5431
20. Roetling J (1965) Yield stress behavior of polymethylmethacrylate. Polymer 6:311–317

Chapter 22
Shrinkage Coefficient: Drying Microcrack Indicator

Dragana Jankovic

Abstract In order to investigate the material behaviour in reality (macro-level) during drying shrinkage at early age, the shrinkage coefficient (α_{sh}) is determined in the laboratory at micro-level on specially prepared and dried (1 mm) thick cement paste specimens. Drying shrinkage experiments are performed in Environmental Scanning Electron Microscope (ESEM). For simplicity, a relationship between drying deformations and relative humidity (ε-RH) is assumed linear. It is found out that a certain value of the 'local' drying shrinkage coefficient, determined in the RH range (40–20 %), presents a threshold value. Above the threshold α_{sh} value, microcracks are noticed in specimens.

Keywords Drying shrinkage • ESEM tests • Micro-level • Shrinkage coefficient • Microcrack

22.1 Introduction

Eigenstresses that are invoked by the differential shrinkage due to moisture flow and consequent low tensile strength may initiate a microcrack (less than 1 μm width) in early age concrete. Concrete is considered as a three phase material that consists of shrinkable cement paste, solid aggregate and Interfacial Transition Zone (ITZ) between them [1]. Microcracks in cement paste could later on affect the final (mechanically enlarged) crack path [2–5] and especially endanger structurally heterogeneous, composite systems (concrete–steel composites, precast–prestressed structures) and their bond strength. It is thus necessary to find out under which circumstances, microcracks may appear in order to prevent them, which is a demanding task. It is believed that a certain value of shrinkage coefficient (α_{sh}) could be related to microcracks. The shrinkage coefficient presents the relationship between differential moisture-induced shrinkage strains ($\Delta\varepsilon_{sh}$) and moisture gradient (Δh), which is assumed linear (22.1).

$$\Delta\varepsilon_{sh} = \alpha_{sh}\Delta h \tag{22.1}$$

A relationship between α_{sh} and concrete age, curing conditions and RH values during drying, neglecting of the already present (micro)cracks, was mostly searched in the macro-level analyzes of shrinkage coefficients [6, 7]. In [6] it was concluded that "obviously, the extent of a crack formation increases with shrinkage coefficient, α_{sh}". The approach to these relationships is changed here, as we consider the idea that α_{sh} coefficient value is related to the lack or presence of microcracks as well as their specific size during/after drying at a certain relative humidity, RH. We determine the shrinkage coefficient in cement paste at micro-level based on shrinkage strains invoked by drying. In order to calculate the value of shrinkage coefficient, drying shrinkage displacements and strains in ESEM tests are determined nondestructively by digital image technique. The Vic-2D code (from Correlated Solutions) is used. Further on, statistical regression analysis is applied to determine the coefficients.

The presented α_{sh} coefficients are determined from the strain data from specimens without and with microcracks, for comparisons. Unlike the macro-level shrinkage coefficients that are known from the literature [6–8], current micro-level

D. Jankovic (✉)
CITG, Microlab, Delft University of Technology, Stevinweg 1, 2628 CN Delft, Netherlands;
Gradacacka 37, Sarajevo, Bosnia-Herzegovina
e-mail: dj32826@yahoo.com

H.J. Qi et al. (eds.), *Challenges in Mechanics of Time-Dependent Materials, Volume 2: Proceedings of the 2014 Annual Conference on Experimental and Applied Mechanics*, Conference Proceedings of the Society for Experimental Mechanics Series,
DOI 10.1007/978-3-319-06980-7_22, © The Society for Experimental Mechanics, Inc. 2015

shrinkage coefficients are determined on the dried 1 mm thick cement paste specimens in ESEM XL30. That is to obtain as much as possible 'real' shrinkage in specimens without microcracks, which means to avoid apparent shrinkage [9]. In comparison to the literature suggestions on the mathematical expressions for the shrinkage coefficients [10–12], the shrinkage coefficients in this work are not determined at relative humidity (RH = 0). Technically, it was very difficult if not impossible to obtain such relative humidity in ESEM XL30. There were difficulties in obtaining stable pressure and temperature already during drying to 10 % RH, since pressure in ESEM had to be beneath 2 Torr. Under such a low pressure, clear and sharp imaging was disabled as low contrast and brightness developed.

22.2 Experiments

22.2.1 Specimen Preparations and Drying Procedure in ESEM

The drying tests are performed on thin (10 × 10 × 1 mm^3, Fig. 22.1a), Portland cement paste (cement type I, 32.5 R) with the water/cement mass ratio w/c = 0.5 [13]. The specimens are of young age (from 2 to 52 days old). Prior to testing, the specimens are either wet cured in lime-saturated water (at 95 % RH and 20 °C) or dry (sealed) cured in plastic. The specimens are grind/polished on the day of tests, in order to create a fresh surface for sharp imaging in ESEM. The stainless steel mould and cylindrical tool are specially designed to cast a 2 mm thick specimen and then to grind it and polish to 1 mm thickness [13, 14]. The direct casting of thin and small-sized specimens is performed to avoid: (a) the apparent shrinkage mechanisms and (b) eventual damage and consequential microcracking due to cutting the thin cement paste specimen of very young age, from a lager specimen.

The specimens are dried in ESEM XL30 [13] using a new inbuilt thermo-electric couple (Fig. 22.1b), based on the Seeback effect (the opposite to the Peltier effect). Both effects operate on the principle of forming an electric circuit in two wires of different metals welded together. A thin layer of a conductive paste is used to fix the specimen to the cooling stage (Fig. 22.1b). That is to prevent it from moving on the plate and to control the temperature more accurately. Relative humidity in ESEM chamber is created by controlling (lowering) a pressure, while keeping a temperature constant. The initial (prior to drying) conditions in ESEM chamber are 100 % RH, with a pressure of 9.1 Torr and temperature of app. 10 °C. The stepwise reduction of RH by 10 %, from 100 to 90 %, 80 %, 70 %, . . ., and 20 %, is performed every time the equilibrium is reached. At 20 % RH, the minimum pressure reaches 1.8 Torr.

All drying observations and measurements are focused on the surface of the specimen. In order to gain reasonable results from the image analysis, surface of the specimen must be extremely flat such that any out-of-plane warping must be avoided. Due to very low specimen thickness, the drying of a 3-D cement paste is observed as a 2-D problem.

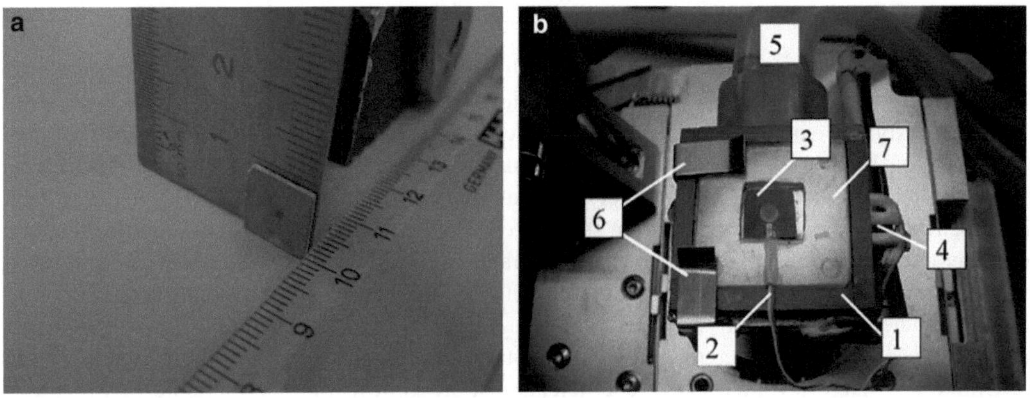

Fig. 22.1 (**a**) Polished specimen (10 × 10 × 1 mm^3), ready for drying in ESEM. (**b**) Improved cooling house (1), with inbuilt thermocouple sensors and conductive paste (2), for temperature measurements on the specimen surface (3), and integrated circuits inside the polyvinyl box (4), (5) pipes for liquid circulation, (6) metal clamps, (7) plastic frame

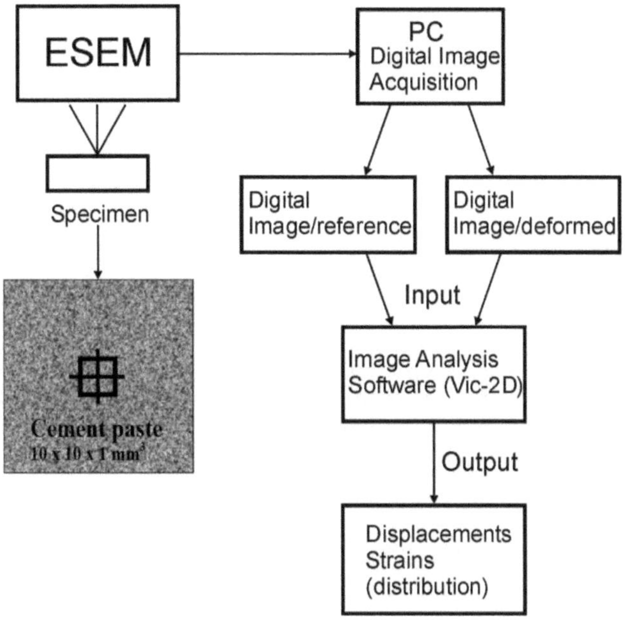

Fig. 22.2 Flow chart: strain determination in dried cement paste in ESEM

22.2.2 Acquiring ESEM Images

A special GSE (Gaseous Secondary Electron) detector for non-vacuum environment is used in order to detect (SE) signals during image acquiring. The signal is digitized during image acquisition. The physical image is divided into a rectangular grid of set of pixels (n × m): arrays of 8-bit numbers (the intensity of a monochromatic light, in the brightness range from 0, black, to 255, white). Beam current varied between 25 and 30 kV, depending on the condition of ESEM filament (used/new). Images are acquired in several magnifications (from 50 to 2 μm) for comparisons and further investigations. The 50 μm magnification images are correlated to obtain shrinkage deformations and shrinkage coefficient. All digital images are acquired with high resolution (1,424 × 968) pixels, and non-squared pixel size ratio (y/x = 0.907). The shape of a pixel is taken into account as an aspect ratio during strain calculation. The size of a pixel in (μm) varies with magnification and direction (x or y). For the magnification of 50 μm, the x-size of the pixel is 0.2 μm and y-size is 0.18 μm. That means that a field of view observed under ESEM is approx. (280 × 180) μm at 50 μm magnification. To minimize the noise, which increases remarkably as soon as the pressure (relative humidity) in the ESEM chamber is lowered, images are taken when the temperature and the pressure equilibrium appear without any fluctuations. The brightness and the contrast of the images during different drying steps are carefully chosen to stay in the closest possible range in order to avoid large differences and to produce more accurate results.

The complete procedure of acquiring images (reference and deformed images) during drying tests in ESEM and further deformation analysis by means of software (Vic-2D) can be summarized into the chart below (Fig. 22.2). In order to determine the deformations of plain cement paste specimen, the middle spot is chosen (Fig. 22.2) as an observational spot, with the coordinates (x = 0) and (y = 0).

22.3 Determination of Micro-Deformations/Strains by Vic-2D

Computation of displacements and strains is performed by mapping and correlation technique, using the commercial code Vic-2D. In order to properly map and correlate digital images, specimen surface must be well grained and polished, to obtain sharp digital images. Usually, in the cases of non-destructive testing with various materials, surface must have a random (speckle) pattern, sometimes applied by paint. In the case of cement paste preparation for ESEM tests, creation of an artificial speckle pattern on the surface was not needed, since the cement surface provides a naturally diverse pattern.

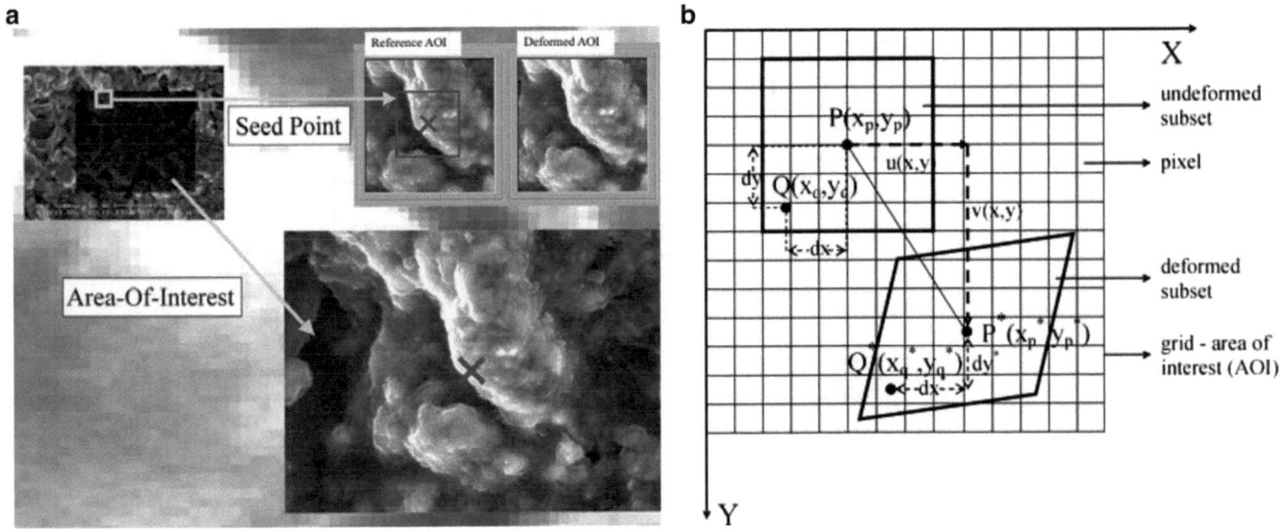

Fig. 22.3 (**a**) Arbitrarily chosen Area-Of-Interest (AOI) or a subset, and a seed point in the digital image analysis with Vic-2D code. Seed point, the point in the reference image where the correlation starts, is chosen at the spots where some recognizable features exist. (**b**) Deformation of a subset in a sampling grid, after [15]

The principle of the 2-D mapping technique is searching for the similar pixel-based intensity pattern in the 'reference' image (taken initially at 100 % RH) and the 'deformed' image (taken at all other reduced RH). Since the intensity pattern is a 2-D projection of the object onto plane [15], it is used for the basis of calculation of displacements and strains. Namely, it is assumed that the mapping of local subsets is equivalent to the mapping that applies to the deformation of the object surface, which reflected the intensity pattern [16].

All deformed images in Vic-2D are correlated to the reference image via chosen subsets, so-called Area-Of-Interest (AOI), Fig. 22.3a. The size of the subsets is (1,334 × 878) pixels, excluding image edges of (90 × 90) pixels. In the AOI, the tracking of translation and rotation of imaginary point P(x, y) related to point Q(x + dx, y + dy), is performed (Fig. 22.3b). The assumed sign convention for deformation in images is determined according to the coordinate system in the upper left corner: x is positive to the right, y is positive downwards (Fig. 22.3b). It is assumed that in-plane arbitrary line segment PQ(dx, dy) remains infinitesimally small P*Q*(dx*, dy*) after the deformation. The adequate choice of a subset- and the step-size is chosen as the subset-size 100 and the step-size 15. Since both translation and rotation are assumed in the specimens, the derivatives of u and v displacements (Fig. 22.3b) become unknowns. The unknowns are initially assumed and then calculated for the center point of the subset during comparing of the intensity values from undeformed subset to all deformed subsets. The comparison among assumptions, results in the closest value. The error made during assumption is calculated by a cross correlation function with a more accurate Quintic Bi-spline interpolation (surface fit) [17]. Introduction of a higher accuracy is necessary, especially when it comes to shrinkage deformations in the complex microstructure of cement paste. The 2-D 'in-situ' displacements (u, v) and Lagrangian strains (ε_{xx}, ε_{yy}, ε_{xy}) are computed as well as principal strains. The components of strain are two normal strains (ε_{xx} and ε_{yy}) and a shear strain (ε_{xy}). The average value of each strain component is given in the strain diagrams (Fig. 22.4). The principal strains do not differ much from the strain components and they are not showed. Besides calculated displacements and strains, temperature changes on the surface and the bottom of the specimens in ESEM chamber, are carefully watched in order to get more details about specimen behaviour during drying.

22.4 Example of Non-cracked and Microcracked Cement Paste Specimens

Two cement paste specimens are showed as examples. The 6-day old specimen is presented as an example for specimens where no microcrack occurred and the 5-day old specimen as an example of microcracked specimens. The drying is performed with drying step of 10 % RH, from 100 % (wet condition) to 20 or 10 % RH (dry condition). The diagrams of strains in three directions are given in Fig. 22.4. The diagrams are based on the strains calculated from the microstructure images taken at ESEM (Figs. 22.5a and 22.6a) and strain distribution in the chosen AOI (analyzed by Vic-2D, Figs. 22.5b

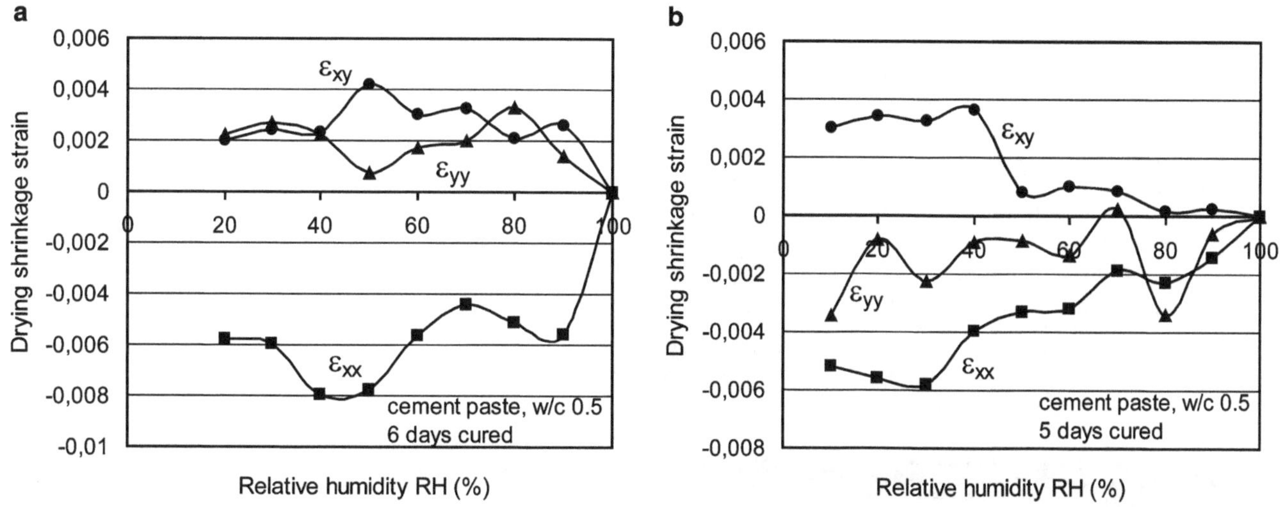

Fig. 22.4 Average drying shrinkage strains as a function of RH, based on analysis of large AOI in cement paste specimens (CEM I 32.5, w/c 0.5, wet cured): (**a**) 6-day old, without microcracks and (**b**) 5-day old, microcracked

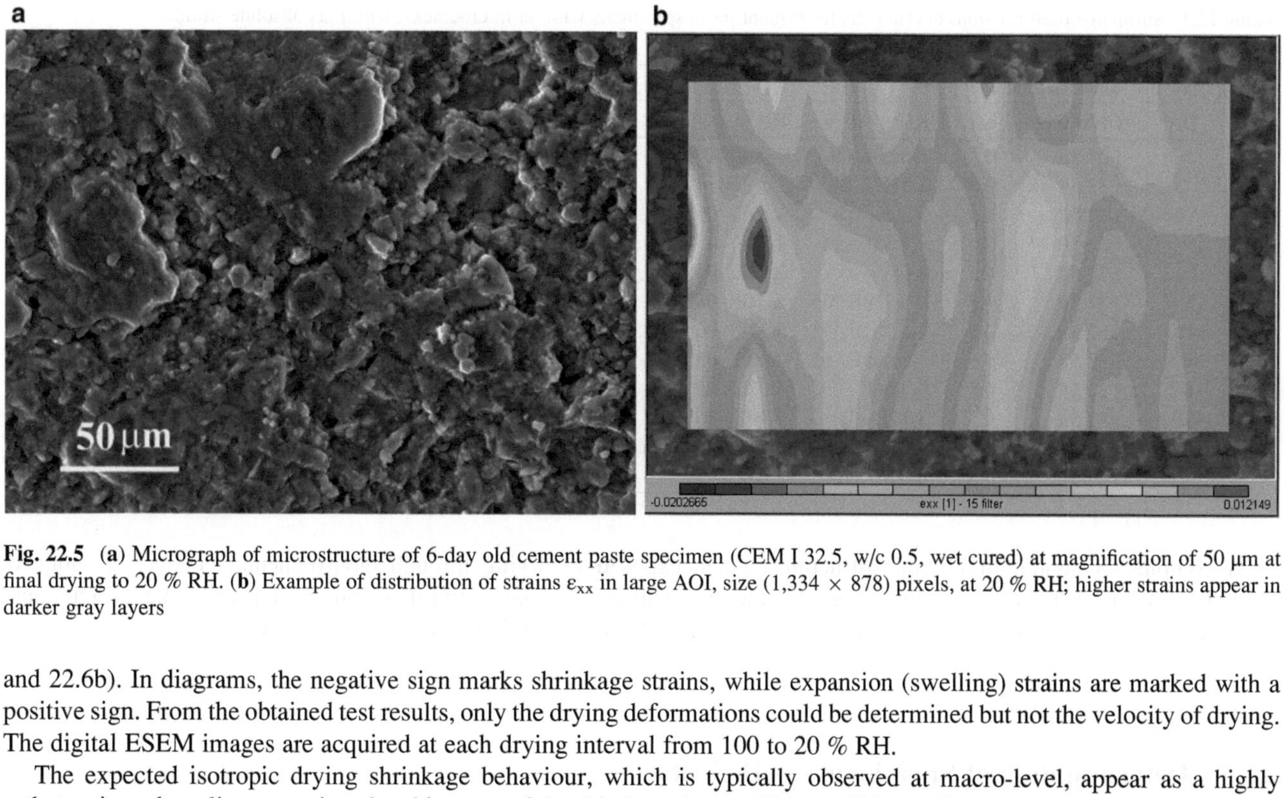

Fig. 22.5 (**a**) Micrograph of microstructure of 6-day old cement paste specimen (CEM I 32.5, w/c 0.5, wet cured) at magnification of 50 μm at final drying to 20 % RH. (**b**) Example of distribution of strains ε_{xx} in large AOI, size (1,334 × 878) pixels, at 20 % RH; higher strains appear in darker gray layers

and 22.6b). In diagrams, the negative sign marks shrinkage strains, while expansion (swelling) strains are marked with a positive sign. From the obtained test results, only the drying deformations could be determined but not the velocity of drying. The digital ESEM images are acquired at each drying interval from 100 to 20 % RH.

The expected isotropic drying shrinkage behaviour, which is typically observed at macro-level, appear as a highly orthotropic and nonlinear at micro-level in most of the dried specimens. This could be attributed to a non-uniform moisture flow and drying of heterogeneous cement paste microstructure (Figs. 22.5 and 22.6) and particularly two-phased C–S–H (at submicro-level). In the average strain-RH diagrams (Fig. 22.4) the main drying intervals could be distinguished from 100 to 40 % RH, and below 40 % RH. Depending on the RH interval and specimen characteristics (age, w/c ratio and curing conditions), cement paste reacts differently. While in cement paste up to 4 days of age, only expansion is present, in specimens over 4 days old (Figs. 22.5 and 22.6) both expansion and shrinkage develop. Expansion and shrinkage occur regardless of the age, which was also confirmed by Neubauer [18]. The domination of expansion or shrinkage depends on the specimen age such that expansion in early aged specimens is mostly 'replaced' by shrinkage in mature specimens. This applies especially for strain components ε_{xx} and ε_{yy}.

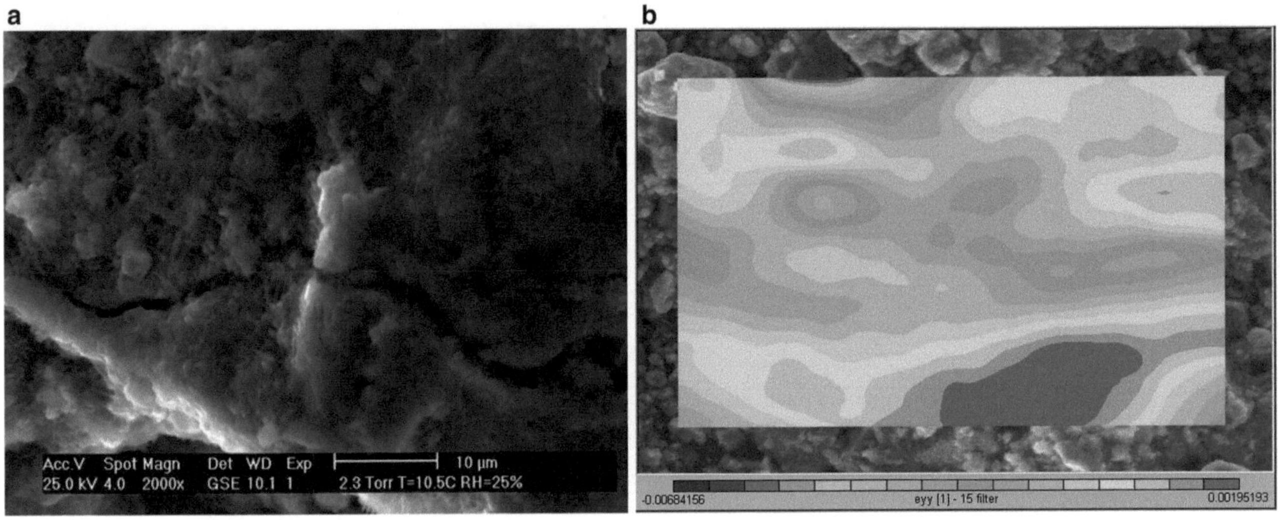

Fig. 22.6 (**a**) Microcrack at 10 µm in 5-day old cement paste (CEM I 32.5, w/c 0.5, wet cured). Image acquired after drying to 25 % RH. (**b**) Example of distribution of strains ε_{yy} in large AOI, size (1,334 × 878) pixels, at 20 % RH; higher strains appear in darker gray layers

Table 22.1 Shrinkage (deformation) coefficients for cement paste specimens without microcracks, with max absolute strains

Curing (days)		Material character		Shrinkage coefficient						Max
				$b_{xx\ L}$		$b_{yy\ L}$		$b_{xy\ L}$		
Wet	Dry	CEM type	w/c	RH ≤ 40 %	RH ≥ 40 %	RH ≤ 40 %	RH ≥ 40 %	RH ≤ 40 %	RH ≥ 40 %	strain (%)
6		I 32.5R	0.5	−0.000109	0.000102	−1.496E−06	−1.348E−05	1.73E−05	−3.98E−05	0.0080
7		I 32.5R	0.5	2.53E−05	−3.92E−05	7.65E−05	−3.25E−05	8.15E−06	1.08E−05	0.0025
29		I 32.5R	0.5	5.50E−05	4.89E−05	−1.76E−05	4.68E−05	−1.69E−05	3.10E−05	0.0042
30		I 32.5R	0.5	−1.55E−05	2.72E−05	3.30E−05	3.06E−05	−8.15E−06	−1.73E−05	0.0022
30		I 32.5R	0.5	4.08E−05	2.58E−05	3.04E−05	1.23E−05	7.54E−07	−1.06E−05	0.0023
	2	I 32.5R	0.5	−2.259E−05	5.07E−05	−1.89E−05	5.60E−05	4.13E−06	4.87E−06	0.0028
	4	I 32.5R	0.5	−6.701E−05	8.37E−06	4.68E−05	−3.95E−06	−1.79E−05	−3.72E−05	0.0036
	4	I 32.5R	0.5	−8.685E−05	2.80E−05	−0.000113	3.68E−05	−1.482E−05	1.91E−07	0.0024
	4	I 32.5R	0.5	−4.66E−05	5.07E−05	−1.894E−05	5.60E−05	4.13E−06	4.87E−06	0.0015
	5	I 32.5R	0.5	3.12E−05	2.99E−05	7.79E−05	3.44E−05	6.72E−05	−4.04E−07	0.0040
	7	I 32.5R	0.5	−2.145E−05	1.46E−05	6.19E−05	1.06E−05	3.64E−05	−1.52E−05	0.0020
	9	I 32.5R	0.5	4.09E−05	2.04E−05	3.01E−05	1.84E−05	−9.10E−06	−4.36E−07	0.0035

A complex behaviour of all microstructures, when exposed to drying through environmental changes, is noticed. Actually: the lower the scale, the more complex the observations. At some point it appears as unpredictable, which could be expected since cement-based matrix behaves more like a 'living tissue' [19] and less like a mechanism.

22.5 Determination of Shrinkage Coefficients

The coefficient of shrinkage deformation is calculated in a straightforward manner with the statistical method of regression, based on the obtained strain results from the Vic-2D correlation analysis given in the diagrams (Fig. 22.4). The ESEM tests showed the nonlinear relationship in a higher or lesser degree in all directions (x–x, y–y and x–y) already from the start of drying (at 100 % RH). The correlation of nonlinear (ε-RH) relationship with a polynomial approximation of fourth or fifth order, is replaced by a simple linear approximation. That means that H-function = (1-H) [10–12] is an approximation. The details about coefficient calculation are given in [13].

Due to the observed difference in cement paste specimens behaviour during drying, the shrinkage coefficient is determined as 'global' and 'local' in specimens without and with microcracks. The global shrinkage coefficient is defined and determined in the full drying range from 100 % RH to 20 or 10 % RH, when the obtained ESEM images appear sharp. The local shrinkage coefficients (presented in Tables 22.1 and 22.2) are determined in two ranges: from 100 RH to 40 % RH and below 40 % RH (to 20 or 10 % RH). The 40 % RH is taken as a sort of drying threshold, since the changes in brightness and contrast were noticed during drying tests at 40 % RH. The shrinkage coefficients in Table 22.1 are termed b_{xx}, b_{yy} and b_{xy} in order to follow the statistical regression terms [13]. They are listed as local (b_{xxL}... etc.) and global (b_{xxG}... etc.). In the further text, terms local (b_{xxL}... etc.) and global (b_{xxG}... etc.) are replaced by a standard shrinkage coefficient symbol: "α_{shL}" (local) or "α_{shG}" (global). In the given test results, we concentrate on α_{shL}.

The shrinkage coefficient presents a slope of the regression line, marked with (+) or (−) sign in the given Tables 22.1 and 22.2. The (+) sign shows tendency of shrinkage in the examined RH range. The (−) sign shows shrinkage reduction. Thus (+) sign, ascribed to shrinkage coefficient in the tables, do not necessarily shows swelling as it would be expected according to the (ε-RH) diagrams (Fig. 22.4). Both coefficient signs in the table can result from the shrinkage deformations, which are marked with sign (−) in the (ε-RH) diagrams.

22.6 Discussion on Shrinkage Coefficients

22.6.1 Shrinkage Coefficients and Strains in Non-cracked Specimens

Only local shrinkage coefficients are presented in Table 22.1. Both local and global shrinkage coefficients for the cement paste specimens without microcracks reach the order of magnitude of 10^{-5} (rarely 10^{-6} or 10^{-7}). However, the absolute values of local α_{shL} in (x–x) and (y–y) direction appear always higher in comparison to the absolute values of global α_{shG} in any direction. That is the reason for separating coefficients into local and global. Local α_{shL} below 40 % RH appears typically higher when compared to α_{shL} above 40 % RH as an absolute value.

It is visible that a higher number of young sealed (dry cured) specimens were found without microcracks in comparison to the wet cured non-cracked specimens. That could be related to the softer microstructure of wet cured specimens. The highest (absolute) coefficient value is around 0.00011, seen in a local α_{shL}, below 40 % RH, regardless of the specimen age or curing conditions. Based on the comparisons of all the coefficients, this value could be treated as a sort of a threshold between the coefficients in specimens in which a microcrack occurred and those without a microcrack. The strains ε_{xx} or ε_{yy} vary around 0.002 and 0.004 (absolute values) as given in Table 22.1. Since no microcracks are traced we could say that specimens could reach the higher strain values only in one of the directions (either ε_{xx} or ε_{yy}) without having any microcracks. When compared, slightly higher values of ε are reached by the specimens that are dried by RH reduction of 10 % (around 0.0035 with an exception of 0.008); strain in specimens that are dried by RH reduction of 20 % varies around 0.002. In many cases the similar drying trend of ε_{xx} and ε_{yy} curves occur and the curves have a similar shape.

22.6.2 Shrinkage Coefficients and Strains in Micro-Cracked Specimens

The values of the local coefficients α_{shL} (marked as b_{xxL}) in Table 22.2 again appear higher in their absolute value for specimens with microcracks, when RH drops below 40 %. The local coefficients at RH below 40 % vary typically between min (10^{-4}) or 0.00012 and higher order of magnitude (10^{-3}). The coefficient values increase before a microcrack occurs at 20 or 10 %.

Local coefficients are higher in all directions compared to the coefficients in Table 22.1, but a bit higher in (x–x) and (y–y) direction compared to (x–y). The questions remain: does that mean microcracks would appear only at a certain specimen curing age, material characteristics and at these particular coefficient values? Or the calculated coefficient has its minimum value when microcracks occur and that value could be higher and how is that related? Actually, we should know the lowest coefficient at which the microcrack would appear. If the coefficient crosses the threshold value of 0.00011 as in Table 22.2, the microcrack should appear in any case, regardless of the way of curing, specimen age or drying steps (20 or 10 % RH). It seems that the value of shrinkage coefficient is sufficient to define if the specimen would microcrack, but when accompanied by a certain strain value. However, specimen age, curing way and ε values define RH at which a microcrack may initiate (Table 22.2).

In the dry cured (sealed) specimens α_{shL} have similar values. They range from the highest 10^{-3} to 10^{-5} in young specimens (2–4 days of age). These specimens experienced microcracks and even cracks at the end of drying tests (at 10 %

Table 22.2 Shrinkage (deformation) coefficients with maximum absolute strains for cement paste specimens with microcracks

	Curing (days)		Material character		Shrinkage coefficient						
					$b_{xx\ L}$		$b_{yy\ L}$		$b_{xy\ L}$		
Microcrack initiates (%)	Wet	Dry	CEM type	w/c	RH ≤ 40 %	RH ≥ 40 %	RH ≤ 40 %	RH ≥ 40 %	RH ≤ 40 %	RH ≥ 40 %	Max strain (%)
Between 70–60 % and 30–20 %	3		I 32.5R	0.5	−6.539E−05	−0.0001041	−0.000154	−0.0001014	−5.539E−06	−6.881E−07	0.009
Between 60–50 % and 20–10 %	3		I 32.5R	0.5	−0.007306	−0.0001255	0.004129	2.03E−05	−0.001573	−2.66E−05	0.200
Between 70–60 % and 30–20 %	4		I 32.5R	0.5	0.001518	−0.0008417	0.007571	0.000536	−0.002620	0.000127	0.200
Between 70–60 % and 20–10 %	4		I 32.5R	0.5	0.003308	−7.044E−05	0.007455	−3.745E−05	−0.002381	−2.064E−05	0.200
At start and between 30 and 20 %	5		I 32.5R	0.5	3.38E−05	0.000137	6.15E−05	9.76E−06	1.67E−05	−0.000108	0.003
At start, at test end (at 20 %)	6		I 32.5R	0.5	−0.011299	−3.816E−05	0.006715	4.27E−05	−0.001772	−9.617E−06	0.200
At test end (20 %)	7		I 32.5R	0.5	−0.000127	1.21E−05	0.000559	−3.45E−05	5.63E−05	3.28E−06	0.049
At test end (20 %)	32		I 32.5R	0.5	−0.001744	5.88E−05	0.001228	−2.06E−05	0.000602	−5.36E−05	0.030
At test end (10 %)	40		I 32.5R	0.5	−0.002081	2.15E−05	0.001848	3.64E−05	0.000738	2.79E−06	0.067
At test end (20 %)	52		I 32.5R	0.5	0.001010	4.47E−05	−0.001449	9.51E−06	0.000552	−1.27E−05	0.027
At test end (10 %) Crack		2	I 32.5R	0.5	0.004743	−2.82E−05	0.013769	−3.29E−05	0.000180	5.90E−06	0.400
At test end (at 10 %)		3	I 32.5R	0.5	0.005302	6.20E−06	0.007350	−1.31E−05	0.000255	−1.49E−06	0.250
Between 70–60 % and 30–20 %		8	I 32.5R	0.5	0.000368	0.000481	−0.000570	−0.0002023	−0.0003453	0.000346	0.050
Between 50–40 % and 20–10 %		8	I 32.5R	0.5	−0.000272	3.06E−05	0.000358	4.15E−05	2.58E−05	3.11E−06	0.014
Between 40 and 20 %		37	I 32.5R	0.5	−0.000785	3.82E−05	0.001717	1.64E−07	1.06E−05	1.90E−05	0.010

RH) with coefficient 10^{-2}. Average max strain reached 0.3, similar as in young, wet cured specimens. In specimens 8–37 days old, microcracks are found also during drying (at 70–60 % RH and 30–20 % RH) with coefficients 10^{-4} and lower. Average max strain is 0.03 (again similar as in wet cured specimens of the similar age). This could be expected. Sealed specimens become more brittle while maturing and microcrack earlier.

In the specimens where tiny microcracks are found at the end of drying tests, microcracks initiated during the drying test. Only in one case microcracks were found at the beginning of the test but that did not influence the strain. In some cases they are found after rewetting was performed. The location of the microcracks was much more often present further away from the examined area. Only in a few cases not showed here, the microcrack was located in the observed dried area under ESEM. The drying strains ε_{xx} or ε_{yy} reached much higher values (up to 10^{-1}) compared to specimens without microcracks. Based on them we can recognize when a microcrack occurred (Table 22.2). Higher ε_{xx} or ε_{yy} was reached by young specimens (wet or dry cured). In specimens older than 7 days, ε was typically lower. Higher strain was reached in sealed (dry) cured specimens of 2–3 days of age.

22.7 Conclusions

Calculation of shrinkage coefficients in cement paste specimens without and with microcracks based on nondestructively determined deformations, shows the existence of a coefficient threshold value 0.00011, which appears in a local α_{shL}, below 40 % RH, regardless of the specimen age or curing conditions. When the coefficient value is above 0.00011, microcracking should be expected. This is important to know when performing a coupled numerical analysis of moisture flow (drying) and microcracking in the specimens, based on a certain shrinkage coefficient value. The micro-level drying tests show also all the complexity of cementitious materials behaviour during drying (swelling, shrinkage and microcracking), which are not seen or they are neglected when drying tests at macro-level are performed. Hence, the coefficient can be called not only a shrinkage coefficient but a coefficient of drying deformations. It seems meaningful to consider only the order of the magnitude of the calculated coefficients as well as their absolute values due to all the complexity.

References

1. Jankovic D (2000) Moisture flow and microstructure development of the Interface Zone, Report, Delft University of Technology, Faculty of Civil Engineering and Geo-science, CM2000-010, Netherlands
2. Bazant ZP, Prat PC (1988) Effect of temperature and humidity on fracture energy of concrete. ACI Mater J 85:262–271
3. Van Mier JGM (1990) Mode I behaviour of concrete: influence of the rotational stiffness outside the crack zone. In: Elfgren L, Shah SP (eds) Analysis of concrete structures by fracture mechanics. Chapman and Hall, London, p 19
4. Sadouki H, Van Mier JGM (1997) Simulation of hygral crack growth in concrete repair systems. Mater Struct 30:518–526
5. Sadouki H, Van Mier JGM (1997) Meso-level analysis of moisture flow in cement composites using a lattice-type approach. Mater Struct 30:579–587
6. Alvaredo AM (1994) Drying shrinkage and crack formation, Ph.D. thesis, ETH, No. 10754
7. Alvaredo AM, Helbling A, Wittmann FH (1995) Shrinkage data of drying concrete, building materials Reports No.4, Aedificio Verlag GmbH, Freiburg
8. Martinola G, Wittmann FH (1995) Application of fracture mechanics to optimize repair mortar systems. In: Wittmann FH (ed) Proceedings of fracture mechanics of concrete structures (FraMCoS-2). AEDIFICATIO Publishers, D-79104 Freiburg, pp 1481-1492
9. Wittmann FH (1982) Creep and shrinkage mechanisms. In: Bazant ZP, Wittmann FH (eds) Creep and shrinkage in concrete structures. Wiley, New York, pp 129–161
10. Bazant ZP, Panula LP (1982) New model for practical prediction of creep and shrinkage, designing for creep and shrinkage in concrete structures, SP 76, ACI, Detroit, pp 7-23
11. Bazant ZP, Chern JC (1985) Concrete creep at variable humidity: constitutive laws and mechanisms. Mater Struct 18(103):1–20
12. Bazant ZP, Xi Y (1994) Drying creep of concrete: constitutive model and new experiments separating its mechanisms. Mater Struct 27:3–14
13. Jankovic D (2008) Nondestructive determination of drying deformations in cement paste by means of ESEM and digital image analysis, Optical Micro- and Nanometrology in Microsystems Technology II. In: Gorecki C, Asundi AK, Osten W (eds) Proceedings of SPIE Europe photonics Europe 7–10 April 2008, vol 6995. Strasbourg, 69950G1-G12
14. Jankovic D, Van Mier JGM (2002) Preliminary investigation of drying shrinkage cement paste specimens. In: Pyrz R, Schjodt-Thomsen J, Rauche JC, Thomsen T, Jensen LR (eds) International conference on new challenges in mesomechanics. Aalborg University, pp 265-271, vol 1 August 2002
15. Chu TC, Ranson WF, Sutton MA, Peters WH (1985) Application of digital image correlation technique to experimental mechanics. Exp Mech 25:232–244

16. Sutton MA, Wolters WJ, Peters WH, Ranson WF, McNeill SR (1983) Determination of displacements using an improved digital correlation method. Image Vision Comput 1(3):133–139
17. Bruck HA, McNeill SR, Sutton MA, Peters WH III (1989) Digital image correlation using Newton–Raphson method of partial differential correlation. Exp Mech 28(3):261–267
18. Neubauer CM (1997) On the chemistry, microstructure, and deformation properties of cement pastes: towards a new strategy for controlling drying shrinkage, Ph.D. thesis, Northwestern University, Illinois
19. Freyssinet E (1951) The deformation of concrete. Mag Concrete Res 3(8):49–56

Chapter 23
Thermo-Fluid Modeling of the Friction Extrusion Process

H. Zhang, X. Deng, X. Li, W. Tang, A.P. Reynolds, and M.A. Sutton

Abstract Friction extrusion is a novel manufacturing process for producing high-value materials (e.g. metal wires) from low-cost precursors (e.g. powders) or through recycling of machining wastes (e.g. chips). In the friction extrusion process, the material experiences high temperature and severe plastic deformation before it forms the final product in the form of a wire. The present work is focused on understanding the heat transfer and material flow phenomena in the friction extrusion process. A numerical thermo-fluid model has been developed and validated by experimental measurements. In the model, the experimentally measured mechanical power is used as the heat input to predict the temperature field in the experiment. The processing material is treated as a non-Newtonian fluid with a viscosity that is temperature and strain rate dependent. Select marker particles in the material are followed and their motions observed in the simulation to study material flow patterns. It is found that predictions of the temperature field and marker particle trajectories match reasonably well with experimental measurements. Results of this study suggest that the proposed thermo-fluid model can capture the main features of the thermo-fluid phenomena in the friction extrusion process and can be used to provide reasonable predictions of the temperature and material flow fields in the friction extrusion process.

Keywords Friction extrusion process • Particle tracking • Heat transfer • Thermo-fluid modeling • Numerical simulation

23.1 Introduction

The friction extrusion process (FEP) is a novel friction based extrusion method, invented and patented in 1993 [1], for producing high value wire or rod from low-cost precursors (e.g. powders and chips). Figure 23.1 shows a schematic diagram of the FEP. At the beginning, the extrusion die with a central extrusion hole is pushed down into the process chamber by a vertical force to consolidate the billets loaded in the chamber. Then the die starts to rotate under an increasing force while heat is generated through the friction at the die-billet interface. After the billets are consolidated and softened by the vertical force and frictional heating, they are extruded out from the extrusion hole in the die and form a solid wire.

The FEP represents a step forward in recycling metal wastes, such as metal chips, scraps, and small blocks produced during manufacturing, because of its attractive merits. Generally there are two categories of recycling methods of metal wastes: conventional recycling methods and direct recycling methods. The most striking difference between the two categories is that conventional recycling methods require more energy and labor and has more material loss than direct recycling methods. In the case of recycling of aluminum waste, it has been estimated by Gronostajski and Matuszak [2] that direct recycling can save about 40 % of material, 26-31 % of energy consumption, and 16-60 % of labor costs. The FEP falls into the category of direct recycling methods but it is different from traditional direct recycling methods in that fewer process steps, less energy, and simpler equipment are needed. In addition to recycling metal wastes, the FEP can also serve as a potentially energy efficient method for producing high-value materials (e.g. titanium wires) from low-cost precursors (e.g. titanium powders).

Until recently the FEP has received little attention and development since and little scientific literature exists for FEP [3–6]. However, some lessons can be learned from investigations of a similar but well-known process, the friction stir

H. Zhang • X. Deng (✉) • X. Li • W. Tang • A.P. Reynolds • M.A. Sutton
Department of Mechanical Engineering, University of South Carolina, Columbia, SC 29208, USA
e-mail: deng@cec.sc.edu

H.J. Qi et al. (eds.), *Challenges in Mechanics of Time-Dependent Materials, Volume 2: Proceedings of the 2014 Annual Conference on Experimental and Applied Mechanics*, Conference Proceedings of the Society for Experimental Mechanics Series,
DOI 10.1007/978-3-319-06980-7_23,

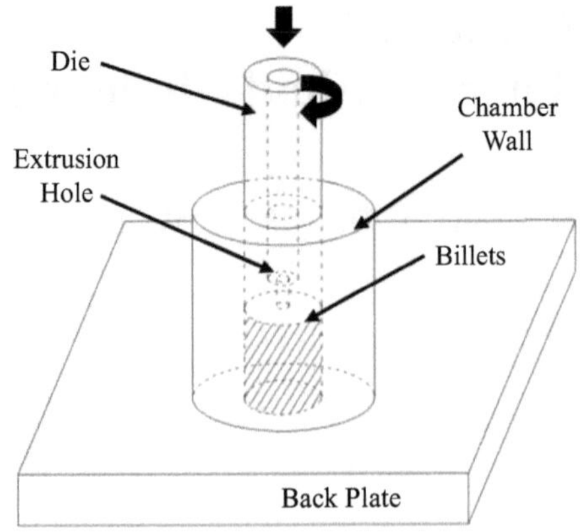

Fig. 23.1 A schematic diagram of the friction extrusion process

welding (FSW) process, which has been studied numerically using a fluid or solid mechanics based approach. In fluid based approaches, the workpiece material is treated as a fluid. Although frictional heating is not directly modeled, heating is treated either in the form of viscous dissipation in the fluid volume (e.g. [7]) or as a heat flux at the tool–workpiece interface (e.g. [8, 9]). In solid mechanics based approaches, the workpiece material is treated as a solid, and frictional heating at the tool–workpiece interface can be directly modeled (e.g. [10–13]). The current work uses a fluid based method to model the friction extrusion process.

23.2 Experimental Details

23.2.1 Extrusion Materials and Devices

In order to visualize the material flow pattern in the friction extrusion process, a maker insert technique is employed. In this study, a Φ25.4 mm × 19.1 mm Aluminum Alloy 6061 solid cylinder was adopted as the extrusion billet charge (so that the consolidation process can be omitted). For the marker material, a Φ2.54 mm Aluminum Alloy 2195 wire was inserted into a through hole in the AA6061 cylinder at 1/3 radius from the center. A top view of the extrusion cylinder is shown in Fig. 23.2. Since the mechanical and thermal properties of the two alloys are very similar and the volume fraction of the marker material is relatively small, it is believed that insertion of a thin marker wire will not greatly affect the friction extrusion process.

The FEP device used in this experiment consists of four parts: a cylindrical die with a flat tip (see Fig. 23.3 for a bottom view of the die, where the central hole is the extrusion exit hole), an O1 tool steel chamber, a H13 steel back plate, and a friction stir welding (FSW) machine used to apply the extrusion force and torque during FEP. The dimensions of the extrusion die with a flat contact surface are Φ25 mm × 114.3 mm with a Φ2.54 mm through hole at the center. Eight thermocouples were inserted in the system at different positions as shown in Fig. 23.4 to measure temperature variations during FEP.

The inner and outer diameters of the stationary 'billet' chamber are 26.4 and 54.6 mm, respectively, and the height of the chamber is 29.2 mm. The chamber also has a 70.0 mm diameter shoulder with two semicircle breaches for convenience of clamping and location.

The FSW PDS (Process Development System) at the University of South Carolina (USC) was used to control the die movement in the extrusion process, and the Z-axis load control mode was adopted in the operation. Except at the initial stage, the entire extrusion process occurred under a constant Z-axis load (the vertical force or pressure) but the extrusion rate and die advance rate may vary during the process. This is in contrast to a "normal" extrusion process wherein the extrusion rate is held constant and the pressure may vary to maintain the specified rate. A stainless steel plate was used as the back plate to support the chamber and cylinder sample from below. In order to prevent slipping between the back plate and the billet charge, a 4 mm height small pin was set on the back plate inside the 'billet' chamber area.

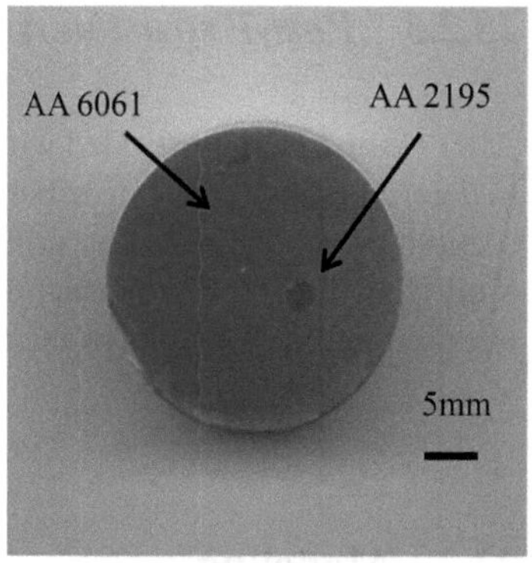

Fig. 23.2 An AA6061 billet cylinder with an AA 2195 marker wire inserted at 1/3 radius from the center

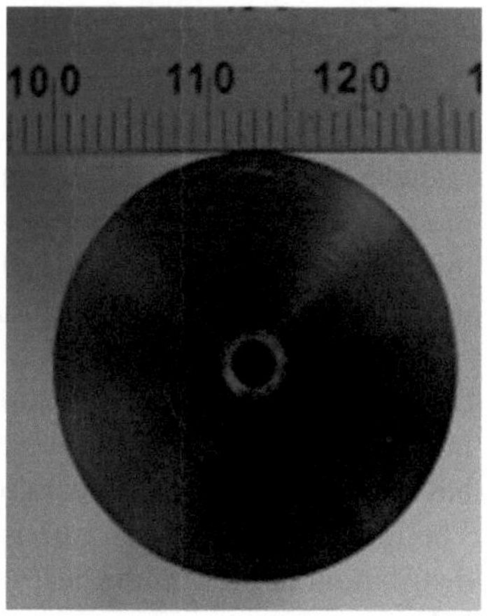

Fig. 23.3 A friction extrusion die with a flat contact surface

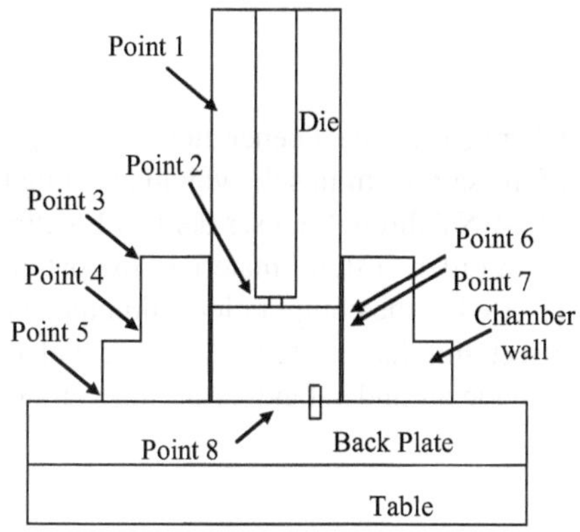

Fig. 23.4 Thermocouple locations

23.2.2 *Friction Wire Extrusion Process*

The wire extrusion process as performed using the FSW PDS is best described as a back extrusion wherein the rotating die is pressed into the billet chamber under load control and the wire is extruded up through the central hole in the die. During the process, the die is pushed downward until it contacts the charge in the billet chamber and then the die starts to rotate. Under downward pressure and with frictional heating, the aluminum billet sample expands towards the chamber wall to fill in the small gap (~0.4 mm) between the sample and chamber, and when a sufficient temperature is achieved, wire extrusion begins. Because of the limited headspace above the die in the FSW PDS, only a 150 mm long wire was extruded in the experiment.

During the friction extrusion process, the die rotation rate, die position, Z-force (extrusion force) and torque were sampled at a rate of 10 Hz. For this study, all extrusion runs were performed using a constant die rotation rate of 300 rpm and a constant Z force of 44.45 KN.

23.2.3 *Post-Extrusion Analysis Procedures*

After extrusion, the extruded wire and the remaining billet disk (in the billet chamber) were serially sectioned and examined to determine the spatial distribution of the marker material, which provides a visualization of the flow pattern in FEP. Wire sections were elucidated by grinding, polishing, and etching with Keller's reagent (190 ml water, 2 ml HF, 3 ml HCl, and 5 ml HNO3). The distribution of AA2195 marker material in the AA6061 wire cross-sections was determined by optical microscopy. Six sections of the extruded wire were examined ranging from near the first material extruded to near the last.

23.3 Modeling

23.3.1 *Model Description*

A three-dimensional thermo-fluid model has been developed to simulate the friction extrusion process. The model domain includes the extrusion die, the process chamber, the billet sample, the back plate, and the support table. The finite volume method was used to discretize the computational domain. A converged computational grid is shown in Fig. 23.5. The minimum cell size in the grid is 0.05 mm and the maximum cell size is 1.2 mm. The commercial code ANSYS FLUENT was used to carry out the computation.

The current work employs a volume heating model in which frictional heating at the die-sample interface and plastic dissipation underneath the interface are represented by a volume heat source in a material layer in the sample near the interface that experiences severe plastic deformation and material flow. A uniform height of 4 mm for the layer is taken based on experimental observation of marker material positions in the remnant sample at the end of extrusion experiment. The total heat generation rate is equal to the mechanical power input recorded in the experiment. It is assumed that the distribution of the volume heat source in the layer is linear along both the radial and vertical loading directions, as described below

$$q = \frac{3Q_{total}}{\pi(H^2 - h^2)R^3} z\sqrt{x^2 + y^2} \tag{23.1}$$

where q is the heat generation rate, Q_{total} is the mechanical power in experiment, H is the height of the sample, h is the height of the sample minus the volume heating layer, and R is the radius of the sample. This heat source is implemented in ANSYS FLUENT through a user defined function.

In order to study material flow patterns during extrusion, several massless solid particles with a 10^{-3} mm diameter are inserted in the computational domain at different positions as markers. Since they are massless, these particle markers don't affect simulation predictions (e.g. velocity and temperature fields). The trajectory paths of these particles can be used to provide an understanding of material flow patterns during extrusion.

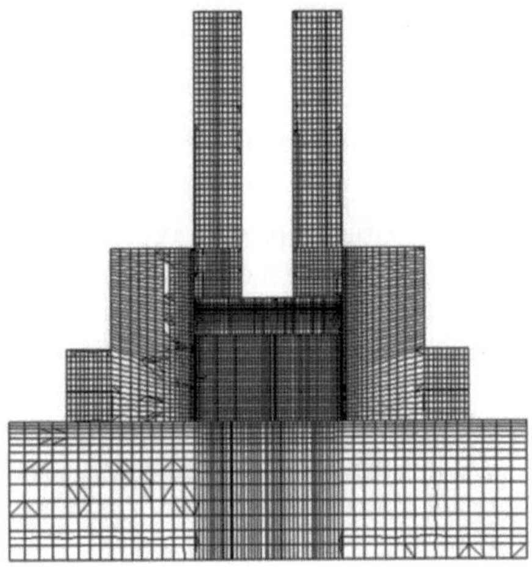

Fig. 23.5 A cross-section of a converged grid

23.3.2 Material Properties

In the simulation model, the billet sample is treated as a fluid with relevant fluid and thermal properties, and other parts of the model domain are taken to be rigid solids that only require thermal properties. The sample in the experiment is made of aluminum alloy 6061, which is widely used in the industry and its properties are available from the literature. Following a study of FSW [7], in the current work AA6061 is treated as a highly viscous non-Newtonian fluid with a viscosity given by

$$\mu = \frac{1}{3\alpha\bar{\dot{\varepsilon}}} \ln\left\{ \left(\frac{Z}{A}\right)^{\frac{1}{n}} + \left[\left(\frac{Z}{A}\right)^{\frac{2}{n}} + 1\right]^{\frac{1}{2}} \right\} \tag{23.2}$$

where

$$Z = \bar{\dot{\varepsilon}} \exp\left(\frac{Q}{GT}\right) = A[shinh(\alpha\sigma)]^n \tag{23.3}$$

and $\bar{\dot{\varepsilon}}$ is the effective strain rate defied by strain components ε_{ij}

$$\bar{\dot{\varepsilon}} = \left(\frac{2}{3}\varepsilon_{ij}\varepsilon_{ij}\right)^{\frac{1}{2}} \tag{23.4}$$

and Q is the activation energy, T is temperature, G is the gas constant, and A, α and n are constants whose values are obtained from reference [14]. Other material properties for AA6061 and for the solids are given in [15].

23.3.3 Governing Equations and Conditions

The governing equations involved are the continuity equation, Navier–Stokes equations, and the energy equation. The continuity equation for a incompressible flow in indicial notation ($i = 1$, 2, and 3) is given by

$$\frac{\partial u_i}{\partial x_i} = 0 \tag{23.5}$$

where u is velocity. The Navier–Stokes equations for incompressible non-Newtonian transient fluid flow are

$$\rho\frac{\partial u_j}{\partial t} + \rho u_i \frac{\partial u_j}{\partial x_i} = -\frac{\partial P}{\partial x_j} + \frac{\partial}{\partial x_i} + \left(\mu\frac{\partial u_j}{\partial x_i}\right) + F_j \tag{23.6}$$

where ρ is the density, P is the pressure, μ is the viscosity given by Eq. (23.2), and F is the body force. The energy equation is

$$\rho c_p \frac{\partial T}{\partial t} + \rho c_p u_i \frac{\partial T}{\partial x_i} = \frac{\partial}{\partial x_i}\left(K\frac{\partial T}{\partial x_i}\right) + q \tag{23.7}$$

where c_p is specific heat, T is temperature, K is conductivity, and q is the heat generation rate given by Eq. (23.1).

The initial temperature is set at the lab temperature of 295 K. Convection heat transfer boundary conditions are used for all the outer surfaces of the model. The existence of an initial gap between the billet sample and process chamber wall is also modeled and it disappears gradually as the sample deforms to fill it. Based on experimental data, an approximation is made that the gap is filled from time t = 4.7 s and to the beginning of extrusion at t = 10 s. The gap is modeled by changing the conductivity of the gap layer in the numerical simulation.

Studies have shown that there is slipping between the workpiece and the rotating tool in FSW [11] and it is expected that in the friction extrusion process such a phenomenon exists as well. A contact state variable, $\delta = 0.3$, is employed in the current work based on findings from reference [16], so that the top surface of the sample has a rotation speed of 90 rpm, which is 30 % of the die rotation speed of 300 rpm.

23.4 Results and Discussions

23.4.1 *Thermal Validation*

To validate simulation predictions of the temperature field, temperature variations with time were measured using thermal couples at several positions in the friction extrusion experiment. Figure 23.6 shows both the measured temperature variations with time and temperature predictions at the same positions. In general, the predictions have a reasonably good agreement with experimental measurements.

23.4.2 *Temperature Field Features*

Figure 23.7 shows the predicted contour graphs of the temperature field of the entire cross-section of the model system at some instants. Figure 23.7a depicts the temperature field at t = 7.3 s. Obviously, the heat source region has the highest temperature since this region experiences the most severe heating. At this moment, only a part of the initial gap between the aluminum alloy sample and the chamber wall is filled with hot aluminum alloy. As a result, heat transfer from the aluminum alloy sample to the chamber wall is affected and the sample generally has much higher temperatures than the chamber wall. Figure 23.7b gives the temperature field at t = 17.1 s. At this moment, the gap has been filled completely and heat flows from the aluminum sample to the chamber wall without the obstruction of a gap. Even so, the sample material still has much higher temperatures due to the fact that it contains the heat source layer and has a higher thermal conductivity. Figure 23.7c presents the temperature field at the end of extrusion. The chamber wall has a pronounced temperature rise while the sample does not. A small region near the die-sample interface has a temperature close to the solidus of aluminum alloy 6061.

23.4.3 *Velocity Field Features*

Figure 23.8 shows the resultant velocity contours of the cross-section of the sample in the chamber at different times. It can be seen that a small region under the die-sample interface has a high velocity magnitude and the area of this region hardly shows any change during the friction extrusion process. At the beginning of extrusion, as shown in Fig. 23.8a, there is only a tangential velocity and the velocity is almost zero near the extrusion hole. Along the vertical direction, the velocity drops

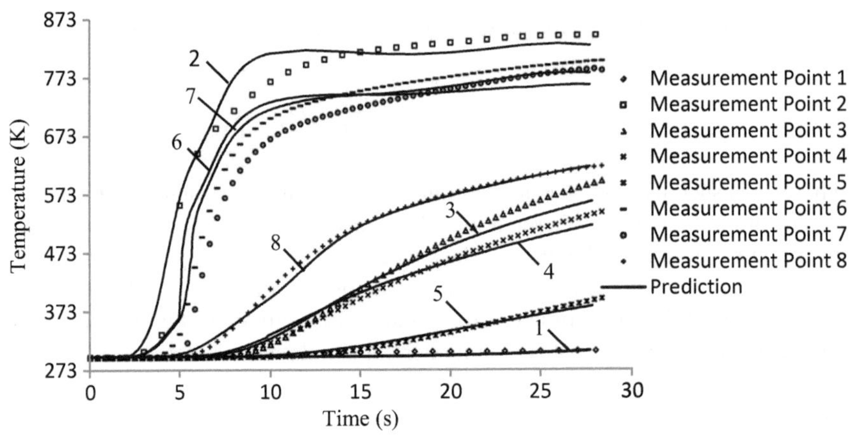

Fig. 23.6 Comparisons between temperature predictions and experimental measurements

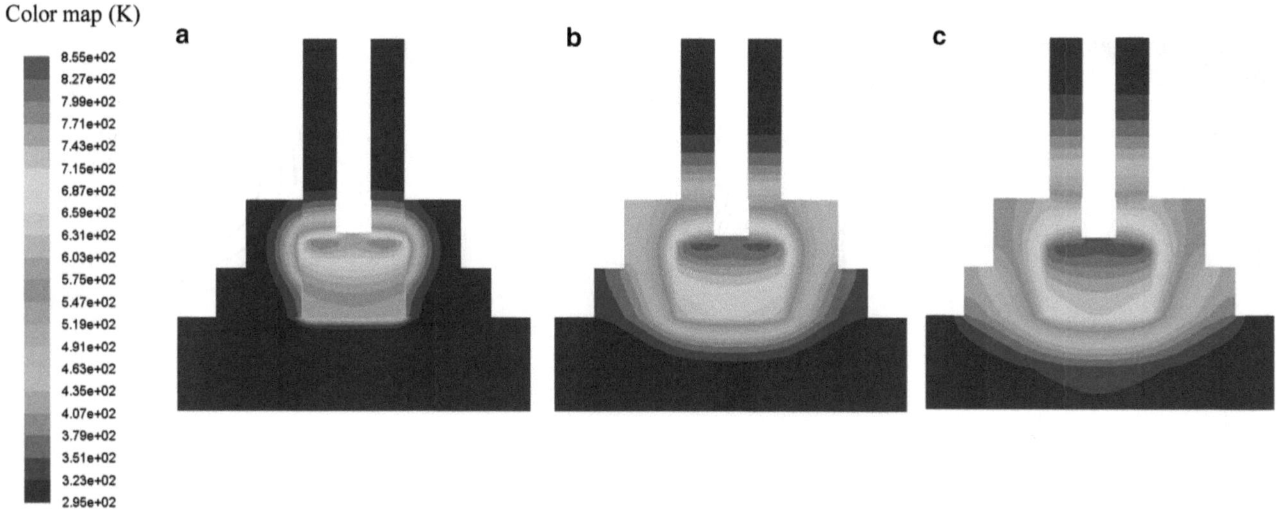

Fig. 23.7 Predicted temperature contours at (**a**) t = 7.3 s; (**b**) t = 17.1 s; (**c**) t = 27.7 s

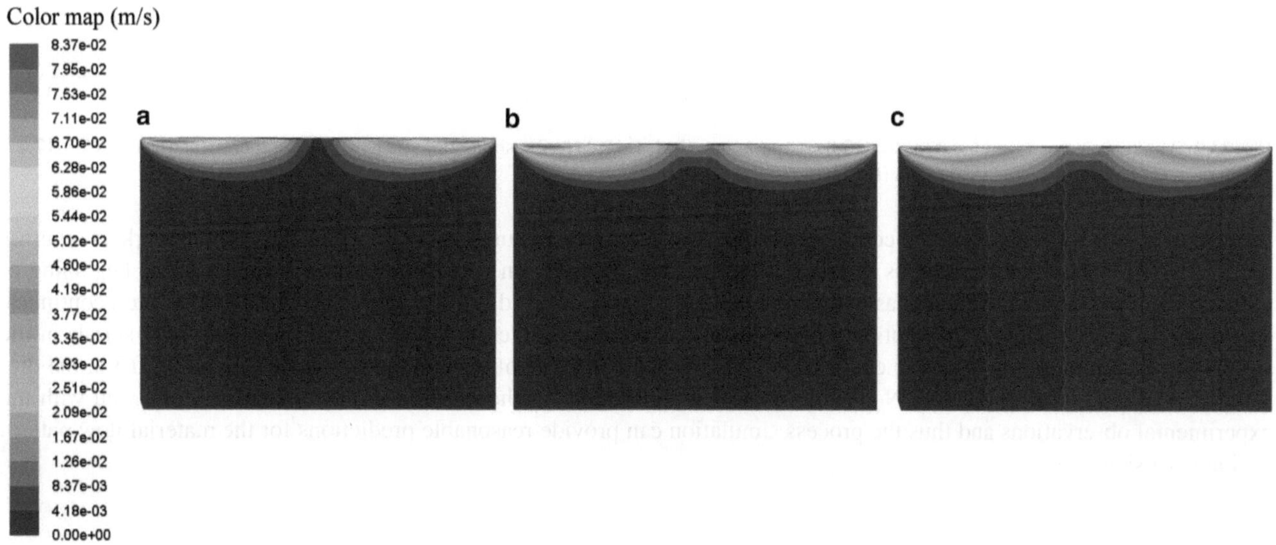

Fig. 23.8 Predicted velocity contours at (**a**) t = 10 s; (**b**) t = 18 s; (**c**) t = 24 s

drastically when the distance from the interface increases except near the extrusion hole. The velocity has a peak near the center of the radius at the same height. During extrusion, as shown in Fig. 23.8b and c, compared with the beginning of extrusion shown in Fig. 23.8a, the contours almost have no change except in the region under the extrusion hole, indicating that the vertical velocity dominates only in the small region under the extrusion hole.

23.4.4 Marker Distribution in the Extruded Wire

Marker particles released in the simulation in the sample material is observed to flow with the sample material and leave behind continuous trajectory paths, which reflect the material flow patterns. Because the number of marker particles used in the simulation is finite and is not enough to provide spatially continuous information of the flow, some interpolations are made to obtain more information. Figure 23.9 shows a qualitative comparison of the positions of marker particles on extruded wire cross-sections at various times from both experimental measurements and simulation predictions (during an

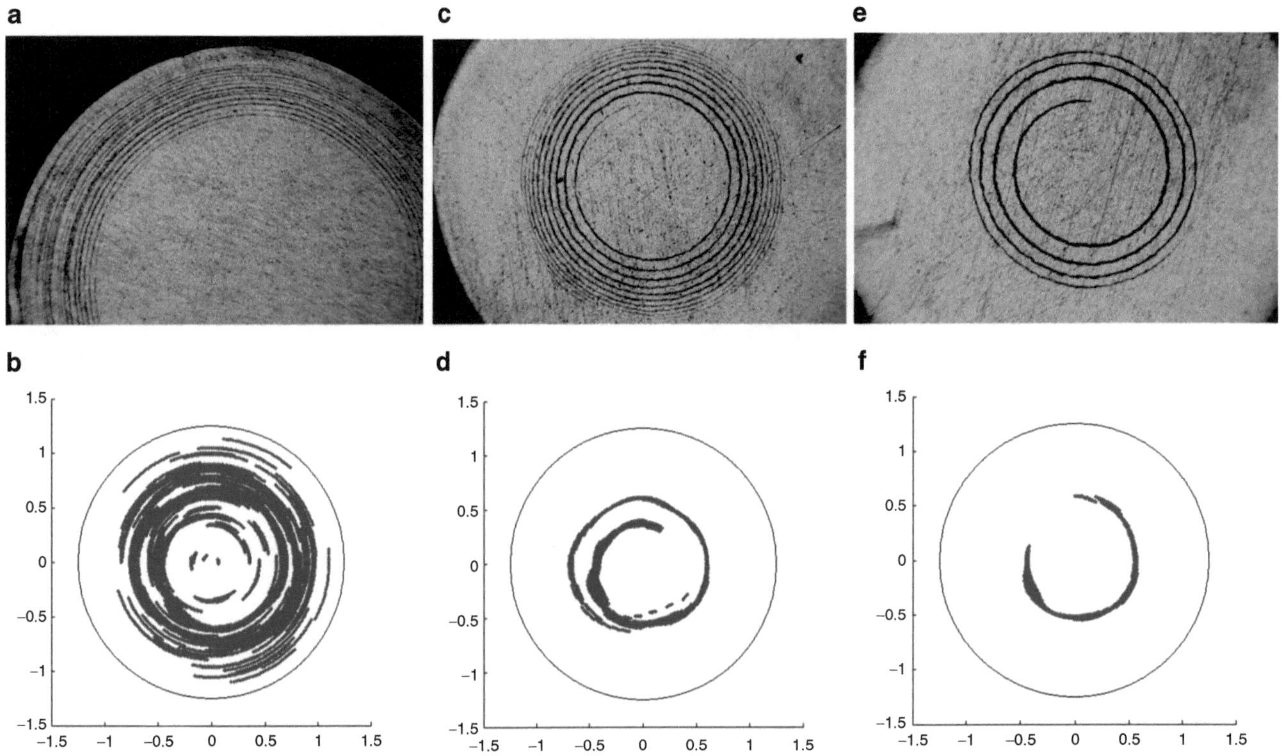

Fig. 23.9 Markers on cross-sections at (**a**) t = 13.0 s, experiment; (**b**) t = 12.8–13.2 s, simulation; (**c**) t = 16.9 s, experiment; (**d**) t = 16.7–17.1 s, simulation; (**e**) t = 18.8 s, experiment; and (**f**) t = 18.6–19.0 s, simulation

interval of 0.4 s). It is seen from the comparisons that there are some features in common, such as (a) the markers on cross sections generally formed continuous spirals, (b) the number of spirals on wire cross section is higher at the beginning of extrusion and the number decreases as extrusion continues, and (c) the radii of the spirals decrease as extrusion continues. The main difference between simulation predictions and experimental measurements is that there are fewer spirals in the simulation predictions. One possible cause for this difference is the use of a small contact state variable of 0.3 at the die-sample interface in the simulation. Nonetheless, the variation trends of the markers in the simulation agree well with the experimental observations and thus the process simulation can provide reasonable predictions for the material flow pattern during extrusion.

23.4.5 Trajectory Paths of Particle Markers

Since the marker particles in the simulation are massless, they flow with the nearby material and thus their trajectory paths represent the material flow paths. Figure 23.10 depicts the path lines of the particle markers with different initial positions in the simulation. The initial heights of these path lines range from 15 to 18 mm. Every path line has a unique color. Figure 23.10a shows a group of path lines which originate at a radial location that is one third of sample radius away from the center. All of these path lines go through the extrusion hole, indicating that the material near the initial path line positions is extruded out of the chamber to form the wire. Before reaching the extrusion hole, all path lines have spiral shapes with decreasing radii and increasing heights as they flow towards the extrusion hole. This means that the material there is pushed both inwards and upwards. Figure 23.10b shows path lines of particles which are released at positions with a distance 7.2 mm from the center. Only some of these path lines reach the extrusion hole and so only some of the material at the path line location flows out of the chamber to become part of the extrusion wire. Path lines with initial positions at 10.4 mm from the sample center are shown in Fig. 23.10c. Apparently, none of them passes through the extrusion hole and thus no material at this location goes out of the chamber and become part of the extruded wire. The path lines in Fig. 23.10b and c suggest that there exists a dead zone in the process chamber as in traditional extrusion where the material can never be extruded out of the chamber.

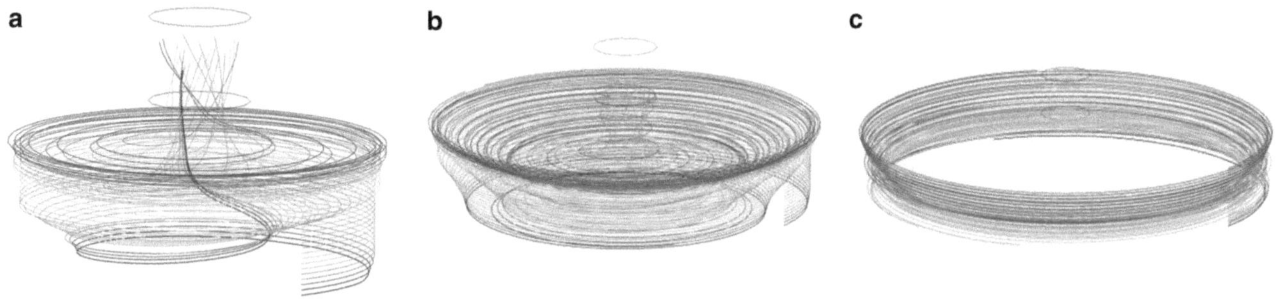

Fig. 23.10 Path lines with initial positions at (**a**) r = 4.3 mm; (**b**) r = 7.2 mm; (**c**) r = 10.4 mm from the center

23.5 Summary and Conclusions

A thermo-fluid simulation model of the friction extrusion process has been developed and validated by experimental measurements. The model treats the processing material, the billet sample, as a highly viscous non-Newtonian fluid with a viscosity that is temperature and strain rate dependent. A volume heating method was employed to approximate the heat source distribution in the sample due to friction at the die-sample interface and due to plastic dissipation. The mechanical power in the experiment is used as the heat source input in the model. Small massless solid particles are inserted in the simulation domain and are released in the fluid as markers, so that their trajectory paths are observed to study the material flow patterns. It is found that temperature predictions from the model agree well with experimental measurements and the marker particles extruded into the wire in the simulation show the same features as in experimental observations. It is concluded that (a) the volume heating model can be used to provide a good approximation of the heat source in the friction extrusion process, (b) the initial gap between the chamber wall and the sample needs to be modeled for good temperature predictions, and (c) there exists a dead zone in the friction extrusion process chamber.

Acknowledgment The financial support provided in part by NASA Consortium Agreement NNX10AN36A and by the National Science Foundation through NSF-CMMI-1266043 is gratefully acknowledged.

References

1. Thomas WM, Nicholas ED, Jones SB (1993) US Patent 5,262,123
2. Gronostajski J, Matuszak A (1999) The recycling of metals by plastic deformation: an example of recycling of aluminum and its alloys chips. J Mater Process Technol 92–93:35–41
3. Tang W, Reynolds AP (2010) Production of wire via friction extrusion of aluminum alloy machining chips. J Mater Process Technol 210:2231–2237
4. Li X, Tang W, Reynolds AP (2012) Visualization of material flow in friction extrusion. In: ICAA13: 13th International Conference on Aluminum Alloys, Hoboken, NJ, pp 1659–1664
5. Zhang H, Zhao X, Deng X, Sutton MA, Reynolds AP, McNeill SR, Ke K (accepted) Investigation of material flow during friction extrusion process. Int J Mech Sci
6. Zhang H, Li X, Tang W, Deng X, Sutton MA, Reynolds AP (in review) Heat transfer modeling of the friction extrusion process. J Mater Process Technol
7. Seidel TU, Reynolds AP (2001) Visualization of the material flow in AA2195 friction-stir welds using a marker insert technique. Metallurgical Mater Trans A 32A:2879–2884
8. Schmidt H, Hattel J, Wert J (2004) An analytical model for the heat generation in friction stir welding. Model Simul Mater Sci Eng 12:143–157
9. Nandan R, Roy GG, Debroy T (2006) Numerical simulation of three dimensional heat transfer and plastic flow during friction stir welding. Metallurgical Mater Trans A 37:1247–1259
10. Xu S, Deng X, Reynolds AP, Seidel TU (2001) Finite element simulation of material flow in friction stir welding. Sci Technol Welding Joining 6:191–193
11. Deng X, Xu S (2004) Two-dimensional finite element simulation of material flow in the friction stir welding process. J Manuf Process 6:125–133
12. Assidi M, Fourment L, Guerdoux S (2010) Tracy Nelson, friction model for friction stir welding process simulation: calibrations from welding experiments. Int J Mach Tool Manuf 50:143–155

13. Chiumenti M, Cervera M, Agelet C, Dialami N (2013) Numerical modeling of friction stir welding processes. Comput Methods Appl Mech Eng 254:353–369
14. Sheppard T, Wright DS (1979) Determination of flow stress I constitutive equation for aluminum alloys at elevated temperatures: II-radial and axial temperature distribution during torsion testing. Met Technol 6:215–229
15. Seidel TU (2002) The development of a friction stir welding process model using computational fluid dynamics. Ph.D. Dissertation, University of South Carolina
16. Schmidt HNB, Dickerson TL, Hattel JH (2055) Material flow in butt friction stir welds in AA2024-T3. Acta Materialia 54:1199–1209

MIX
Papier aus verantwortungsvollen Quellen
Paper from responsible sources
FSC® C105338
FSC
www.fsc.org

If you have any concerns about our products, you can contact us on
ProductSafety@springernature.com

In case Publisher is established outside the EU, the EU authorized representative is:
Springer Nature Customer Service Center GmbH
Europaplatz 3, 69115 Heidelberg, Germany

Printed by Libri Plureos GmbH
in Hamburg, Germany